ATOMIC PROCESSES IN PLASMAS

ATOMIC PROCESSES IN PLASMAS

Eleventh APS Topical Conference

Auburn, Alabama March 1998

EDITORS
Eugene Oks
Michael S. Pindzola
Auburn University, Alabama

American Institute of Physics

AIP CONFERENCE
PROCEEDINGS 443

Woodbury, New York

Editors:

Eugene Oks and Michael S. Pindzola
Department of Physics
Auburn University
Auburn, AL 36849

Email: goks@physics.auburn.edu

Email: pindzola@physics.auburn.edu

Authorization to photocopy items for internal or personal use, beyond the free copying permitted under the 1978 U.S. Copyright Law (see statement below), is granted by the American Institute of Physics for users registered with the Copyright Clearance Center (CCC) Transactional Reporting Service, provided that the base fee of $15.00 per copy is paid directly to CCC, 222 Rosewood Drive, Danvers, MA 01923. For those organizations that have been granted a photocopy license by CCC, a separate system of payment has been arranged. The fee code for users of the Transactional Reporting Service is: 1-56396-802-9/ 98 /$15.00.

© 1998 American Institute of Physics

Individual readers of this volume and nonprofit libraries, acting for them, are permitted to make fair use of the material in it, such as copying an article for use in teaching or research. Permission is granted to quote from this volume in scientific work with the customary acknowledgment of the source. To reprint a figure, table, or other excerpt requires the consent of one of the original authors and notification to AIP. Republication or systematic or multiple reproduction of any material in this volume is permitted only under license from AIP. Address inquiries to Office of Rights and Permissions, 500 Sunnyside Boulevard, Woodbury, NY 11797-2999; phone: 516-576-2268; fax: 516-576-2499; e-mail: rights@aip.org.

L.C. Catalog Card No. 98-86803
ISBN 1-56396-802-9
ISSN 0094-243X
DOE CONF- 980365

Printed in the United States of America

CONTENTS

Preface ix
Previous Meetings. x
Program and Local Committees. xi

ATOMIC STRUCTURE AND PHOTOIONIZATION

All-Order Methods in Relativistic Atomic Structure Theory. 3
    W. R. Johnson, M. S. Safronova, and A. Derevianko
Photoionization of Atomic Ions and Related Electron Scattering Processes ... 19
    S. T. Manson, Z. Altun, H. Chakraborty, E. Dias, P. C. Deshmukh,
    and C. S. Turner
**Angle-Resolved 2D Imaging of Electron Emission Processes in Atoms
and Molecules** 29
    E. Kukk, A. A. Wills, B. Langer, J. D. Bozek, and N. Berrah

ATOMIC PROCESSES IN MAGNETIC FUSION PLASMAS

**Study of Volume Recombination and Radiation Opacity Effects
in Alcator C-Mod** 43
    J. L. Terry, B. Lipschultz, A. Yu. Pigarov, C. Boswell, S. I. Karsheninnikov,
    B. LaBombard, and D. A. Pappas
Monte Carlo Impurity Transport Modeling in the DIII-D Tokamak. 58
    T. E. Evans and D. F. Finkenthal
**Calculated Radiative Power Losses from Mid- and High-Z Impurities
in Tokamak Plasmas** 73
    K. B. Fournier, M. J. May, D. Pacella, B. C. Gregory, J. E. Rice,
    J. L. Terry, M. Finkenthal, and W. H. Goldstein

ATOMIC PROCESSES IN LASER PLASMAS

Intense and Ultrashort Pulse Laser Interactions with Matter 91
    R. W. Falcone
Studies of Ultrafast Laser-Produced X-ray Sources 92
    J. C. Gauthier, J. P. Geindre, P. Audebert, S. Bastiani, Th. Schlegel,
    C. Quoix, A. Rousse, and A. Antonetti
High Gain X-ray Lasers Pumped by Transient Collisional Excitation. 106
    J. Dunn, A. L. Osterheld, V. N. Shlyaptsev, J. R. Hunter, R. Shepherd,
    R. E. Stewart, and W. E. White

ELECTRON EXCITATION OF ATOMIC IONS

Recent Developments in the Theory of Electron-Ion Collisions 121
    K. Bartschat
Low-Energy Electron Collisions with Multiply-Charged Positive Ions 134
    A. Chutjian, J. B. Greenwood, and S. J. Smith
Recent Experiments on Near-Threshold Electron-Impact Excitation
of Multiply Charged Ions 149
    M. E. Bannister, N. Djurić, O. Woitre, G. H. Dunn, Y.-S. Chung,
    A. C. H. Smith, and B. Wallbank

ATOMIC PROCESSES IN ASTROPHYSICAL PLASMAS

Quasars, the Early Universe, and Plasma Simulations..................... 163
    G. J. Ferland and D. A. Verner
SOHO: Atomic Physics and the Solar Atmosphere......................... 173
    T. A. Kucera
Heavy Particle Atomic Collisions in Astrophysics:
Beyond H and He Targets... 185
    P. C. Stancil, P. S. Krstić, and D. R. Schultz

SPECTROSCOPY AND DIAGNOSTICS I

Plasma Spectroscopy of Pulsed Power Driven Z-Pinch Plasmas.............. 199
    J. Davis, R. W. Clark, J. L. Giuliani, Jr., and J. W. Thornhill
Recent Advances in Spectroscopy of Strongly Correlated Plasmas 216
    E. Leboucher-Dalimier, P. Sauvan, P. Gauthier, P. Angelo, H. Derfoul,
    S. Alexiou, A. Poquerusse, T. Ceccotti, and A. Calisti
Interaction of Intense Laser Pulses with Neutral Gases and
Preformed Plasmas... 229
    A. J. Mackinnon, M. Borghesi, A. Iwase, M. W. Jones, and O. Willi

IONIZATION AND RECOMBINATION I

Electron-Impact Ionization and Recombination of Atomic Ions Studied
at Storage Rings 241
    A. Müller, T. Bartsch, C. Brandau, G. Gwinner, A. Hoffknecht,
    C. Kozhuharov, A. A. Saghiri, S. Schippers, M. Schmitt, and A. Wolf
Dielectronic Recombination and Excitation-Autoionization in
Highly Ionized Heavy Elements... 256
    P. Mendelbaum, E. Behar, R. Doron, M. Cohen, A. Peleg, and J. L. Schwob
Non-Perturbative vs. Perturbative Methods for Photorecombination 265
    T. W. Gorczyca, N. R. Badnell, F. Robicheaux, and M. S. Pindzola

ATOMIC PROCESSES IN INERTIAL FUSION PLASMAS

Atomic Physics in Inertial Confinement Fusion (ICF) 281
    C. J. Keane

Atomic Processes in Inertial Fusion Plasmas 282
    C. A. Bach, N. C. Woolsey, O. L. Landen, S. B. Libby, and R. W. Lee

**Atomic and Nuclear Processes Produced in Ultra-High Intensity
Laser Irradiation of Solid Targets** 283
    M. H. Key, M. D. Cable, T. E. Cowan, K. G. Estabrook, B. A. Hammel,
    S. P. Hatchett, E. A. Henry, D. E. Hinkel, J. D. Kilkenny, J. A. Koch,
    W. L. Kruer, A. B. Langdon, B. F. Lasinski, R. W. Lee, B. J. MacGowan,
    A. J. MacKinnon, J. D. Moody, M. J. Moran, A. A. Offenberger, D. M. Pennington,
    M. D. Perry, T. J. Phillips, T. C. Sangster, M. S. Singh, M. A. Stoyer,
    M. Tabak, G. L. Tietbohl, M. Tsukamoto, K. Wharton, and S. C. Wilks

SPECTROSCOPY AND DIAGNOSTICS II

Spectroscopic Studies on Well-Diagnosed Pinch Columns 287
    Th. Wrubel, S. Büscher, I. Ahmad, and H.-J. Kunze

Models for Stark Broadening Applied to Plasma Diagnostics 299
    R. Stamm, A. Calisti, S. Ferri, M. Koubiti, T. Meftah, L. Mouret, C. Mossé,
    F. Reva, and B. Talin

IONIZATION AND RECOMBINATION II

Recombination at Ultra-Low Energies 317
    M. R. Flannery and D. Vrinceanu

**Recoil Ion Momentum Spectroscopy: A 'Momentum Microscope'
to View Atomic Collision Dynamics** 334
    R. Dörner, V. Mergel, H. Bräuning, M. Achler, T. Weber, Kh. Khayyat,
    O. Jagutzki, L. Spielberger, J. Ullrich, R. Moshammer, Y. Azuma,
    M. H. Prior, C. L. Cocke, and H. Schmidt-Böcking

Projectile and Recoil Ion Momentum Spectroscopy in p+He Collisions 347
    M. Schulz, L. An, R. E. Olson, J. Ullrich, and H. Schmidt-Böcking

ATOMIC PROCESSES IN INDUSTRIAL PLASMAS

UV/VUV High Sensitivity Absorption Spectroscopy 359
    A. N. Goyette, L. W. Anderson, K. L. Mullman, and J. E. Lawler

Electron Collision Frequency in Inductively Coupled Plasmas 365
    V. A. Godyak

**Spectroscopic Modeling and Interpretation of Industrial Plasmas
in Materials Processing** 366
    J. L. Giuliani, Jr., J. P. Apruzese, A. Dasgupta, P. Kepple, J. Rogerson,
    V. Shamamian, D. Counts, R. Bicknell-Tassius, and F. J. Grunthaner

List of Attendees..367
Author Index..379

PREFACE

On March 23-26, 1998 atomic and plasma physicists from around the world gathered in Auburn, Alabama to participate in the Eleventh APS Topical Conference on "Atomic Processes in Plasmas". This volume contains the papers and abstracts of the 33 invited talks presented at the meeting. The order of presentation in this volume is the same as that followed at the conference.

On the first day of the conference there were invited talk sessions on atomic structure and photoionization, atomic processes in magnetic fusion plasmas, and atomic processes in laser plasmas. On the second day there were sessions on electron excitation of atomic ions, atomic processes in astrophysical plasmas, and spectroscopy and diagnostics. On the third day there were sessions on ionization and recombination, atomic processes in inertial fusion plasmas, and again spectroscopy and diagnostics. On the morning of the fourth day there was another session on ionization and recombination and then a concluding session on atomic processes in industrial plasmas. Although not included here, the atomic and plasma physics research projects discussed at the three afternoon poster sessions were vital to the overall success of the conference.

We hope that all the conference participants enjoyed the lively interdisciplinary interaction found at the invited talk and poster sessions and the unusually beautiful spring time weather in the deep South. The surface and electronic mail addresses of all participants are included at the back of this volume. We are grateful to the Office of Fusion Energy of the U.S. Department of Energy for providing financial support towards the publication of these proceedings.

                                                    Eugene Oks
                                                    Michael S. Pindzola
                                                    Auburn, Alabama
                                                    Spring 1998

Previous Meeting Sites

1. Knoxville, Tennessee   Feb. 16-18, 1977

2. Boulder, Colorado   Jan. 17-19, 1979

3. Baton Rouge, Louisiana   Feb. 25-27, 1981

4. Princeton, New Jersey   April 13-15, 1983

5. Pacific Grove, California   Feb. 25-28, 1985

6. Santa Fe, New Mexico   Sept. 28-Oct. 2, 1987

7. Gaithersburg, Maryland   Oct. 2-5, 1989

8. Portland, Maine   August 25-29, 1991

9. San Antonio, Texas   Sept. 19-23, 1993

10. San Francisco, California   Jan. 14-18, 1995

Program Committee

Dr. James Cohen
Los Alamos National Laboratory

Dr. Michael Crisp
U.S. Department of Energy

Dr. William Goldstein (Chairman)
Lawrence Livermore National Laboratory

Prof. Donald C. Griffin
Rollins College

Dr. Bruce Hammel
Lawrence Livermore National Laboratory

Prof. Charles F. Hooper, Jr.
University of Florida

Prof. G. Kalman
Boston College

Dr. Yong-Ki Kim
National Institute of Standards
 and Technology

Dr. Marcel Klapisch
Naval Research Laboratory

Dr. Richard W. Lee
Lawrence Livermore National Laboratory

Prof. Eugene Oks (Co-Secretary)
Auburn University

Prof. Michael S. Pindzola (Co-Secretary)
Auburn University

Prof. Ronald A. Phaneuf
University of Nevada-Reno

Dr. Douglas E. Post
ITER - San Diego

Dr. John Raymond
Center for Astrophysics-Harvard

Dr. William L. Rowan
University of Texas

Prof. Jon C. Weisheit
Rice University

Local Committee

Prof. Eugene Oks (Co-Chairman)
Prof. Michael S. Pindzola (Co-Chairman)
Prof. Francis Robicheaux
Dr. John Shaw
Dr. Dario Mitnik
Dr. Alice Kolakowska
Mr. Jimmy Touma
Ms. Cheryl Matheny
Ms. Susan Benson

ATOMIC STRUCTURE AND PHOTOIONIZATION

All-Order Methods in Relativistic Atomic Structure Theory

W.R. Johnson, M.S. Safronova, and A. Derevianko

Department of Physics, Notre Dame University, Notre Dame, IN 46556

Abstract. The singles-doubles (SD) method, in which single and double excitations of the Hartree-Fock wave function are summed to all-orders in perturbation theory, is discussed and applied to valence removal energies, hyperfine constants, and transition matrix elements.

INTRODUCTION

In this paper, we review recent applications of the singles-doubles (SD) approximation [1–3] to obtain highly accurate wave functions and valence electron removal energies for atoms and ions with one valence electron. To make our discussion as specific as possible, we restrict our attention to sodiumlike ions with nuclear charges in the range $Z = 11$–16. These systems are simple enough to make extensive calculations feasible, yet they are significantly more complicated than lithiumlike ions, to which SD equations [2], and the closely related coupled-cluster singles-doubles (CCSD) equations [4], have also been applied successfully. We formulate the SD equations relativistically in order to give an *ab-initio* account of fine structure. Our discussion begins with a brief review of relativistic many-body perturbation theory (MBPT). We show that only 90% of the correlation correction to removal energies is recovered by third-order MBPT for $3s$ and $3p$ states in neutral sodium. The all-order SD approximation accounts for a major part of the remaining 10%.

The SD equations can be separated into core equations, which are the same for every state, and valence equations. It is shown that the core correlation energy, when expanded in a perturbation series, agrees with MBPT through third-order. The valence correlation energy, however, does not agree with third-order MBPT exactly. The missing third-order energy is from omitted triple excitations. By including these triples to lowest nonvanishing order, the complete third-order valence correlation energy is recovered.

The SD equations are solved numerically for low-lying states of sodium and sodiumlike ions with $Z < 16$. The resulting theoretical energies agree with the measured energies to 1-2 cm^{-1} (5 parts in 10^5) for neutral sodium. As Z increases along the sodium isoelectronic sequence, differences between experimental and theoretical energies for 3s states increase. These differences are shown to be consistent with the 3s Lamb-shift. Indeed, adding any of the available estimates of the 3s Lamb-shift to the theoretical 3s energy, reduces differences between experiment and theory below 20 cm^{-1} for the entire interval $Z < 16$. Theoretical 3p fine-structure intervals agree with experiment to better than 0.3% over the same range of nuclear charge.

Dipole matrix elements and hyperfine constants are also evaluated using the SD wave functions. The resulting hyperfine constants agree with measurements to about 0.3%, whereas the dipole matrix elements agree with recent measurements to about 0.1%. Other calculated atomic parameters are also found to be in excellent agreement with experiment.

To summarize, the relativistic SD approximation provides an accurate method for summing important terms in MBPT to all orders. Methods such as this are of obvious importance for PNC measurements [5], where atomic theory is combined with experiment to determine "atomic" values of weak-interaction coupling constants.

REVIEW OF MBPT

The Dirac equation
$$h_0 \phi_i = \epsilon_i \phi_i,$$
with
$$h_0 = c\boldsymbol{\alpha} \cdot \boldsymbol{p} + \beta mc^2 + U(r) + V_{\text{nuc}},$$
describes the motion of an electron in a potential $U(r) + V_{\text{nuc}}$, where V_{nuc} is the nuclear potential and $U(r)$ is a central potential describing the atomic cloud. It is often convenient to choose $U(r) = V_{\text{HF}}$, the self-consistent Hartree-Fock potential. The spectrum of h_0 consists of electron bound states ($0 < \epsilon_i \leq mc^2$), electron scattering states ($\epsilon_i > mc^2$), and positron states ($\epsilon_i < -mc^2$). A useful many-electron generalization of the Dirac Hamiltonian is the *no-pair* Hamiltonian; an approximation to the field-theoretic QED Hamiltonian obtained by performing a contact transformation to eliminate the electron-photon interaction to order e^2 [6]. In second-quantized form, the *no-pair* Hamiltonian is given by $H = H_0 + V$, where
$$H_0 = \sum_i \epsilon_i a_i^\dagger a_i,$$
$$V = \frac{1}{2} \sum_{ijkl} v_{ijkl} a_i^\dagger a_j^\dagger a_l a_k - \sum_{ij} U_{ij} a_i^\dagger a_j.$$

In these equations, $a_i^\dagger$ and a_i are creation and annihilation operators, respectively, for the state i. The sums over i, j, k, and l are restricted to electron states only, and v_{ijkl} is a two-particle matrix element of the sum of the Coulomb and Breit interactions. The term $U_{ij} = \langle i|U|j\rangle$ compensates for including $U(r)$ in h_0. In the applications considered here, Coulomb contributions are far more important than those from the Breit interaction. We evaluate the Coulomb contributions to V first, then linearize V in the Breit interaction to obtain the Breit corrections. In the equations to follow v_{ijkl} is understood to be a two-particle Coulomb matrix element. We introduce $\tilde{v}_{ijk} = v_{ijkl} - v_{ijlk}$ to represent an antisymmetrized Coulomb matrix element.

To investigate alkali atoms or ions, it is convenient to choose $U = V_{HF}$, the HF potential of the closed-shell ionic core. The atomic wave function Ψ and the energy E can be expanded in powers of the potential V as

$$\Psi = \Psi^{(0)} + \Psi^{(1)} + \ldots$$
$$E = E^{(0)} + E^{(1)} + \ldots,$$

where the lowest-order wave function is just the HF wave function $\Psi^{(0)} = \Psi^{HF}$. In each order of perturbation theory, $k = 0, 1, \cdots$, the energy can be separated into a core contribution and a valence contribution, $E^{(k)} = E_C^{(k)} + E_v^{(k)}$. In the first few orders of perturbation theory [7], the contributions to the valence removal energy $E_v^{(k)}$ are:

$$E_v^{(0)} = \epsilon_v, \tag{1a}$$

$$E_v^{(1)} = 0, \tag{1b}$$

$$E_v^{(2)} = -\sum_{bmn} \frac{v_{mnvb}\tilde{v}_{vbmn}}{\epsilon_{mn} - \epsilon_{vb}} + \sum_{abn} \frac{v_{vnab}\tilde{v}_{abvn}}{\epsilon_{vn} - \epsilon_{ab}}, \tag{1c}$$

$$\begin{aligned}E_v^{(3)} =& \sum_{mabcd} \frac{\tilde{v}_{abvm}v_{cdab}v_{mvcd}}{(\epsilon_{ab}-\epsilon_{vm})(\epsilon_{cd}-\epsilon_{mv})} + \sum_{mabcr} \frac{\tilde{v}_{abvm}\tilde{v}_{cmra}\tilde{v}_{vrbc}}{(\epsilon_{ab}-\epsilon_{vm})(\epsilon_{bc}-\epsilon_{vr})} \\ &+ \sum_{mnrbc} \frac{\tilde{v}_{vbmn}\tilde{v}_{cnrb}\tilde{v}_{mrvc}}{(\epsilon_{vb}-\epsilon_{mn})(\epsilon_{vc}-\epsilon_{mr})} + \sum_{mnbrs} \frac{\tilde{v}_{vbmn}v_{mnrs}v_{rsvb}}{(\epsilon_{vb}-\epsilon_{mn})(\epsilon_{vb}-\epsilon_{rs})} \\ &+ \sum_{mnabc} \frac{\tilde{v}_{camn}v_{nmba}\tilde{v}_{vbvc}}{(\epsilon_{mn}-\epsilon_{ab})(\epsilon_{mn}-\epsilon_{ac})} + \sum_{mnrab} \frac{\tilde{v}_{abrn}v_{nmba}\tilde{v}_{vrvm}}{(\epsilon_{mn}-\epsilon_{ab})(\epsilon_{rn}-\epsilon_{ab})} \\ &+ \left[\sum_{mabrs} \frac{\tilde{v}_{abvm}v_{mvrs}v_{rsab}}{(\epsilon_{ab}-\epsilon_{vm})(\epsilon_{ab}-\epsilon_{rs})} + \sum_{mabcr} \frac{\tilde{v}_{abvm}\tilde{v}_{cvrb}\tilde{v}_{mrac}}{(\epsilon_{ab}-\epsilon_{vm})(\epsilon_{ac}-\epsilon_{mr})} \right. \\ &+ \sum_{mnbcd} \frac{\tilde{v}_{vbmn}v_{cdvb}v_{mncd}}{(\epsilon_{vb}-\epsilon_{mn})(\epsilon_{cd}-\epsilon_{mn})} + \sum_{mnrbc} \frac{\tilde{v}_{vbmn}\tilde{v}_{cmrv}\tilde{v}_{nrbc}}{(\epsilon_{vb}-\epsilon_{mn})(\epsilon_{bc}-\epsilon_{nr})} \\ &+ \left. \sum_{mabnr} \frac{\tilde{v}_{vavm}v_{mbnr}\tilde{v}_{nrab}}{(\epsilon_a-\epsilon_m)(\epsilon_{ab}-\epsilon_{nr})} - \sum_{mabcn} \frac{\tilde{v}_{vavm}v_{bcan}\tilde{v}_{mnbc}}{(\epsilon_a-\epsilon_m)(\epsilon_{bc}-\epsilon_{mn})} + \text{c.c.} \right]. \end{aligned} \tag{1d}$$

Here and later, we use the notation $\epsilon_{ij} = \epsilon_i + \epsilon_j$. We adopt the convention that indices at the beginning of the alphabet, $a, b, \cdots$, refer to occupied core states; those in the middle of the alphabet $m, n, \cdots$, refer to excited states; and v and w refer to valence orbitals. An angular momentum decomposition of the two-particle

TABLE 1. Na removal energies (a.u.)

Term	$3s_{1/2}$	$3p_{1/2}$	$3p_{3/2}$
$E^{(0)}$	-0.18203	-0.10949	-0.10942
$E^{(2)}$	-0.00589	-0.00179	-0.00178
$E^{(3)}$	-0.00039	-0.00015	-0.00015
$B^{(1)}$	0.00003	0.00001	0.00001
Total	-0.18828	-0.11142	-0.11134
Expt.	-0.18886	-0.11160	-0.11152
$E^{(4)+} \approx$	-0.00058	-0.00018	-0.00018

integrals v_{ijkl} is carried out to simplify the above sums. The radial wave functions needed to evaluate the resulting integrals are taken from a B-spline basis set [8].

In Table 1, we show the results of a third-order MBPT calculation [9] of removal energies of $3s$ and $3p$ states in sodium. In this table, $E^{(k)}$, $k = 0 \cdots 3$, are Coulomb contributions to the removal energy, and $B^{(1)}$ is the first-order Breit interaction. It can be seen from this table that third-order MBPT accounts for about 90% of the correlation energy; the remaining 10% is from fourth and higher orders. Third-order calculations of transition matrix elements and hyperfine constants have also been carried out, leading to theoretical values that differ from precise measurements by 1–3%. To achieve higher accuracy, one must include the missing fourth- and higher-order terms.

SINGLES-DOUBLES EQUATIONS

One approach to the correlation problem that goes beyond perturbation theory is the SD approximation in which single and double excitations of the HF wave function are evaluated exactly. The SD approximation is a linearized version of the CCSD approximation discussed, for example, in Ref. [10]. We assume that the wave function for a state v of an alkali atom such as sodium can be represented as $\Psi_v = \Psi_v^{\rm HF} + \delta\Psi_v$, where $\delta\Psi_v$ is a linear combination of single and double excitations of the HF wave function:

$$\delta\Psi_v = \left\{ \sum_{am} \rho_{ma} a_m^\dagger a_a + \frac{1}{2} \sum_{abmn} \rho_{mnab} a_m^\dagger a_n^\dagger a_b a_a \right.$$
$$\left. + \sum_{m \neq v} \rho_{mv} a_m^\dagger a_v + \sum_{bmn} \rho_{mnvb} a_m^\dagger a_n^\dagger a_b a_v \right\} \Psi_{\rm HF}. \quad (2)$$

Later, we will consider the effects of triple excitations also.

Substituting this equation into the many-electron Schrödinger equation one finds that the single- and double-excitation coefficients ρ_{ma} and ρ_{mnab} for atomic core orbitals satisfy the equations [2],

$$(\epsilon_a - \epsilon_m)\rho_{ma} = \sum_{bn} \tilde{v}_{mban}\rho_{nb} + \sum_{bnr} v_{mbnr}\tilde{\rho}_{nrab} - \sum_{bcn} v_{bcan}\tilde{\rho}_{mnbc}, \qquad (3a)$$

$$(\epsilon_a + \epsilon_b - \epsilon_m - \epsilon_n)\rho_{mnab} = v_{mnab} + \sum_{cd} v_{cdab}\rho_{mncd} + \sum_{rs} v_{mnrs}\rho_{rsab}$$

$$+ \left[\sum_r v_{mnrb}\rho_{ra} - \sum_c v_{cnab}\rho_{mc} + \sum_{rc} \tilde{v}_{cnrb}\tilde{\rho}_{mrac}\right] + \left[\begin{array}{c} a \leftrightarrow b \\ m \leftrightarrow n \end{array}\right]. \qquad (3b)$$

The corresponding core correlation energy is

$$\delta E_C = \frac{1}{2} \sum_{abmn} v_{abmn}\tilde{\rho}_{mnab}. \qquad (4)$$

The valence excitation coefficients ρ_{mv} and ρ_{mnvb} satisfy

$$(\epsilon_v - \epsilon_m + \delta E_v)\rho_{mv} = \sum_{bn} \tilde{v}_{mbvn}\rho_{nb} + \sum_{bnr} v_{mbnr}\tilde{\rho}_{nrvb} - \sum_{bcn} v_{bcvn}\tilde{\rho}_{mnbc}, \qquad (5a)$$

$$(\epsilon_v + \epsilon_b - \epsilon_m - \epsilon_n + \delta E_v)\rho_{mnvb} = v_{mnvb} + \sum_{cd} v_{cdvb}\rho_{mncd} + \sum_{rs} v_{mnrs}\rho_{rsvb}$$

$$+ \left[\sum_r v_{mnrb}\rho_{rv} - \sum_c v_{cnvb}\rho_{mc} + \sum_{rc} \tilde{v}_{cnrb}\tilde{\rho}_{mrvc}\right] + \left[\begin{array}{c} v \leftrightarrow b \\ m \leftrightarrow n \end{array}\right], \qquad (5b)$$

and the valence correlation energy is

$$\delta E_v = \sum_{ma} \tilde{v}_{vavm}\rho_{ma} + \sum_{mab} v_{abvm}\tilde{\rho}_{mvab} + \sum_{mna} v_{vbmn}\tilde{\rho}_{mnvb}. \qquad (6)$$

In numerical applications, the core equations are first solved, then the valence equations are solved successively for each valence state considered.

Perturbation expansion of the SD equations

To help understand the relation of the SD equations to MBPT, we expand the excitation coefficients in powers of the interaction potential. In lowest order, the core excitation coefficients are seen to be

$$\rho^{(1)}_{mnab} = \frac{v_{mnab}}{\epsilon_a + \epsilon_b - \epsilon_m - \epsilon_n} \quad \text{and} \quad \rho^{(1)}_{ma} = 0,$$

from which it follows that the second-order contribution to the core energy is

$$\delta E^{(2)}_C = \frac{1}{2} \sum_{mnab} \frac{v_{abmn}\tilde{v}_{mnab}}{\epsilon_a + \epsilon_b - \epsilon_m - \epsilon_n};$$

a result that is identical to second-order MBPT [7]. Therefore, δE_C agrees with MBPT through second order. Let us consider the next term in a perturbation expansion of the core-excitation coefficients:

$$\rho^{(2)}_{mnab} = \frac{1}{\epsilon_{ab} - \epsilon_{mn}} \left\{ \sum_{cd} \frac{v_{cdab}v_{mncd}}{\epsilon_{cd} - \epsilon_{mn}} + \sum_{rs} \frac{v_{mnrs}v_{rsab}}{\epsilon_{ab} - \epsilon_{rs}} + \left[\sum_{rc} \frac{\tilde{v}_{cnrb}\tilde{v}_{mrac}}{\epsilon_{ac} - \epsilon_{mr}} + \left(\begin{array}{c} a \leftrightarrow b \\ m \leftrightarrow n \end{array}\right)\right]\right\}.$$

Substituting this into the expression for the core energy, we obtain the following contribution to the third-order energy:

$$\delta E_C^{(3)} = \frac{1}{2} \sum_{mnabcd} \frac{\tilde{v}_{abmn} v_{cdab} v_{mncd}}{(\epsilon_{ab} - \epsilon_{mn})(\epsilon_{cd} - \epsilon_{mn})} + \frac{1}{2} \sum_{abmnrs} \frac{\tilde{v}_{abmn} v_{mnrs} v_{rsab}}{(\epsilon_{ab} - \epsilon_{mn})(\epsilon_{ab} - \epsilon_{rs})}$$

$$+ \sum_{abcmnr} \frac{\tilde{v}_{abmn} \tilde{v}_{cmbr} \tilde{v}_{nrac}}{(\epsilon_{ab} - \epsilon_{mn})(\epsilon_{ac} - \epsilon_{nr})}.$$

This result for $\delta E_C^{(3)}$ is identical to the third-order MBPT energy given in Ref. [7]. Thus, a perturbation expansion of the SD core correlation energy agrees with MBPT through third order. Differences occur, however, starting in fourth order.

Now let us turn to the equations for the valence excitation coefficients ρ_{mv} and ρ_{mnvb}. Expanding these coefficients as described above and substituting into the expression for the valence energy, we find in second order,

$$\delta E_v^{(2)} = \sum_{mab} \frac{v_{abvm} \tilde{v}_{mvab}}{\epsilon_{ab} - \epsilon_{mv}} + \sum_{mna} \frac{v_{vbmn} \tilde{v}_{mnvb}}{\epsilon_{vb} - \epsilon_{mn}},$$

a result that agrees with the second-order valence energy given in Eq.(1c). The third-order correction to δE_v is found to be:

$$\delta E_v^{(3)} = \sum_{mabcd} \frac{\tilde{v}_{abvm} v_{cdab} v_{mvcd}}{(\epsilon_{ab} - \epsilon_{vm})(\epsilon_{cd} - \epsilon_{mv})} + \sum_{mabrs} \frac{\tilde{v}_{abvm} v_{mvrs} v_{rsab}}{(\epsilon_{ab} - \epsilon_{vm})(\epsilon_{ab} - \epsilon_{rs})}$$

$$+ \sum_{mabcr} \frac{\tilde{v}_{abvm} \tilde{v}_{cvrb} \tilde{v}_{mrac}}{(\epsilon_{ab} - \epsilon_{vm})(\epsilon_{ac} - \epsilon_{mr})} + \sum_{mabcr} \frac{\tilde{v}_{abvm} \tilde{v}_{cmra} \tilde{v}_{vrbc}}{(\epsilon_{ab} - \epsilon_{vm})(\epsilon_{bc} - \epsilon_{vr})}$$

$$+ \sum_{mabnr} \frac{\tilde{v}_{vavm} v_{mbnr} \tilde{v}_{nrab}}{(\epsilon_a - \epsilon_m)(\epsilon_{ab} - \epsilon_{nr})} - \sum_{mabcn} \frac{\tilde{v}_{vavm} v_{bcan} \tilde{v}_{mnbc}}{(\epsilon_a - \epsilon_m)(\epsilon_{bc} - \epsilon_{mn})}$$

$$+ \sum_{mnbcd} \frac{\tilde{v}_{vbmn} v_{cdvb} v_{mncd}}{(\epsilon_{vb} - \epsilon_{mn})(\epsilon_{cd} - \epsilon_{mn})} + \sum_{mnbrs} \frac{\tilde{v}_{vbmn} v_{mnrs} v_{rsvb}}{(\epsilon_{vb} - \epsilon_{mn})(\epsilon_{vb} - \epsilon_{rs})}$$

$$+ \sum_{mnbrc} \frac{\tilde{v}_{vbmn} \tilde{v}_{cnrb} \tilde{v}_{mrvc}}{(\epsilon_{vb} - \epsilon_{mn})(\epsilon_{vc} - \epsilon_{mr})} + \sum_{mnbrc} \frac{\tilde{v}_{vbmn} \tilde{v}_{cmrv} \tilde{v}_{nrbc}}{(\epsilon_{vb} - \epsilon_{mn})(\epsilon_{bc} - \epsilon_{nr})}, \quad (7)$$

which *disagrees* with the third-order valence energy $E_v^{(3)}$ given in Eq.(1d). To account for the missing third-order energy, it is necessary to add triple excitations to the SD wave function. Let us consider the following subset of triples:

$$\frac{1}{6} \sum_{abmnr} \rho_{mnrvab} a_m^\dagger a_n^\dagger a_r^\dagger a_v a_b a_a \Psi_{\text{HF}}.$$

Adding this term to the wave function leads to the following addition to the valence correlation energy:

$$E_{v\,\text{extra}} = \frac{1}{2} \sum_{mnab} \tilde{v}_{abmn} \rho_{mnvvab}.$$

TABLE 2. 3rd-order energies in Na (a.u.)

Term	$3s_{1/2}$	$3p_{1/2}$	$3p_{3/2}$
$\delta E_v^{(3)}$	-3.446[-4]	-1.454[-4]	-1.449[-4]
$E_{v\,\text{extra}}^{(3)}$	-0.418[-4]	-0.070[-4]	-0.072[-4]
$E_v^{(3)}$	-3.865[-4]	-1.525[-4]	-1.521[-4]

The lowest nonvanishing approximation (third-order) to this term is:

$$E_{v\,\text{extra}}^{(3)} = \sum_{mnabc} \frac{v_{abmn}\tilde{v}_{cmav}\tilde{v}_{nvbc}}{(\epsilon_{ab}-\epsilon_{mn})(\epsilon_{bc}-\epsilon_{nv})} + \sum_{mnabs} \frac{\tilde{v}_{abmn}\tilde{v}_{nvas}\tilde{v}_{msvb}}{(\epsilon_{ab}-\epsilon_{mn})(\epsilon_{vb}-\epsilon_{ms})}$$

$$+ \sum_{mnabc} \frac{v_{abmn}\tilde{v}_{cvbv}\tilde{v}_{mnca}}{(\epsilon_{ab}-\epsilon_{mn})(\epsilon_{ca}-\epsilon_{mn})} + \sum_{mnabs} \frac{v_{abmn}\tilde{v}_{mvsv}\tilde{v}_{nsba}}{(\epsilon_{ab}-\epsilon_{mn})(\epsilon_{ab}-\epsilon_{ns})}$$

$$+ \sum_{mnabs} \frac{v_{abmn}\tilde{v}_{mnvs}v_{vsba}}{(\epsilon_{ab}-\epsilon_{mn})(\epsilon_{ab}-\epsilon_{vs})} + \sum_{mnabc} \frac{v_{abmn}\tilde{v}_{cvba}v_{mnvc}}{(\epsilon_{ab}-\epsilon_{mn})(\epsilon_{vc}-\epsilon_{mn})}$$

$$+ \sum_{mnabc} \frac{v_{abmn}\tilde{v}_{cmab}\tilde{v}_{vnvc}}{(\epsilon_{ab}-\epsilon_{mn})(\epsilon_{c}-\epsilon_{n})} + \sum_{mnabs} \frac{v_{abmn}\tilde{v}_{mnas}\tilde{v}_{vsvb}}{(\epsilon_{ab}-\epsilon_{mn})(\epsilon_{b}-\epsilon_{s})}. \qquad (8)$$

The sum $\delta E_v^{(3)} + E_{v\,\text{extra}}^{(3)}$ gives the entire MBPT expression for the third-order valence correlation energy in Eq.(1d). In determining theoretical SD energies, we add $E_{v\,\text{extra}}^{(3)}$ to the SD correlation energy δE_v to account for the missing third-order terms. In Table 2, we give values of $\delta E_v^{(3)}$, $E_{v\,\text{extra}}^{(3)}$, and $E_v^{(3)}$ for the 3s and 3p states of sodium. The missing third-order energy is about 11% of $E_v^{(3)}$ for the 3s state and about 5% of $E_v^{(3)}$ for 3p states.

Iteration solution to SD equations

As mentioned previously, we first solve the core SD equations completely and then solve the valence equations successively for each state of interest. We retain terms in the angular-momentum decomposition of the equations from single-particle orbitals states with angular momenta $l = 0$ to 6 and extrapolate to obtain the final correlation energies. We limit our basis set to $n = 27$ out of 30 B-spline basis functions for each value of l.

This type of calculation is carried out for 3s and 3p states of sodium and sodiumlike ions with $Z = 12$ to 35. In the upper panel of Fig. 1, we plot the valence correlation energy δE_v for these states. These graphs are dominated by the second-order correlation energy $E_v^{(2)}$ which is 80-90% of δE_v over the range considered. For neutral sodium, the correlation energy is seen to be largest for the 3s state; however, at higher Z the correlation energy is larger for 3p states.

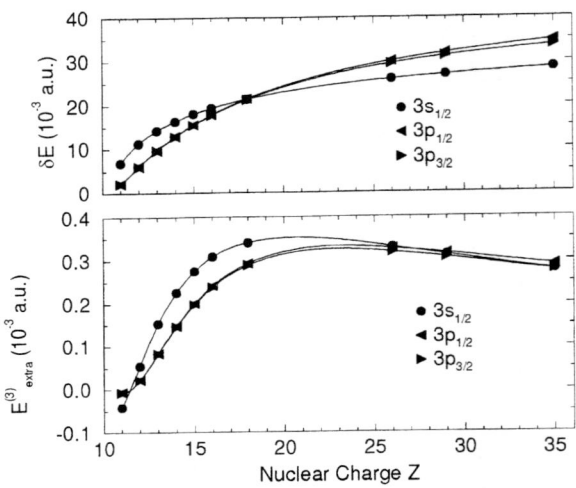

FIGURE 1. All-order correlation energies for sodiumlike ions

In the lower panel of Fig. 1, contributions to the correlation energy from $E^{(3)}_{v\,\text{extra}}$ are plotted. These contributions are small for sodium, but grow very rapidly with Z. For the highest Z considered ($Z = 35$), $E^{(3)}_{v\,\text{extra}}$ is 85% of $E^{(3)}_v$ for the 3s state and of comparable size for 3p states. For all Z, the contribution of $E^{(3)}_{v\,\text{extra}}$ is two orders of magnitude smaller than δE_v.

In Fig. 2, we show $E^{\text{corr}}_v - E^{(2+3)}_v$, where $E^{\text{corr}}_v = \delta E_v + E^{(3)}_{v\,\text{extra}}$. These are the contributions to the correlation energy of fourth and higher orders. The higher-order correlation contributions are seen to decrease by a factor of about 6 over the range $Z = 11 - 35$ for 3s states. For 3p states the higher-order corrections have maxima near $Z = 13$. The relative size of the higher-order correlation corrections decreases from 6-9% for Z=11 to 1% for Z=20. For $Z > 20$, third-order MBPT, therefore, accounts for more than 99% of the correlation energy.

In Table 3, we list the various contribution to the removal energies of 3s, 3$p_{1/2}$, 3$p_{3/2}$ states of sodium. The zeroth-order DHF energy is given in the row labeled DHF. The row labeled δE gives all-order results obtained by calculating partial waves $l \leq 6$ and extrapolating the remainder as discussed previously. The number in parentheses for the 3s entry is an estimate of the extrapolation error. The row labeled $E^{(3)}_{\text{extra}}$ is the "extra" third-order contribution given in Eq.(8). The row labeled "Breit" gives the Breit contribution evaluated as the expectation value of the Breit operator using SD wave functions as discussed in the following paragraph. The row labeled RM+MP contains the sum of the reduced-mass and mass-polarization corrections, which are evaluated to third-order MBPT as described in [9]. The row

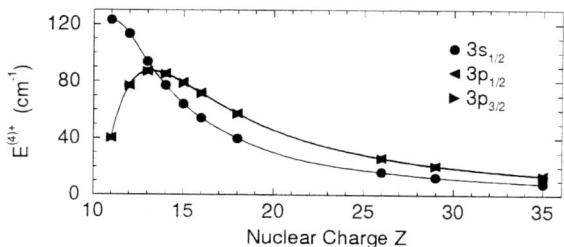

FIGURE 2. Contributions beyond third-order

labeled "Theory" lists the theoretical energy, which is the sum of all of the above contributions. The row labeled "Expt." gives experimental removal energies taken from the NIST database [11].

The Breit correction, as discussed in Ref. [9], is significantly reduced by correlation contributions. The DHF value of the Breit contribution for the $3s$ state $B^{\text{DHF}} = 2.63 \times 10^{-5}$ a.u. is reduced to 0.80×10^{-5} a.u. by the one-body correlations $B^{(1)}$ and further to 0.56×10^{-5} a.u. by the two-body correlations $B^{(2)}$. Moreover, the all-order value (the SD matrix element of the Breit operator) is about two times smaller than the value 1.15×10^{-5} a.u. given by third-order MBPT. Thus,

TABLE 3. Summary for Na (cm^{-1})

Term	$3s$	$3p_{1/2}$	$3p_{3/2}$
DHF	-39951.6	-24030.4	-24014.1
δE	-1488.8(4)	-463.9	-461.6
$E^{(3)}_{\text{extra}}$	-9.2	-1.5	-1.6
Breit	1.2	1.4	0.1
RM+MP	1.0	0.5	0.5
Theory	-41447.3(4)	-24493.9	-24476.7
Expt.	-41449.4	-24493.3	-24476.1

TABLE 4. Removal energies (cm^{-1}) for sodiumlike ions

Z	Term	3s	$3p_{1/2}$	$3p_{3/2}$
11	SD	-41447.3(.4)	-24493.9	-24476.7
	NIST	-41449.4	-24493.3	-24476.1
12	SD	-121264.6(1.0)	-85601.1(.3)	-85509.8(.3)
	NIST	-121267.6	-85598.3	-85506.8
13	SD	-229447.0(1.9)	-175762.6(1.0)	-175529.6(.9)
	NIST	-229445.7	-175762.8	-175529.1
14	SD	-364102.5(2.2)	-292799.8(1.4)	-292340.0(1.6)
	NIST	-364093.1	-292805.6	-292344.5
15	SD	-524488.1(3.3)	-435801.7(2.8)	-435007.7(2.6)
	NIST	-524462.9	-435811.0	-435015.6
16	SD	-710240.5(4.6)	-604310.2(4.2)	-603049.8(4.3)
	NIST	-710194.7	-604321.1	-603057.0

higher-order Coulomb diagrams contribute substantially to the Breit correction for near neutral ions. The MBPT and SD results approach one another as the ionic charge increases.

A comparison of the SD theory (rows labeled SD) and experiment (rows labeled NIST which are taken from Ref. [11]) for 3s and 3p states of sodiumlike ions is given in Table 4. Differences between the theoretical and experiment energies for 3s states of ions with $Z \leq 13$ range from 1 to 3 cm^{-1}, but these differences increase rapidly for $Z > 13$ as shown in Fig. 3. We attribute the major part of the differences for $Z > 13$ to omitted QED corrections, which are dominated by the 3s self-energy. Codes to evaluate the self-energy in a realistic atomic potential, such as the one described in [12], do not converge for the low values of Z considered here, so it is necessary to turn to approximate schemes to estimate QED corrections. In Fig. 3, we show several different estimates. First, we show values of the self-energy obtained by replacing Z by $Z-6$ and $Z-7$ in the Coulomb-field self-energies for 3s states given by Mohr and Kim [13]. Next, we compare with values of the Lamb shift from the MCDF code of Grant et al. [14], which were obtained by scaling Coulomb-field values of the self-energy; these values and vacuum-polarization corrections. Finally, we present values of the Lamb shift from Kim et al. [15] that also include corrections for vacuum polarization. The values from Ref. [14] agree well with the $Z-7$ Coulomb values, while those from Ref. [15] lie between the $Z-5$ and $Z-6$ Coulomb-field values. It can be seen that if we add any of these estimates of the Lamb shift to the theoretical energy, the resulting values will be within 20 cm^{-1} of the experimental energy. In the absence of more reliable QED corrections, it is impossible to reduce the difference between theory and experiment for these states further.

In Table 5, we compare the theoretical $3p_{3/2} - 3p_{1/2}$ fine-structure intervals with

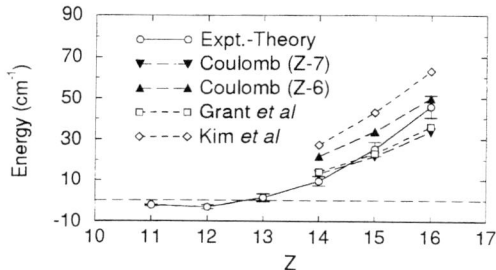

FIGURE 3. Comparison with estimates of $3s$ Lamb-shift

TABLE 5. $3p$ (cm^{-1})

Z	$3p_{3/2} - 3p_{1/2}$	
	Theory	NIST
11	17.15	17.20
12	91.33	91.57
13	233.13	233.67
14	459.93	461.10
15	793.96	795.38
16	1260.53	1264.10

measured intervals. Our uncertainties for the $3p$ intervals range from 0.2cm^{-1} for $Z=11$ to 0.4cm^{-1} for $Z=16$. The third- and higher-order contributions to the splitting are less than 1% and there is substantial cancellation between the second-order Coulomb contribution and the Breit contribution. The all-order calculations are in better agreement with experiment than the third-order calculations, as expected. The overall agreement with experiment is seen to be about 0.3%.

TRANSITION AMPLITUDES AND HYPERFINE CONSTANTS

The formalism for calculating matrix elements of a one-body operator Z in the SD approach is described in Ref. [2], where it was applied to determine transition matrix elements and hyperfine constants in Li and Be$^+$. Here, we apply the SD method to calculations of electric-dipole matrix elements and hyperfine constants along the sodium isoelectronic sequence. A one-body operator is represented in second quantization as

$$Z = \sum_{ij} z_{ij} a_i^\dagger a_j, \qquad (9)$$

where z_{ij} is the matrix element of the operator z between single-particle orbitals. Here, z is the coordinate operator when one is evaluating dipole transition matrix elements or the hyperfine operator written down and discussed in Ref. [16] when one is evaluating hyperfine constants. Substituting SD wave functions into the matrix element $\langle \Psi_w | Z | \Psi_v \rangle$ and correcting for normalization, one obtains the size-consistent expression [2]

$$\langle \Psi_w | Z | \Psi_v \rangle = \frac{Z_{\text{val}}}{[(1+\delta N_w)(1+\delta N_v)]^{1/2}}. \qquad (10)$$

The term Z_{val} consists of the lowest-order (DHF) matrix element z_{wv} corrected by the sum of 20 terms, given together with normalization correction δN_v in Ref. [2]. Two of the important contributions to Z_{val} are the RPA term Z_{wv}^{RPA} and the polarization (Brueckner orbital) correction Z_{wv}^{BO}, given by:

$$Z_{wv}^{\text{RPA}} = \sum_{am} z_{am}\, \tilde{\rho}_{wmva} + \text{c.c.}$$
$$Z_{wv}^{\text{BO}} = \sum_{m} z_{wm}\, \rho_{mv} + \text{c.c.}$$

The remaining eighteen contributions to Z_{val} are linear or quadratic functions of the SD excitation amplitudes that can be evaluated once the SD equations are solved.

E1 transitions

Let us apply the SD wave functions to $E1$ reduced matrix elements for $3p - 3s$ transitions in sodium and sodiumlike ions with Z=12-16. The single- and double-excitation amplitudes discussed above are used to evaluate the twenty terms in the expression for Z_{val}. Since the excitation amplitudes ρ_{mnab}, etc., occurring in these expressions include partial waves with $l \leq 6$, the sums over excited states in Z_{val} are also limited. To estimate the size of the truncation error caused by limiting the number of partial waves, we recalculate the amplitudes using $l \leq 5$, and find that the resulting matrix elements are unchanged to four digits.

The level of agreement between length and velocity forms for electric-dipole transition matrix elements serves to measure the consistency of the theoretical formalism as well as the accuracy of the numerical algorithms. A sample comparison of length and velocity forms of reduced dipole matrix elements in sodium, where the correlation contribution is most important, is given in Table 6. The length and velocity forms are seen to agree to better than 0.05% for these transitions.

TABLE 6. Length vs. velocity for Na (a.u.)

Transition	Length	Velocity
$3p_{1/2} - 3s_{1/2}$	3.531	3.531
$3p_{3/2} - 3s_{1/2}$	4.994	4.991

TABLE 7. Length-form dipole matrix elements (a.u.)

Transition	Mg II	Al III	Si IV	P V	S VI
$3p_{1/2} - 3s_{1/2}$	2.369	1.845	1.523	1.314	1.154
$3p_{3/2} - 3s_{1/2}$	3.351	2.611	2.165	1.859	1.634

The length-velocity agreement improves with increasing nuclear charge Z along the isoelectronic sequence.

In Table 7, we list length-form reduced dipole matrix elements for ions with $Z = 11 - 16$ for the $3p - 3s$ transitions. We estimate that the uncertainty (from truncation and other numerical errors) is less than 0.05% for the data presented in this table.

Comparisons of our results with available experimental data Refs. [17–21] for $3p_{1/2} - 3s_{1/2}$ and $3p_{3/2} - 3s_{1/2}$ transitions in sodiumlike ions are given in Table 8, where we list reduced matrix elements to eliminate the strong dependence of decay rates on photon energy. Our calculations agree with experimental values to within the experimental error bars for transitions in Mg II, Si IV, and S VI, except for the $3p_{1/2} - 3s_{1/2}$ transition in S VI. For Na I, where high precision experiment data are available, our data differ from experiment by about 0.1%.

TABLE 8. Reduced matrix elements (a.u.) for sodiumlike ions.

Ion	$3p_{1/2} - 3s_{1/2}$			$3p_{3/2} - 3s_{1/2}$		
	Present	Expt.	Ref.	Present	Expt.	Ref.
Na I	3.531	3.5267(17)	[18]	4.994	4.9875(25)	[18]
		3.5246(23)	[17]		4.9838(34)	[17]
Mg II	2.369	2.376(9)	[19]	3.351	3.366(16)	[19]
Si IV	1.523	1.53(3)	[20]	2.165	2.17(5)	[20]
S VI	1.154	1.19(2)	[21]	1.634	1.64(3)	[21]

TABLE 9. ^{23}Na hyperfine constants A (MHz)

	$3s_{1/2}$	$3p_{1/2}$	$3p_{3/2}$
DHF	623.5	63.39	12.59
SD	888.1	94.99	18.84
SD[a]	884.5(1.0)	92.4(2)	19.3(1)
CI[b]	882.2	94.04	18.80
CCSD[c]	883.8	93.02	18.318
MBPT[d]	860.9	91.40	19.80
Expt.	885.81[e]	94.42(19)[f]	18.79(12)[g]
		94.44(13)[h]	18.534(15)[i]

[a] Liu [1]
[b] Jönsson et al. [29]
[c] Salomonson and Ynnerman [30]
[d] Johnson et al. [31]
[e] Beckman et al. [23]
[f] Carlsson et al. [24]
[g] Volz et al. [25]
[h] Wijngaarden and Li [26]
[i] Yei et al. [27]

Hyperfine constants

Calculations of hyperfine constants follow the same pattern as the calculations of reduced dipole matrix elements described in the previous subsection. The magnetic moments and nuclear spins used in the present calculations are taken from Ref. [22]. In Table 9, we give the present SD values of the magnetic-dipole hyperfine constants A for ^{23}Na and compare our values with available theoretical [1,29–31] and experimental [23–27] data. The present SD value for A_{3s} disagrees with the very precise experimental value from Ref. [23] by about 0.25%.

The precise experimental value for the $3s$ state in sodiumlike ^{25}Mg, given in Ref. [28], $A(3s) = -596.254$ MHz, differs by 0.2% with the value -597.56 MHz from the present work.

The ratio of the hyperfine constant B to the nuclear quadrupole moment Q for the $3p_{3/2}$ state of ^{23}Na is found to be 26.85 MHz/barn. If we combine this theoretical ratio with the measured value $B=2.724(30)$ MHz from Ref. [27], we obtain $Q = 101.4(11)$ mb for the quadrupole moment of the ^{23}Na nucleus. The error in this value of Q is experimental. The theoretical error is 1% or less, based on the comparisons between theoretical and experimental values for removal energies, dipole matrix elements, and magnetic dipole hyperfine constants A. The "atomic" value of nuclear quadrupole moment obtained in this way is in agreement with the value $Q = 100.6(20)$ mb obtained in muonic experiments [32] and resolves the long-standing disagreement [29,33] between "atomic" and "muonic" values of the nuclear quadrupole moment of ^{23}Na.

Other Applications

The SD equations have also been applied recently to evaluate the Lennard-Jones C_3 coefficient for atom-wall interactions [34]. In this application, one obtains the value $C_3 = 1.8884$ a.u. for sodium from an SD calculation compared to the precise semi-empirical value $C_3 = 1.888$ a.u. found in Ref. [35]. A HF calculation gives $C_3 = 2.2600$ a.u., by contrast.

SUMMARY

As we have seen, the SD equations lead to highly accurate removal energies, fine-structure intervals, transition amplitudes, and hyperfine constants for sodium and sodiumlike ions. For sodiumlike ions, the accuracy of the theoretical energies is limited by the accuracy of Lamb-shift calculations. The SD transition reduced matrix elements from the present calculations agree in length- and velocity-forms to about 0.05%. These matrix elements agree with recent precise experiments to about 0.1%. The hyperfine constants for sodium also agrees with experiments to about 0.3%.

These calculations illustrate the efficacy of the SD approach for atomic-structure calculations in alkali-metal atoms and ions.

ACKNOWLEDGMENTS

This work was supported in part by NSF grant No. PHY95-13179.

REFERENCES

1. Z.W. Liu, Ph. D. thesis, Notre Dame University, (1989).
2. S.A. Blundell, W.R. Johnson, Z.W. Liu and J. Sapirstein, Phys. Rev. A **40**, 2233 (1989).
3. S.A. Blundell, W.R. Johnson, and J. Sapirstein, Phys. Rev. A**43**, 3497 (1991).
4. A.-M. Mårtensson-Pendrill and A. Ynnerman, Phys. Scr. **41**, 329 (1990).
5. C.S. Wood, S.C. Bennett, D. Cho, B.P. Masterson, J.L. Roberts, C.E. Tanner, and C.E. Wieman, Science **275**, 1753 (1997).
6. G.E. Brown and D.G. Ravenhall, Proc. R. Soc. London, Ser. A **208**, 552 (1951).
7. S.A. Blundell, D.S. Guo, W.R. Johnson, and J. Sapirstein, At. Data At. Data and Nucl. Data Tables **37**, 103 (1987).
8. W.R. Johnson, S.A. Blundell, and J. Sapirstein, Phys. Rev. A**37**, 307 (1988).
9. W.R. Johnson, S.A. Blundell, and J. Sapirstein, Phys. Rev. A**38**, 2699 (1988).
10. I. Lindgren and J. Morrison, *Atomic Many-Body Theory*, 2d ed. (Springer, Berlin, 1986), chap. 15.

11. J.R. Fuhr, W.C. Martin, A. Musgrove, J. Sugar, and W.L. Wiese *NIST Atomic Spectroscopic Database* ver. 1.1 January 1996. NIST Physical Reference Data. Online. National Institute of Standards and Technology. Available http://physics.nist.gov/PhysRefData/contents.html
12. K.T. Cheng, W.R. Johnson, and J. Sapirstein, Phys. Rev. Lett. **66**, 2960 (1991).
13. P.J. Mohr and Y.-K. Kim, Phys. Rev. A**45**, 2727 (1991).
14. I.P. Grant, B.J. McKinzie, P.H. Norrington, D.F. Mayers, and N.C. Pyper, Comput. Phys. Commun. **21**, 207 (1980); B.J. McKinzie, I.P. Grant, and P.H. Norrington, Comput. Phys. Commun. **21**, 233 (1980).
15. Y.-K. Kim, D.H. Baik, P. Indelicato, and J.P. Desclaux, Phys. Rev. A**44**, 148 (1991).
16. K. T. Cheng and W. J. Childs, Phys. Rev. A **31**, 2775 (1985).
17. U. Volz and H. Schmoranzer, Phys. Scr. T**65**, 48 (1996).
18. K.M. Jones, P.S. Julienne, P.D. Lett, W.D. Phillips, E. Tiesinga, and C.J. Williams, Europhys. Lett. **35**, 85 (1996).
19. W. Ansbacher, Y. Li and E.H. Pinnington, Phys. Lett. **A 139**, 165, (1989).
20. S.T. Maniak, E. Trabert and L.J. Curtis, Phys. Lett. **A 173**, 407, (1993).
21. J.O. Ekberg, L. Engström, S. Bashkin, B. Denne, S. Huldt, S. Johansson, C. Jupèn, U. Litzèn, A. Trigueiros, and I. Martinson, Phys. Scr. **29**, 226, (1984). Donnelly, Phys. Scr. **19**, 267 (1979).
22. P. Raghavan, At. Data Nucl. Data Tables **42**, 189 (1989).
23. A. Beckman, K. D. Böklen, and D. Elke, Z. Phys. **270**, 173 (1974).
24. J.Carlsson, P. Jönsson, L. Sturesson, and C. Froese Fischer, Phys. Scr. **46**, 394 (1992).
25. U. Volz, M. Majerus, H. Liebel, A. Schmitt, and H. Schmoranzer, Phy. Rev. Lett. **76**, 2862 (1996).
26. W.A. Wijngaarden and J. Li, Z. Phys. D **32**, 67 (1994).
27. W. Yei, A. Sieradzan and M.D. Havey, Phys. Rev. A **48**, 1909 (1993) .
28. W.M. Itano and D.J. Wineland, Phys. Rev. A **24**, 1364 (1981).
29. Per Jönsson, Anders Ynnerman, Charlotte Froese Fischer, Michel R. Godefroid, and Jeppe Olsen, Phys. Rev. A**53**, 4021 (1996).
30. S. Salomonson and A. Ynnerman, Phys. Rev. A**43**, 2233 (1991).
31. W. R. Johnson, M. Idrees, and J. Sapirstein, Phys. Rev. A**35**, 3218 (1987).
32. B. Jeckelmann, W. Beer, I. Beltrami, F.W.N. de Boer, G. de Chambrier, P.F.A. Goudsmit, J. Kern, H.J. Leisi, W. Ruckstuhl, and A. Vacchi, Nucl. Phys. **A408**, 495 (1983).
33. D. Sundholm and J. Olsen, Phys. Rev. Lett. **68**, 927 (1992).
34. A. Derevianko, W.R. Johnson, and Stephan Fritzsche, Phys. Rev. A, to be published (1998).
35. P. Kharchenko, J. F. Babb, and A. Dalgarno, Phys. Rev. A**55**, 3566 (1997).

Photoionization of Atomic Ions and Related Electron Scattering Processes

Steven T. Manson[*], Zikri Altun[+], Himadri Chakraborty[#], Eric Dias[#], Pranawa C. Deshmukh[#], and Clay S. Turner[*]

[*]Department of Physics and Astronomy, Georgia State University, Atlanta, Georgia 30303
[+]Department of Physics, Marmara University, Istanbul, Turkey
[#]Department of Physics, Indian Institute of Technology-Madras, Madras, India 600 036

> A selection of recent theoretical results revealing new phenomenology in the photoionization and electron elastic scattering properties of multicharged atomic ions. Particularly striking is the non-smooth behavior of cross sections along isoelectronic and isonuclear sequences of ions, indicating that, although ionic phenomena become simpler eventually, this does not happen monotonically or quickly. This has importance concerning the possibility of transferring our vast knowledge of neutrals to ionic species

INTRODUCTION

Atomic ions abound in very hot environments such as fusion or astrophysical plasmas. Unfortunately, they do not abound under normal conditions on the surface of the earth. Thus, it is only with considerable difficulty that they can be produced in sufficient quantity for laboratory experiments. Theory is, therefore, of particular importance in this area for two reasons: To provide accurate data for cases where experiment has not been done; and to provide a framework for interpolation/extrapolation of the ionic data that does exist. In a general sense, this represents the continuation of a long-standing project to understand the spectral and collision properties of atomic ions (1-3). In this review, we focus on some recent theoretical results on ionic photoionization and electron-ion elastic scattering which give some insight into new phenomenology and/or deeper understanding of the behavior of ionic properties along isoelectronic or isonuclear sequences.

In the next section, some general theoretical ideas are put forth which serve as a framework for assessing the behavior of ionic properties. The next section presents several illustrative examples which emphasize different aspects of the physics.

THEORETICAL BACKGROUND

The Hamiltonian for an atomic ion with N electron and nuclear charge Z is given by

$$H = \sum_{j=1}^{N} \left[\frac{p_j^2}{2m} - \frac{Ze^2}{r_j} + \sum_{j'<j}^{N} \frac{e^2}{r_{jj'}} \right] \qquad (1)$$

which can be rewritten as

$$H = \sum_{j=1}^{N} \frac{p_j^2}{2m} - Z \sum_{j=1}^{N} \left[\frac{e^2}{r_j} - \frac{1}{Z} \sum_{j'=1}^{j-1} \frac{e^2}{r_{jj'}} \right] \qquad (2)$$

where e and m are the electron charge and mass respectively, r_j is the position vector of the j-th electron, and $\mathbf{p}_j$ is the corresponding momentum operator. Since we cannot solve the Schrödinger equation exactly for a system of more than a single electron, approximations must be made. The last term in the Hamilitonian, the interaction among the electrons, is the term that requires the approximations since we cannot solve a Hamiltonian with such a term exactly. It is evident from Eq.(2) that, owing to the 1/Z factor multiplying the electron-electron interaction term, this term becomes less important with increasing Z, i.e., along an isoelectronic sequence. It is also clear that along an isonuclear sequence, where N is decreased, this term again becomes less important. Thus, eventually the photoionization (or almost any other ionic property) must become simpler, more hydrogenic, along isoelectronic or isonuclear sequences. But, experience has shown that this simplicity comes neither rapidly, nor monotonically.

ILLUSTRATIVE EXAMPLES

To give some idea how the above discussion manifests itself in concrete cases, we present results for the photoionization cross section of the excited 6f state in the Cs (Z = 55) isoelectronic sequence (4), calculated using a simple, but realistic, central-field potential (5), in Figure 1. The outstanding feature of these results is that the cross section for neutral Cs is a simple monotonic decreasing function of energy, generally reminiscent of hydrogenic, at least near threshold. On the other hand, going to Ba^+, where the electron-electron potential should be less important, the cross section is completely different from the neutral case, having a very deep minimum a bit above threshold. Furthermore, this minimum persists to the eighth member of the isoelectronic sequence. And even by Tb^{+10},

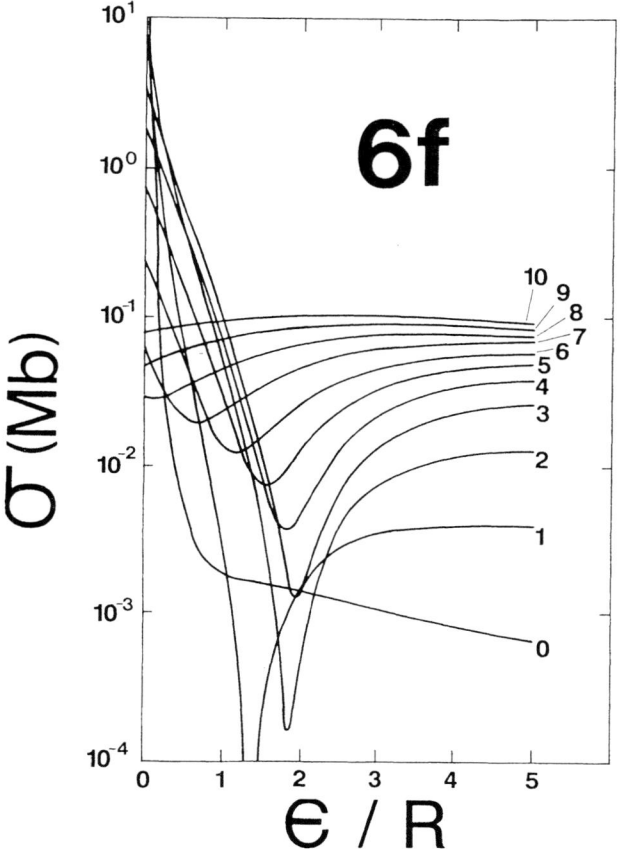

FIGURE 1. Photoionization cross section for the 6f excited states of the first 11 members of the Cs isoelectronic sequence as a function of photoelectron energy in Rydbergs (13.6 eV). Each curve is labeled by the charge of the ion, i.e., 0 refers to neutral Cs, 1 to Ba^+, 2 to La^{+2}, etc.

the highest member of the sequence shown, the cross section is still increasing from threshold. It turns out that this cross section does not become monotone decreasing until about Hg (Z=80) and does not become even vaguely hydrogenic until we reach a Z of about 100. And the effect is even more dramatic for the photoionization of higher nf states (4). This example confirms quite clearly, that the path to simplicity is neither rapid nor monotonic.

Just briefly, the cause of this behavior involves the nature of the 6f bound state wave function. At low energy, in neutral Cs, initial and final states in the dominant 6f→g transition have vanishingly small phase shift (quantum defect) so that the transition is essentially hydrogenic; at higher energies the change in slope is due to the g-wave shape resonance (6). The effective f-wave potential (electrostatic plus centrifugal) is double welled is Cs, but he 6f is bound in the outer well; the 6f does not really "feel" the inner well and are almost hydrogenic as mentioned above. Going to Ba^+, the added electrostatic attraction causes the 6f to "collapse" and to be bound in the inner well, making it decidedly non-hydrogenic and giving a quantum defect close to unity. Then, since the continuum g-wave stall has a near-zero phase shift near threshold, there is a relative phase shift difference between initial and final states in the 6f→εg transitions of about π; from previous work (7-10), this difference implies a Cooper minimum in the cross section, as seen in Fig. 1. With increasing Z, this Cooper minimum still dominates the cross section, even when it moves below threshold. This means that interpolation or extrapolation can on be done meaningfully when the physics of the situation is understood, as Fig. 1 shows dramatically.

As a second example, we scrutinize the near-threshold behavior of the photoionization cross sections of the Ne isoelectronic sequence, calculated using the relativistic-random-phase approximation, RRPA (11,12), which includes significant correlation. Furthermore, the RRPA results have been found to be quite reliable for the neutral noble gases (13), and there is no reason to expect that are less reliable for the ions of the isoelectronic sequence. The results of the calculations for the outer $2p_{3/2}$ and $2p_{1/2}$ in Al^{+3} and Si^{+4} are shown in Fig.2 for the open continuum energy region above the $2p_{1/2}$ threshold. The cross section for Al^{+3} is seen to be monotone decreasing from the threshold and is, in general, similar to the case of neutral Ne (not shown) with the slight delayed maximum moved into the discrete. In addition, the situation with and without coupling to the 2s channels shows no major differences. On the other hand, the Si^{+4} results are completely different at threshold. Both $2p_j$ cross sections are close to zero at threshold and rise dramatically above threshold to a maximum before decreasing at the higher energies. Note that P^{+5}, the next member of the sequence (not shown) follows the pattern established from Ne to Al^{+3}; Si^{+4} is an anomaly. Why does this occur? An important clue comes from Fig.2 where it is seen that the Si^{+4} results without coupling to the 2s channels are completely in keeping with the rest of the sequence. Thus, the anomaly must have something to do with the coupling with the 2s channels. What actually occurs is that autoionizing resonances, whose energies are well above the 2p thresholds in neutral Ne, move closer to the 2p thresholds with increasing Z along the isoelectronic sequence. This is because these Rydberg states move lower below the 2s ionization threshold as the square of the effective charge (roughly z+1, where z is the ionic charge), while the 2p thresholds increase their separation below the 2s threshold only as the first power of z+1. Thus, eventually, each 2s→np resonance moves below the 2p thresholds and becomes a true bound state. At Si^{+4}, the 2s→3p resonance occurs just at the threshold, leading to the

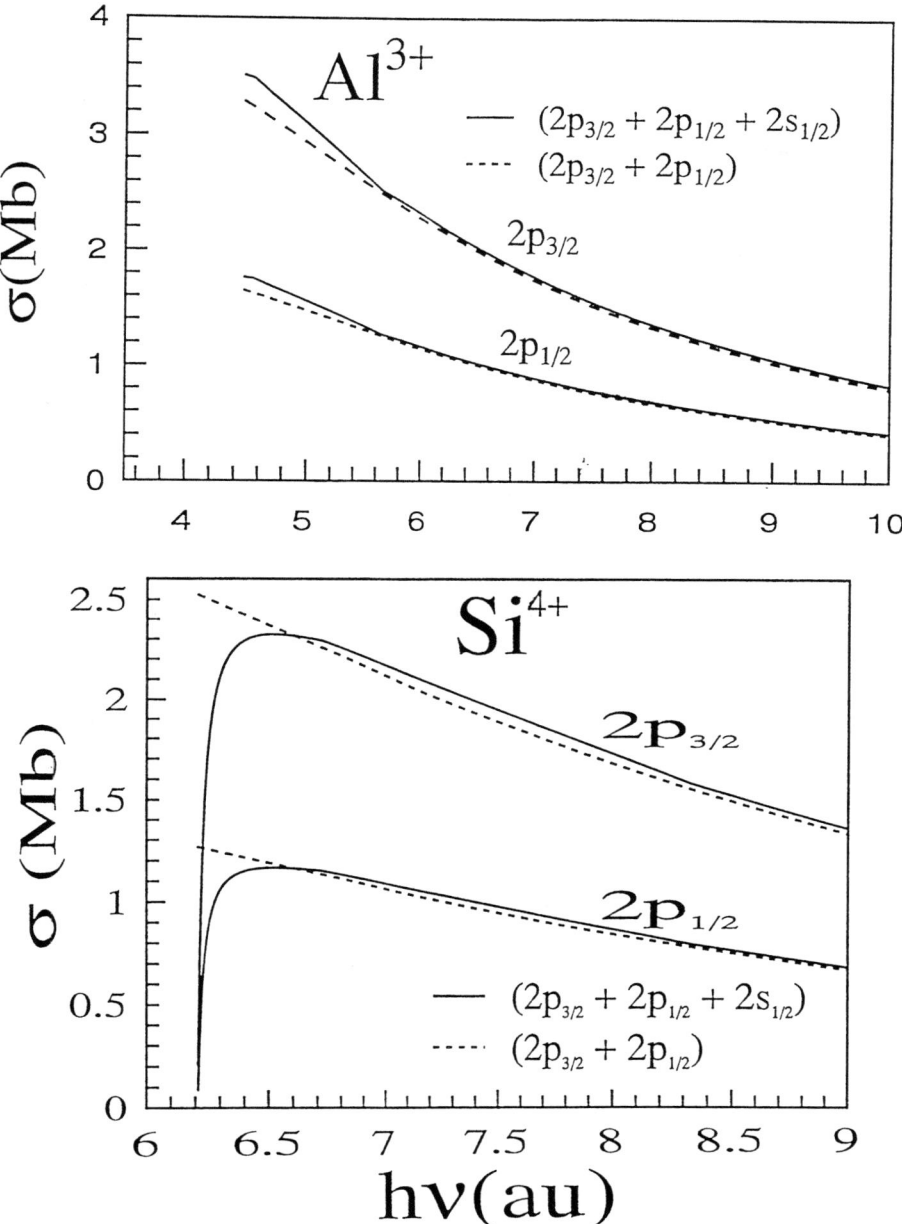

FIGURE 2. Photoionization cross sections for the outer $2p_{3/2}$ and $2p_{1/2}$ subshells of Al^{+3} and Si^{+4} above the $2p_{1/2}$ threshold, as a function of photon energy in au (27.21 Ev) calculated in RRPA. The solid lines are the full calculations and the dashed lines are with coupling to the 2s ionization channels omitted.

phenomenology seen in Fig.2. Similar phenomenolgy is seen as the higher 2s→np states move below the 2p thresholds at higher Z in the sequence, but none of the other cases are quite so dramatic as the Si^{+4} case. The importance of this phenomenology is that there is nothing special about the Ne isoelectronic sequence case; these anomalies should occur in the photoionization cross sections of any outer non-s subshell. Note finally, that there are preliminary experimental indications of this anomaly in Si^{+4} (14).

The next case involves the photoionization of the ground state of Sc^{++}, an ion of the first of the 3d transition metals. It is, in some sense, the third member of the K isoelectronic sequence, an alkali atom, but with a very important proviso; although the ground state structure of Sc^{++} is a single electron outside of a closed Ar-like core, it is not an s-electron like the alkali atoms, but a 3d electron. This is because as one moves from the neutrals to ions, the filling order of subshells becomes more hydrogen-like owing to the dominance of the nuclear attraction as compared to the interelectron repulsion (2). Since the ground state structure changes along the K isoelectronic sequence, it is evident that the photoionization of Sc^{++} will be radically different from K.

Calculations of the photoionization cross section of Sc^{++} have been carried out using many-body-perturbation theory (MBPT) for ground state correlations and a matrix diagonalization coupled channel technique, augmented by MBPT, to deal with the degenerate continuum and autoionizing channels (15). In addition, threshold energies and the energies of the important 3p→3d autoionizing resonances were obtained from extensive (500+ configurations) multiconfiguration Hartree-Fock (MCHF) calculation (16). Of course, the coupled channel technique had to be modified to insure that various interactions between channels were not double-counted (17). In the calculation, all single excitation channels from 3d, 3p and 3s were included.

A sample of the results is presented in Fig.3, where the complexity of the spectrum is evident. The cross section in the 40-60 eV region is dominated by autoionizing resonances leading up to the six $3p^53d$ thresholds; both $3p^53dnd$ and the $3p^53dns$ types contribute significantly to the spectrum. The $3p^53d^2$ resonances dominate the lower part of this spectrum, e.g., the resonance just below 42 Ev, as well as in the region below 40 eV (not shown). Although there is no experiment, the agreement between "length" and "velocity" formulations argue for the accuracy of these calculations. Further, note that the resonance positions are in substantial agreement with a calculation of the inverse process, electron - Sc^{+3} recombination (18), despite the fact that the two calculations were performed using completely different methodologies; the recombination calculation was performed using a modified version of the R-matrix technique (19). In any case, from Fig.3, it is quite clear that the cross section has not become particularly simple, even though a doubly ionized system has been considered.

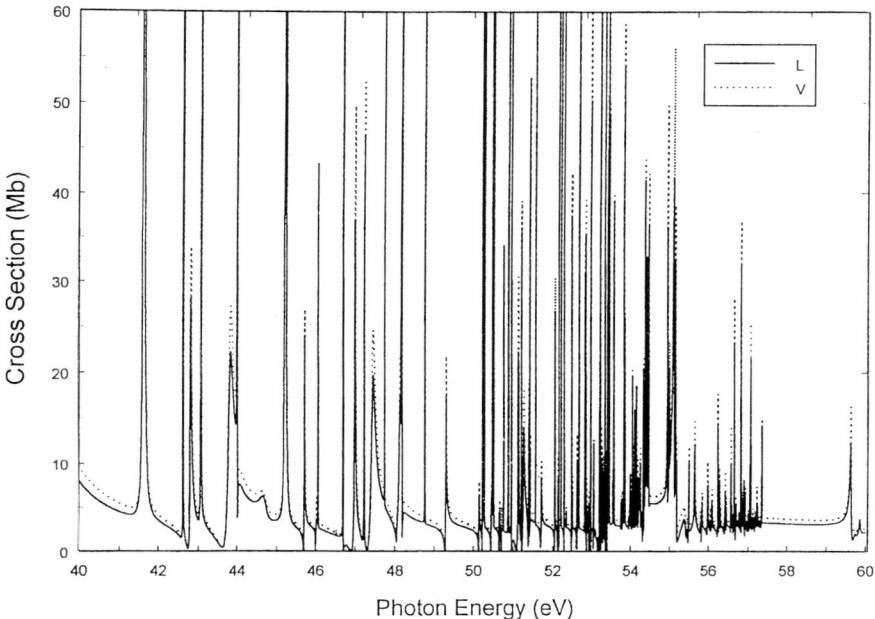

FIGURE 3. Photoionization cross section for the ground state of Sc^{++} in the 40-60 Ev region showing both "length" (solid line) and "velocity" dashed line results calculated as described in the text.

As a final example, the differential cross section for elastic electron-ion scattering is considered. The differential cross section for this process is important in plasmas for several reasons. Although the collision is elastic in the center-of-mass frame, there is energy transfer between the electron and the ion in the laboratory frame. The details of the energy transfer depends upon the relative velocity of the electron and ion, and the scattering angle--the larger the scattering angle the more energy is transferred. Thus, elastic scattering can be a very important energy transfer mechanism in dense plasmas. In addition, in laboratory measurements of differential *inelastic* electron-ion cross sections, it is often the case that it is difficult to separate the small inelastic signal from the large elastic component, so reliable estimates of the differential elastic cross section are quite useful (20).

Present models generally assume the Rutherford cross section, with the asymptotic charge is a reasonable approximation; detailed calculation shows that it is not. To deal with the situation, we have initiated a calculational program for differential elastic electron-ion scattering for the ground states of all ions up to Zn, Z=30 (21). The preliminary calculations, done at the central-field Hartree-Slater (22) level, are surprisingly accurate. As an example, the situation

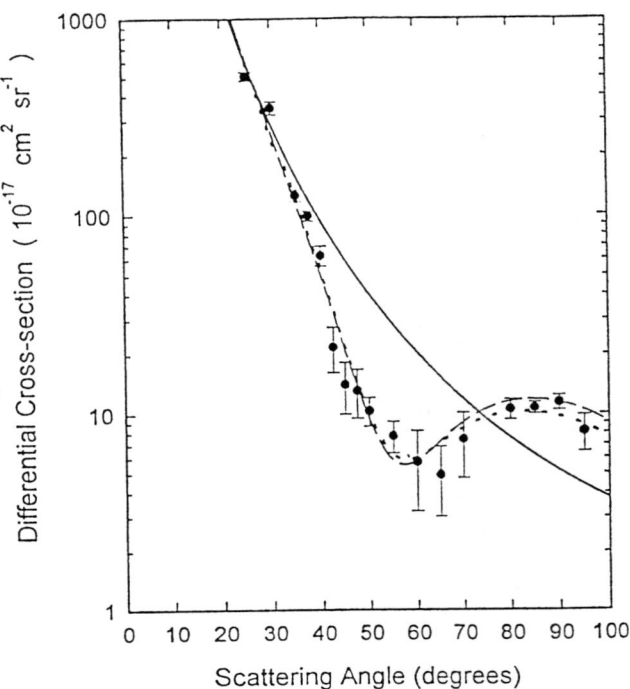

FIGURE 4. Differential elastic scattering cross section for electrons on Na^+. The points are the experimental data of [23], the dashed line is our theoretical result, and the solid line is the Rutherford cross section.

for electrons scattering from Na^+ is shown in Fig.4 where it is seen that the calculation and the experimental results (23) agree extremely well. Of equal importance is that the Rutherford prediction, with the asymptotic charge of unity, is in very poor agreement at the larger angles; it is off by almost an order of magnitude. In fact, comparing our results with the Rutherford predictions with the appropriate charge, it is found that Rutherford can differ by as much as *three orders of magnitude*, particularly in the backward direction (22). It is, thus, quite evident, that Rutherford, with the asymptotic charge, is not valid at the larger scattering angles.

It might be thought that since the electron is attracted to the ion, and penetrates the electron cloud, that the use of the Rutherford cross section with some effective charge between the asymptotic charge and the full nuclear charge, dependent upon angle and energy might be adequate. But even this is generally poor; it has been found through measurements of the "quasi-elastic" scattering which produced the so-called binary peak in ion-atom collisions, that at the backward angles, that the differential elastic scattering cross sections from partially stripped ions can be larger than cross sections arising from fully stripped ions, by factors of three or more (24). This completely negates the idea of an effective charge in the Rutherford formula. The phenomenon arises because it is the scattering amplitudes, not the cross sections, of the electron-nucleus and electron-electron interactions so that interference, both constructive and destructive, can occur. As a matter of fact, the oscillations in the cross section, seen in Fig.4, show that both do occur. Our results for backward scattering show a complex dependence upon nuclear charge, net charge, and energy with the largest cross section often being for the singly-charged ion. For example, for 68 eV electrons scattering from O^+, the back-scattering cross section is a factor of 3.29 larger than the Rutherford cross section for fully stripped O^{+8}. For O^{+7}, on the other hand, which presents a much larger effective charge than O^+, the cross section at 68 Ev is 0.91 times the stripped cross section. Thus, adding electrons, although it decreases the effective charge, increases the back-scattering cross section, emphasizing the quantum mechanical nature of the interferences which lead to the cross section. It is evident, then, that while a picture of the phenomenology of elastic electron-ion scattering is emerging, the deeper physical understanding of the interferences leading to the phenomenology is still developing.

ACKNOWLEDGEMENTS

This work was supported by the National Science Foundation and NASA.

REFERENCES

1. Theodosiou, C.E., Manson, S.T., and Inokuti, M., *Phys. Rev. A* **34**, 943 (1986).
2. Manson, S.T., Theodosiou, C.E., and Inokuti, M., *Phys. Rev. A* **43**, 4688 (1991).
3. Manson, S.T., in *The Physics of Electronic and Atomic Collisions. AIP Conference Proceedings 205*, ed. by Dalgarno, A., Freund, R.S., Koch, P.M., Lubell, M.S., and Lucatorto, T.B., (American Institute of Physics, New York, 1990), pp. 189-200.
4. Lahiri, J., and Manson, S.T., *Phys. Rev. A* **37**, 1047 (1988).
5. Fano, U., and Cooper, J.W., *Rev. Mod. Phys.* **40**, 441 (1968).
6. Rao, A.R.P., and Fano, U., *Phys. Rev.* **167**, 7 (1970).
7. Lahiri, J., and Manson, S.T., *Phys. Rev. A* **33**, 3151 (1986).
8. Wane, S., and Aymar, M., *J. Phys. B* **20**, 2657 (1987).
9. Msezane, A.Z., and Manson, S.T., *Phys. Rev. Lett.* **48**, 473 (1984).
10. Lahiri, J., and Manson, S.T., *Phys. Rev. Lett.* **48**, 614 (1984).
11. Johnson, W.R., and Lin, C.D., *Phys. Rev. A* **20**, 964 (1979).
12. Johnson, W.R., Lin, C.D. Cheng, K.T., and Lee, C.M., *Phys. Scripta* **21**, 409 (1980).
13. Johnson, W.R., and Cheng, K.T., *Phys. Rev. A* **20**, 978 (1979).
14. Kennedy, E.T., private communication (1998).
15. Kelly, H.P., *Adv. Chem Phys.* **14**, 129 (1969).

16. Fischer, C.F., *Computer Phys. Commun.* **64**, 145 (1991).
17. Altun, Z., and Manson, S.T., *Europhys. Lett.* **33**, 17 (1996).
18. Gorczyca, T.W., Pindzola, M.S., Robicheaux, F., and Badnell, N.R., *Phys. Rev. A* **56**, 4742 (1997).
19. Berrington, K.A., Eissner, W.B., and Norrington, P.H., *Computer Phys. Commun.* **92**, 290 (1995).
20. Chutjian, A., private communication, 1997.
21. Turner, C.S., and Manson, S.T., in *XX. ICPEAC. Scientific program and Abstracts of Contributed Papers*, ed. by Aumayr, F., Betz, G., and Winter, HP., (ICPEAC, Vienna, 1997), Vol. II, p. MO155.
22. Herman, F., and Skillman, S., *Atomic Structure Calculations*, (Prentice-Hall, Englewood Cliffs, NJ, 1963).
23. Srigengan, B, Williams, I.D., and Newell, W.R., *Phys. Rev. A* **54**, R2450 (1996).
24. Hidmi, H., Richard, P., Sandfers, J.M., Schone, H., Giese, I.P., Lee, D.H., Zourous, T.J.M., and Varghese, S.L., *Phys. Rev. A* **48**, 4421 (1993).

Angle-resolved 2D Imaging of Electron Emission Processes in Atoms and Molecules

E. Kukk*,[†], A.A. Wills*, B. Langer[‡], J.D. Bozek[†], and N. Berrah*

* *Department of Physics, Western Michigan University, Kalamazoo, MI 49008-5151*
[†] *Lawrence Berkeley National Laboratory, University of California, Berkeley, CA 94720*
[‡] *Fritz-Haber-Institut der Max-Planck-Gesellschaft, Faradayweg 4-6, D-14195, Berlin, Germany*

Abstract. A variety of electron emission processes have been studied in detail for both atomic and molecular systems, using a highly efficient experimental system comprising two time-of-flight (TOF) rotatable electron energy analyzers and a 3rd generation synchrotron light source. Two examples are used here to illustrate the obtained results. Firstly, electron emission in the HCl molecule have been mapped over a 14 eV wide photon energy range over the Cl 2p ionization thresholds. Particular attention is paid to the dissociative core-excited states, for which the Auger electron emission shows photon energy dependent features. Also, the evolution of resonant Auger to the normal Auger decay distorted by post collision interaction has been observed and the resonating beahavior of the valence photoelectron lines studied. Secondly, an atomic system, neon, in which excitation of doubly excited states and their subsequent decay to various accessible ionic states has been studied. Since these processes only occur via inter-electron correlations, the many-body dynamics of an atom can be probed. Electron angular distributions following the decay of certain resonances to a parity-unfavored continuum exhibits significant deviation from LS coupling predicitons, which is surprising for such a light atom.

INTRODUCTION

The electron emission processes induced by VUV radiation have been studied extensively for relatively simple systems such as rare gas atoms and diatomic molecules. Electron spectroscopy of free atoms and molecules has proven to be a valuable tool for testing theoretical models of electronic structure, dynamics and of many-body effects. Rapid development of experimental systems and excitation sources also continues to reveal novel and often surprising effects and details in atomic and molecular electron spectra. The results presented here benefitted from the combined use of 3rd generation synchrotron radiation source (Advanced Light Source at Lawrence Berkeley National Laboratory), highly efficient electron time-of-flight analyzers capable of recording angularly resolved data, and the use of sophisticated two-dimensional (2D) data acquisition and analysis software [1]. The 2D imaging technique - electron emission intensity recorded as a function of the energies of both the exciting photons and emitted electrons - has proven to be an excellent tool for studying processes that are spread over broad ranges of excitation energies and/or kinetic energies of the emitted electrons. Two examples of such processes will be discussed below: (i) Auger decay of $2p$ core excitations in the HCl molecule, especially in competition with molecular dissociation and (ii) excitation of doubly excited states in neon and their subsequent decay to various ionic states, revealing breakdown of LS coupling.

EXPERIMENT

The electron time-of-flight (TOF) spectra were measured at beamline 9.0.1 of the Advanced Light Source synchrotron radiation facility, which uses radiation from a 8-cm 55-period undulator, monochromatized by a spherical grating monochromator. Two TOF analyzers [2] were mounted in a plane perpendicular to the photon beam propagation direction and were housed in a chamber rotatable around the photon beam (see Fig. 1). This setup allowed angle-resolved (with respect to the electric vector of the light) measurements simultaneously at two angles, 125.3° apart, and at a different pair of angles after rotating the chamber.

Each TOF analyzer consists of a retardation/acceleration stage and a long drift tube with a pair of microchannel plates and a detector plate mounted at the end.

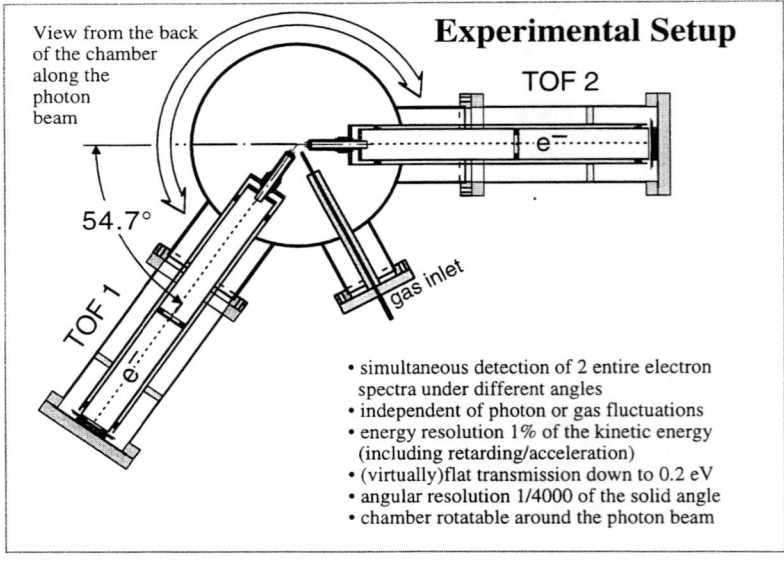

FIGURE 1. Schematic diagram of the experimental setup.

To measure flight times of the electrons in the drift tube, the synchrotron source operated in a 'double-bunch' mode, delivering short bursts of photons in 328 ns intervals, during which the emitted electrons had time to reach the detectors. In order to increase the flight times and thus improve the energy resolution, the electrons can be retarded prior to entering the field-free drift tube. The time scale of the original spectra was then converted to the kinetic energy scale using a procedure described in Ref. [3].

AUGER DECAY OF CORE-EXCITED STATES IN THE HCL MOLECULE

Overview

There have been numerous studies of inner shell processes in small molecules during recent years (*e.g.* Refs. [4–11]. These studies have, however, been performed at only a few selected photon energies. The results presented here give a more complete picture, since the electron emission spectra from HCl following photoexcitation/ionization of the Cl $2p$ core levels were investigated by recording angle-resolved 2D data sets over a 14 eV wide photon energy range in a kinetic energy window covering the majority of the ejected electrons. The photon energy was scanned in 20-meV steps using 75-meV resolution. The resulting 2D images illustrate the evolution of the electron emission processes as the photon energy is scanned towards and across the Cl $2p$ ionization thresholds.

The absorption spectrum of HCl around the Cl $2p$ ionization threshold [12] can be divided into three regions: i) excitation to the antibonding σ^* orbital ($2p \to \sigma^*$), ii) excitation to the Rydberg orbitals ($2p \to nl$) and iii) ionization into the continuum ($2p \to \epsilon l$). These three regions are indicated on the total electron yield curve along the right side of Fig. 2. The 2D electron emission map, shown in the main panel of Fig. 2, was obtained at the angle of 90° relative to the polarization plane. In this map, intensity of the electron signal is represented as different shades of gray. Electrons emitted with the same kinetic energy at different photon energies are aligned vertically in the 2D map. Those emitted with constant binding energy form diagonal lines across the map. The features in the different regions of the 2D map are discussed in more detail below.

Auger decay of dissociating states

Broad bands in the total electron yield between 199.0 and 203.5 eV correspond to excitation of the $2p$ electrons to the first unoccupied molecular orbital, σ^*. When occupied, this antibonding orbital causes the core-excited molecule to dissociate. The dissociation,

$HCl^*(2p^5\sigma^*) \to H(1s) + Cl^*(2p^5 3p^6)$,

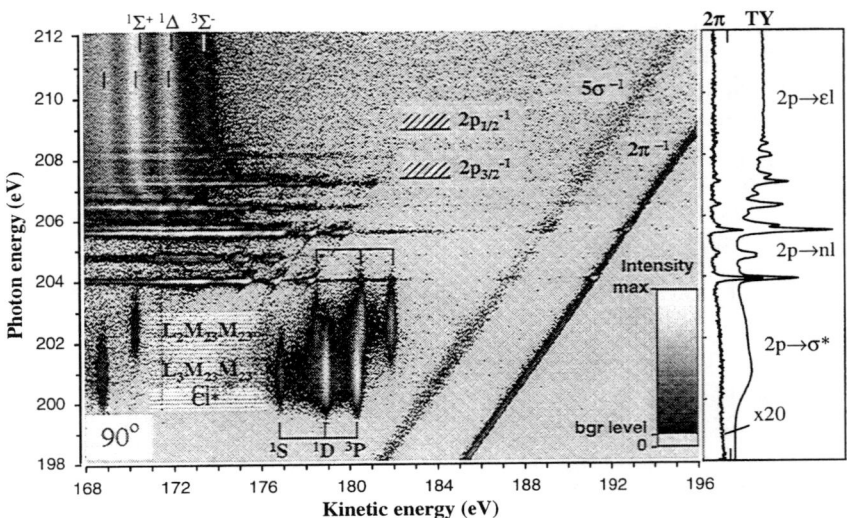

FIGURE 2. Two-dimensional (2D) map of electron emission from the HCl molecule across the Cl 2p ionization threshold, taken at 90° relative to the polarization plane. The total electron and partial 2π photoelectron yields (marked as TY ans 2π, respectively) are shown on the right.

produces neutral hydrogen and core-excited chlorine atoms and proceeds on a time scale ($\approx 10^{-14}$ s) that competes with the Auger decay of the core hole [13,14]. The light hydrogen atom moves quickly away from the Cl atom, so that a large part of the Auger decay occurs in the atomic chlorine as $2p^5 3p^6 \rightarrow 3p^4$ transitions, forming intense vertical lines in the 2D map at the photon energies of the $2p \rightarrow \sigma^*$ excitation, as can be seen in Fig. 2 and in more detail in Fig. 3. These sharp atomic lines have pronounced tails at their low kinetic energy side arising from Auger transitions in the molecular environment prior to the complete dissociation, which form a 'molecular background' to the atomic Auger spectrum [14].

As the photon energy is increased across the breadth of the σ^* band, HCl molecules with decreasing internuclear separation are selected due to the projection of the ground state population onto the dissociative potential energy curve. This shift may influence the time scale of the early stage of the dissociation, which should be reflected in the balance between the molecular background and the sharp atomic peaks in the Auger spectra. In order to investigate this possibility, the intensity of one atomic line and its associated background can be monitored in the 2D maps as a function of photon energy. The $2p_{1/2}^{-1} 3p^6 \rightarrow 3p^4(^3P)$ atomic line at 182.0-eV kinetic energy was selected, since it is well separated from neighboring peaks. The molecular tail of this line is assigned to transitions to the $^4\Sigma^-$, $^2\Sigma^-$ and $^4\Pi$ states of HCl^+ [14]. The distinction between the atomic (marked 'A' in Fig. 3 and molecu-

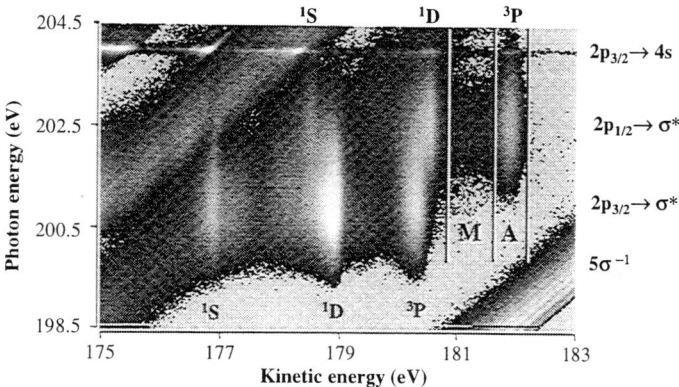

FIGURE 3. 2D map of Auger electron emission at the $2p^{-1}_{1/2,3/2}\sigma^*$ resonances, taken at 35°.

lar ('M') regions is somewhat arbitrary. Here, we represent the latter by choosing the part of the spectrum between the ^1D and ^3P lines, from 180.8 to 181.8 eV. The comparison of the two regions shows that the molecular part of the Auger spectrum dominates at photon energies well below the absorption peak maximum. With increasing photon energy, the internuclear distance at which the excitation can take place shifts to smaller values, where the energy of the core excited state is higher. Here the atomic peak gains intensity over the molecular background, inferring a higher dissociation rate. This effect can be related to the slope of the potential energy curve of the excited state, as discussed in Ref. [15]. Another novel effect connected to the Auger decay of the σ^* resonaces can be observed in Fig. 3. The tail of the $2p^{-1}_{1/2}\sigma^*$ resonance overlaps with the first Rydberg resonance, $2p^{-1}_{3/2}4s$, at the 204.0 eV photon energy. The intensity of the Auger lines from the decay of the dissociating $2p^{-1}_{1/2}\sigma^*$ state shows a sharp increase at the $2p^{-1}_{3/2}4s$ resonance. At the magic angle (54.7°), these Auger lines account for about 6% of the total Auger intensity at this photon energy. About half of their intensity comes from the overlap with the Gaussian-shaped tail of the $2p^{-1}_{1/2}\sigma^*$ resonance. The remainder is a resonant enhancement indicating the possibility of predissociation on the first Rydberg resonance. The transition from the bound $2p^{-1}_{3/2}4s$ to dissociative $2p^{-1}_{1/2}\sigma^*$ potential energy curve must involve some interaction between the two spin-orbit components of the $2p$ orbital.

Auger decay of Rydberg resonances and core-ionized states

A quite different pattern of Auger decay can be seen in the region of the sharp core-to-Rydberg excitations starting at the 204.0 eV photon energy. The spectral structures here arise from molecular Auger decay and are assigned within the simplest spectator model as the $2p^{-1}nl \to (4\sigma 5\sigma 2\pi)^{-2}n'l'$ transitions, where nl and $n'l'$ describe the Rydberg electron in the Auger initial and final state, respectively [16]. The intensity of the Auger spectra of the higher Rydberg states gradually shifts towards lower kinetic energies with the increasing photon energy. This is due to the weakening of the influence of the spectator electron and higher shake-up probability. Finally, the $2\pi^{-2}n'l'$ resonant Auger lines merge into the $2\pi^{-2}(^3\Sigma^-,^1\Delta,^1\Sigma^+)$ normal Auger lines. There is no sudden change in the Auger decay pattern at the $2p$ ionization thresholds (marked in Fig. 2, from Ref. [12]). Rather there is a smooth evolution from the spectator to normal Auger spectra through a strong post-collision interaction (PCI). The energy shift of the normal Auger lines, caused by the PCI between the outgoing slow $2p$ photoelectron and Auger electron, can be seen through threshold and down to the Rydberg states. The shift below the threshold is caused by the screening effect of the bound, core-excited electron.

Anisotropy of the electron emission

Within the dipole approximation, the differential cross-section of the electron emission at an angle θ with respect to the polarization plane is given by

$$\frac{d\sigma_{i \to f}(\theta)}{d\Omega} = \frac{\sigma^T_{i \to f}}{4\pi}(1 + \beta P_2(\cos\theta)), \qquad (1)$$

where $\sigma^T_{i \to f}$ is the total cross-section, Ω is the solid angle and $P_2(\cos\theta)$ is the second-order Legendre polynomial. The anisotropy parameter β id derived from the shape of the initial and final state wavefunctions of the photoelectron in the case of direct ionization. The β-parameter for the Auger electrons is given as a product of a molecular alignment parameter β_m and an intrinsic anisotropy parameter c_a for the Auger decay [17]:

$$\beta = \beta_m c_a. \qquad (2)$$

2D maps similar to Fig. 2 were also measured at 0, 35.3 and 54.7 degrees. From these data sets the normalized intensities and β-parameters of the 2π and 5σ photoelectron lines can be followed continuously as a function of photon energy. Fig. 4 shows the results for the 2π photoelectron line. Below the Rydberg resonances, it has an almost constant β-value of 1.6, but it drops to 0.90 at the first Rydberg resonance. However, β of the 5σ line shows no such pronounced change. At the next strong (overlapping) resonances $2p^{-1}_{3/2}3d$ and $2p^{-1}_{1/2}4s$, the value of β for the 2π line drops again to 1.15. These changes in the value of β indicate that a resonant valence ionization channel becomes available. This is consistent with the sudden increase in the photoelectron yield at these Rydberg resonances, probably due to participator Auger transitions to these ionic states. Such an effect at Rydberg resonances has not been observed for other diatomic molecules like CO [11].

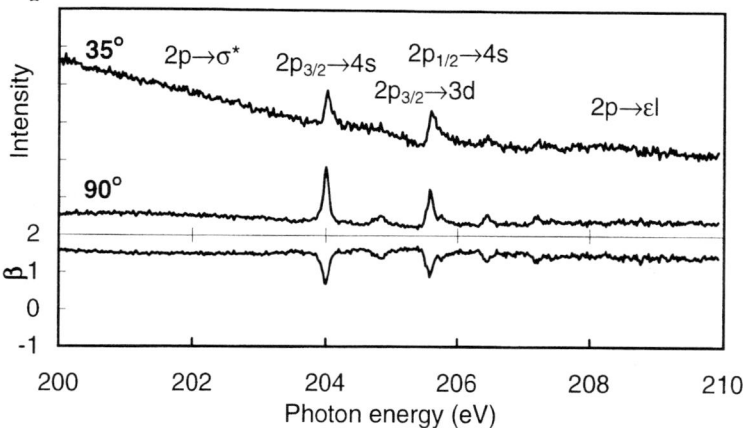

FIGURE 4. 2π photoelectron yield at different angles (upper panel) and the anisotropy parameter (lower panel) as a function of photon energy.

The β-values of the individual Auger lines display large variations (between -0.7 and 0.7) in the spectra measured at the $2p^{-1}\sigma^*$ resonances, but show a markedly more uniform behavior at the Rydberg resonances. At the $2p_{3/2}^{-1}4s$ resonance, all Auger lines have negative β-values. The $2\pi^{-2}(^1\Delta)4s$ line at $E_b=27.3$ eV has the largest anisotropy with $\beta=-0.7(2)$, which infers that the molecule is strongly aligned after the $2p_{3/2} \to 4s$ excitation. We also observed negative β-values at the overlapping $2p_{3/2}^{-1}3d$, $2p_{1/2}^{-1}4s$ resonances. All of the molecular Auger lines have similar β-values and for the the normal Auger lines, β-parameters are equal within the error limits, since there is no change in the branching ratios of the Auger lines at different angles.

SPIN-ORBIT EFFECTS IN PARITY UNFAVORED PHOTOIONIZATION OF NEON

The purpose of this study was to use high-precision measurements in order to reveal details in the dynamics of doubly-excited states. We were able to demonstrate that for certain resonances, the angular distribution of photoelectrons and the ratio of partial cross sections to individual fine structure levels both exhibit behavior that deviates markedly from LS coupling predictions. These observations provide further evidence for the breakdown of LS coupling. Particularly interesting from this viewpoint are the $2p^43pns, nd$ doubly-excited resonances in the photoionization-excitation to the $2p^4(^3P)3s(^2P)$ Ne$^+$ ionic state. These were examined using high-resolution 2D imaging of photoionization cross sections that are differential in photon energy, photoelectron energy, and photoemission angle.

The $2p^43pns, nd$ resonances are particularly strong because the ~10% mixing of the $2p^53p$ configuration in the $2p^6$ ground state leads to the one-electron photoexcitation $2p^53p \to 2p^43pns, nd$ [18]. Breakdown of LS coupling for this light system is due to the spin-orbit interaction, which most likely causes mixing between certain LS-allowed $2p^43pns, nd(^1P_1)$ resonances and their $^3S_1, ^3P_1$, and $^3D_1$ LS-forbidden counterparts. Angle-resolved and level-resolved measurements have provided an insight into such spin-orbit effects.

Fig. 5 displays the photoelectron spectra at two angles (0 and 54.7 degrees). For this measurement, the photon energy resolution was set to 20 meV, which is a significant improvement over the previous resolution of 100 meV in earlier differential measurements [19].

LS-forbidden features in the angular distribution of the $2p^43s(^2P, ^4P)$ photoionization satellites

A useful way of viewing the underlying dynamics of the differential cross-section (Eq. 1) is to consider the angular momentum transfer $\vec{j}_t$ [20]. The differential cross section reduces to an incoherent sum of terms associated with the allowed values of j_t, which are $\ell - 1, \ell$ and $\ell + 1$. While the *parity-favored* contributions $j_t = \ell \pm 1$ have complicated, energy-dependent angular distributions in general, the partial differential cross sections for all *parity unfavored* transitions $j_t = \ell$ have the analytic property $\left(\frac{d\sigma}{d\Omega}\right)_{j_t=\ell} \propto \sin^2\theta$ independent of energy, or equivalently, $\beta_{unf} = -1$ [21]. One important consequence is that the parity-unfavored contribution to the differential cross section vanishes at $\theta = 0°$.

For photoionization from the $2p^6(^1S)$ ground state to the $2p^43s(^2P)\epsilon p(^1P)$ continuum, the angular momentum transfer is restricted to a single, parity-unfavored value $j_t = \ell = 1$. This means that, in LS coupling, this transition is completely parity unfavored and the differential cross section vanishes at $\theta = 0°$, or equivalently, detection of photoelectrons at $0°$ is an unmistakable indication of spin-orbit effects.

It is easy to observe the above-mentioned LS-forbidden features in Fig. 5. Firstly, the vertical lines (b) at 48.8 eV binding energy (corresponding to the $2p^43s(^4P)$ satellite) show, on certain resonances, noticeable photoelectron yields at both angles. Secondly, by comparing the photoelectron yields at 54.7° and 0° along the vertical lines (c) (corresponding to the $2p^43s(^2P)$ satellite at 49.4 eV), it can be seen that several resonances appear at both angles while the others disappear in the 0° spectrum. For instance, the $2p^4(^3P)3p(^2P)3d$ resonance at (photon energy) 51.3 eV shows a photoelectron yield that is almost equally strong at 0° as it is at 54.7°, whereas the $2p^4(^1D)3s(^2D)4p$ resonance at 50.6 eV disappears almost completely in the 0° spectrum. Since this is a parity unfavored transition in LS coupling, the appearance of any signal at 0° is an immediate indication of the breakdown of LS coupling. The appearance of only certain resonances in the photoelectron yield at 0° suggests that spin-orbit mixing between singlet and triplet resonances, not between $^1P_1$ continuum states, is responsible for these LS-forbidden observations.

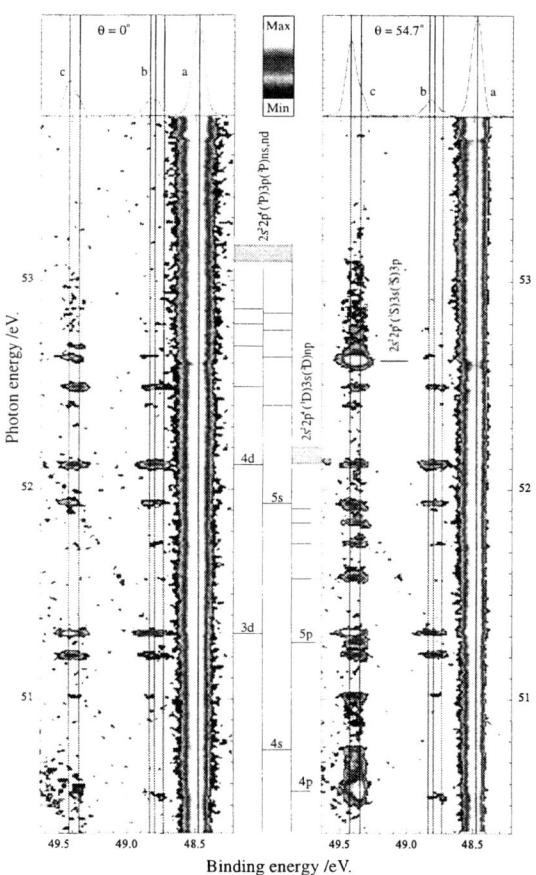

FIGURE 5. Photoelectron yield as a function of photon energy and binding energy at 0° and 54.7°. The upper graphs show the spectrum at 51.3 eV photon energy, and the vertical bars indicate the positions of the a) $2s2p^6(^2S)$, b) $2s^22p^4(^3P)3s(^4P)$, and c) $2s^22p^4(^3P)3s(^2P)$ fine-structure levels.

Deviations of the ratio of the $2p^43s(^2P_{3/2})$ and $2p^43s(^2P_{1/2})$ satellite lines from its LS coupling value

The high resolution in photoelectron energy allows easy identification of one additional LS-forbidden feature. By observing the photoelectron intensity along the two vertical lines (c) in Fig. 5 corresponding to the $2p^43s(^2P_{3/2})$ and $2p^43s(^2P_{1/2})$ satellites (at 49.35 eV and 49.42 eV binding energies, respectively), it can clearly be seen that some resonances such as $2p^4(^1D)3s(^2D)4p$ preferentially autoionize to the $^2P_{3/2}$ ionic state, yielding a ratio $r > 2$, while others such as $2p^4(^3P)3p(^2P)3d$ preferentially autoionize to the $^2P_{1/2}$ state, yielding a ratio $r < 1$. As a more quantitative diagnostic, the photoelectron yield along each of the two fine-structure-split vertical lines (c) in Fig. 5 can be summed to give partial cross sections to the 3/2 and 1/2 levels. These indicate clearly, which resonances exhibit LS-forbidden ratios $r \neq 2$. In particular, the $2p^4(^1D)3s(^2D)4p$ and $2p^4(^3P)3p(^2P)3d$ resonances show ratios of $r = 4.3$ and $r = 0.2$, respectively. This is additional evidence for the breakdown of LS-coupling.

CONCLUSIONS

High-resolution and high-brightness VUV radiation, combined with high performance experimental systems, has allowed detailed studies of some aspects of the dynamics of a molecule, HCl, and of an atom, Ne. We hope that our measurements will stimulate further theoretical work in these areas.

ACKNOWLEDGEMENTS

This work was supported by the US Department of Energy, Office of Basic Energy Science, Division of Chemical Science under contract No. DE-FG02-95ER14299.

REFERENCES

1. A.A. Wills, D. Čubrić, M. Ukai, F. Currell, B.J. Goodwin, T. Reddish, and J. Comer, J. Phys. B **26**, 2601 (1993).
2. B. Langer, A. Farhat, B. Nessar, N. Berrah, O. Hemmers, and J.D. Bozek, in *Proceedings of the Atomic Physics with Hard X-rays from High Brilliance Synchrotron Light Sources Workshop*, ANL/APS/TM-16 (Argonne National Laboratory, Argonne, IL, 1996), p. 245; A. Farhat, M. Humphrey, B. Langer, N. Berrah, J.D. Bozek, D. Cubaynes, Phys. Rev. A **56**, 501 (1997).
3. N. Berrah, B. Langer, J.D. Bozek, T.W. Gorczyca, O. Hemmers, D.W. Lindle, and O. Toader, J. Phys. B **29**, 1 (1996).
4. U. Becker, R. Hölzel, H.G. Kerkhoff, B. Langer, D. Szostak, and R. Wehlitz, Phys. Rev. Lett. **56**, 1455 (1986).

5. O. Hemmers, F. Heiser, J. Eiben, R. Wehlitz, and U. Becker, Phys. Rev. Lett. **71**, 987 (1993).
6. U. Hergerhahn and U. Becker, J. Electron Spectrosc. **72**, 243 (1995).
7. U. Becker and A. Menzel, Nucl. Instr. Meth. in Physics B **99**, 68 (1995).
8. H. Aksela, E. Kukk, S. Aksela, O.-P. Sairanen, A. Kivimäki, E. Nõmmiste, A. Ausmees, S.J. Osborne, S. Svensson, J. Phys. B **28**, 4259 (1995).
9. M. Neeb, M. Biermann, and W. Eberhardt, J. Electron Spectrosc. **69**, 239 (1994).
10. A. Kivimäki, M. Neeb, B. Kempgens, H.M. Köppe, A.M. Bradshaw, Phys. Rev. A **54**, 2137 (1996).
11. S. Sundin, S.J. Osborne, A. Ausmees, O. Björneholm, S.L. Sorensen, A. Kikas, and S. Svensson, Phys. Rev. A **56**, 480 (1997).
12. D.A. Shaw, D. Cvejanović, G.C. King, and F.D. Read, J. Phys. B **17**, 1984 (1984).
13. P. Morin and I. Nenner, Phys. Rev. Lett. **56**, 1913 (1986).
14. E. Kukk, H. Aksela, O.-P. Sairanen, S. Aksela, A. Kivimäki, E. Nõmmiste, A. Ausmees, A. Kikas, S.J. Osborne, and S. Svensson, J. Chem. Phys. **104**, 4475 (1996).
15. E. Kukk, A. Wills, N. Berrah, B. Langer, J.D. Bozek, O. Nayadin, M. Alserhi, A. Farhat, and D. Cubaynes, Phys. Rev. A **57**, R1485 (1998).
16. E. Kukk, H. Aksela, O.-P. Sairanen, E. Nõmmiste, S. Aksela, S.J. Osborne, A. Ausmees, and S. Svensson, Phys. Rev. A **54**, 2121 (1996).
17. D. Dill, J.R. Swanson, S. Wallace, and J.L. Dehmer, Phys. Rev. Lett. **45**, 1393 (1980).
18. K. Schulz et al., Phys. Rev. A **54**, 3095 (1996).
19. A. A. Wills et al., J. Phys. B **23**, 2013 (1990).
20. U. Fano and D. Dill, Phys. Rev. A **6**, 185 (1972).
21. D. Dill and U. Fano, Phys. Rev. Lett. **29**, 1203 (1972).

ATOMIC PROCESSES IN MAGNETIC FUSION PLASMAS

Study of Volume Recombination and Radiation Opacity Effects in Alcator C-Mod

J.L. Terry, B. Lipschultz, A.Yu. Pigarov[+,*], C. Boswell, S.I. Krasheninnikov[+], B. LaBombard, and D.A. Pappas

Plasma Science and Fusion Center, Massachusetts Institute of Technology, Cambridge, Massachusetts, 02139
[+] *Also at I.V. Kurchatov Institute of Atomic Energy, Moscow, Russian Federation*
[*] *Also at The College of William and Mary, Williamsburg, VA*

Abstract. Observations of significant volume recombination within the Alcator C-Mod divertor plasma and in the edge plasma (MARFE) are described. The recombination occurs in regions where $T_e \lesssim 1$ eV and $n_e \gtrsim 1 \times 10^{21}$ m^{-3}. The determinations of the recombination rates are made by measuring the D^0 Lyman and/or Balmer spectra and by using a collisional radiative model describing the level populations, ionization and recombination of D^0. In regions of strong recombination the upper levels ($n \gtrsim 4$) populations are close to those determined by Saha-Boltzmann distribution and are independent of the ground state density. Thus the intensities of lines from these levels are related to the recombination rate, and curves determining the number of 'recombinations per photon' are calculated. Ly$_\beta$ line emission is shown to be trapped in some cases, meaning that Ly$_\alpha$ can be strongly trapped. Since opacity affects the recombination rates, the effects of the trapping of Ly$_{\alpha,\beta}$ photons on the 'recombinations per photon' curves are calculated and considered in the recombination rate determinations. Total recombination rates in the detached divertor plasma and in MARFEs located at the periphery of the main plasma are determined. Recombination can be a significant sink for ions.

INTRODUCTION

Volume recombination is that process by which ions and electrons recombine to form atoms while within the plasma. Tokamaks are devices which create plasmas hot enough (multi-keV) for fusion reactions to occur. It is therefore somewhat surprising that significant volume recombination occurs at the peripheries of these devices. For some time it was generally thought that the ions and electrons in the plasma edge or Scrape Off Layer (SOL) which reach 'open' field lines (those ending on material surfaces, e.g. the divertor plates) recombine only after reaching those surfaces. The heat and particle loads on the divertor plates due to these flows can be enormous in such a scenario. In the Alcator C-Mod Tokamak heat flows in the SOL as high as 500 MW/m^2 are measured. This is roughly only an order of magnitude less than the heat flux from the tip of an arc welder! Thus the approach described above, where the heat is not dissipated before reaching the plates, is simply not tenable for a magnetic fusion reactor, and the search to find ways of dissipating the SOL heat flux is a major part of magnetic fusion research. Dissipative schemes, such as the radiative divertor and the detached divertor [1,2], have been successful in lowering the temperatures at the divertor plates to $\lesssim 3$ eV [3,4]. Stimulated by the achievement of these low temperatures there has been much interest in studying experimentally whether volume recombination is occurring and, if so, in studying its significance.

In a very general sense, if volume recombination fully removed the ion flux to divertor plates, a condition of complete insulation of plasma from the material walls of its containment vessel would have been achieved. Such a condition is significant, since it means that a multi-keV plasma has been effectively insulated from the walls which surround it, and its enormous heat efflux has been radiated away and distributed more uniformly on the surrounding surfaces. In a more specific sense, the 'insulation' provided by volume recombination results in a reduced ion flux to the divertor plates, which is accompanied by an even greater reduction in energy flux, since the 13.6 eV of ion-electron potential energy is released away from the plates. The reduced ion flux also leads to a reduction in the impurity source from these surfaces through reduced sputtering and evaporation.

Volume recombination has now been observed in a number of tokamaks. Computer modelling of these plasmas has also shown the importance of recombination and in many cases has motivated the experimental investigations. (See Ref. [5] for a more extensive listing of experimental and modelling references.) The volume recombination typically occurs in the divertor region of a tokamak. However, recent measurements on Alcator C-Mod[6] have shown that very large volume recombination rates can occur in the edge of the main chamber plasma under some circumstances when a phenomenon called a MARFE[7] occurs. This phenomenon will be described briefly later.

It is the purpose of this paper to describe some of the observations of volume recombination which have been made on Alcator C-Mod and to describe the atomic processes which both influence the phenomenon and allow us to diagnose it. In particular this paper will demonstrate that recombination can be a significant sink for ions, will present an analysis technique which simplifies the evaluation of the total 3-body and radiative recombination rate by relating the number of recombinations to the number of photons in a given Balmer/Lyman series line, and will present measurements of electron temperature (T_e), electron density (n_e), and the opacity of Ly_β.

VOLUME RECOMBINATION PATHWAYS AND THE ALCATOR C-MOD PLASMAS

Volume recombination of deuterons (or protons) can occur in three ways:

- radiative recombination $D^+ + e \rightarrow D^0 + h\nu$,

- 3-body recombination: $D^+ + e + e \rightarrow D^{0*} + e$ (where D^{0*} means that the atom is typically in an excited state), and

- Molecular Activated Recombination (MAR) [8,9], whose multiple pathways require the presence of vibrationally excited hydrogenic molecules, $D_2(\nu)$.

The first process is important for diagnostic purposes, but does not represent a significant ion sink. The latter two processes are predicted to be important at temperatures below ~2 eV and in the 1-3 eV range, respectively. Both 3-body recombination and MAR typically leave a recombined atom in an excited state, which can subsequently either decay to the ground state or re-ionize. In most of the cases of relevance to tokamaks, the decay to the ground state will occur by photon emission. For this reason, the study of recombination is possible through the spectroscopy of hydrogenic line emission. It also requires that a distinction be made between those recombinations which end with an atom in the ground state, something we will call a 'complete' recombination, and those which re-ionize. Only 'complete' recombinations will be significant in the ionization/recombination balance in tokamaks. MAR will not be discussed further in this paper. For further details about MAR see Refs. [9,5].

The discharges created in the Alcator C-Mod Tokamak typically have high (for tokamaks) core densities, $\bar{n}_e \sim 3 \cdot 10^{20}$ m^{-3}, plasma currents up to 1.5 MA, and central temperatures up to 5 keV. They are heated ohmically in combination with up to 4 MW of ICRF heating power. (A description of the general machine operation can be found elsewhere [10].) The edge and divertor regions of non-circular tokamaks like Alcator C-Mod are typically toroidally symmetric, but are otherwise asymmetric. As a result of the toroidal symmetry, the relevant features of the device can be shown in cross-section as in Fig. 1. The Molybdenum armor tiles, a magnetic reconstruction of a typical diverted, lower single-null discharge, some of the different diagnostic views used in this study, and the measured 2D distribution of D_γ emission during divertor detachment are shown. An image of the inside of the tokamak's vacuum vessel, illuminated by D_α light from the plasma, is shown in Fig. 2. For the determination of the 'complete' recombination rates in Alcator C-Mod, measurements of the Balmer and/or Lyman series lines of D^0 are used. Because the Balmer lines occur in the visible and the Lyman lines appear in the vacuum-ultraviolet, both visible and VUV spectrometers have been used. The visible spectrometer accepts multiple views using fibers, while the field-of-view of the VUV spectrometer (shown in Fig. 1) can be scanned vertically. One of the visible views has been aligned to view along the VUV instrument's sight line and to have nearly the same footprint[11]. Additional D_α measurements are made using two interference-filter-based diode arrays[12], one viewing (from the outside midplane) the vertical extent of the plasma with $\sim$ 4 cm spatial resolution, the other viewing the divertor region from above.

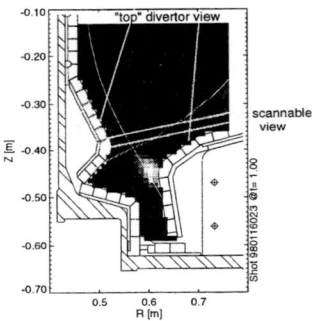
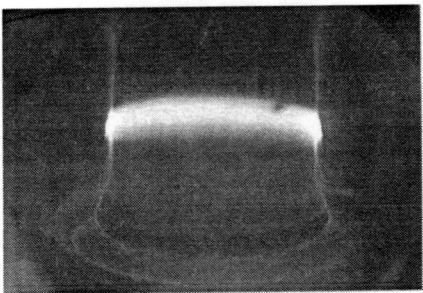

FIGURE 1. A cross section of the lower part of Alcator C-Mod, showing the measured emissivity of D_γ (434 nm). White has an emissivity of $\approx$4 kW/m^3; black is $\approx$0. Also shown: the equilibrium reconstruction of the magnetic separatrix, the closed divertor, and the various views used in this study. Spatially resolved D_α measurements are made within the "top view". The "scannable view" is shared for both Lyman series (VUV) measurements and Balmer series (visible) measurements. Probes embedded in the divertor plates are used to measure the ion current to the plates.

FIGURE 2. A TV image of the inside of the tokamak in D_α light. The bright band of emission ringing the central column is the MARFE. Emission from the MARFE is much greater than emission from the rest of the plasma. The non-MARFE light has been digitally enhanced so that some of the internal components can be seen. The divertor structure is at the bottom of the image.

SIGNATURES OF VOLUME RECOMBINATION

A typical VUV spectrum of the D^0 Lyman series from a region where a large amount of recombination is occurring is shown as the solid line in Fig. 3. Shown by the dashed line is the spectrum characteristic of that produced by electron impact excitation. For both spectra the VUV spectrometer viewed the inner wall of the vacuum vessel through the center of the core plasma, along an approximately horizontal line-of-sight. The 'recombining' spectrum was measured when a MARFE, a radially-thin, toroidal belt of cold, high density plasma, located at the inboard plasma edge, was within the field-of-view. The MARFE geometry is shown by the image in Fig. 2. For the 'excitation' spectrum there was no MARFE in the field-of-view, and it is more typical of the spectrum from the Alcator C-Mod edge (non-divertor region) plasma. A number of differences is evident:

- The decrease in line intensity with principle quantum number, n, is much faster for the 'excitation' spectrum.

- An emission continuum from radiative recombination is evident at wavelengths $\lesssim$ 920 Å in the recombining spectrum.

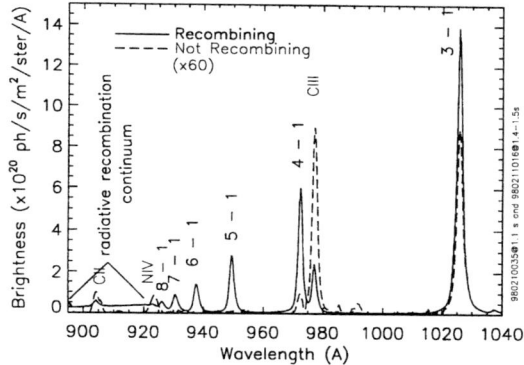

FIGURE 3. The thick solid line shows a D^0 Lyman series spectrum (containing lines n=9, 8, ...3 → 1) produced by volume recombination. The dashed line shows a spectrum produced primarily from electron impact excitation (only the n=4 and 3 → 1 are observed). In the recombining spectrum, continuum emission from radiative recombination is also evident at wavelengths below about 920 Å. The recombining spectrum is characteristic of a 0.7-1.0 eV plasma of density $2.5 - 3 \times 10^{21}$ m^{-3}. The recombining spectrum is from a view of a MARFE.

The difference in the intensity scaling between these two spectra arises from the fact that the upper levels of the transitions are being populated by different processes, one by recombination ($\propto n_i$ (ion density) $\times n_e$) and by excitation from the ground state ($\propto N_0$ (neutral density) $\times n_e$). The two population processes are illustrated by the schematic 'electron flow' diagrams[13] shown in Fig. 4. Population of the excited levels in the recombining case is primarily a result of recombination, linking the levels to the free electron 'continuum'. This leads to significantly greater population of the higher lying levels than does population by electron impact excitation from the D^0 ground state. These schematic representations have been made quantitative using a collisional-radiative (CR) model for population of D^0 within a plasma [14,15,9]. The 'recombining' spectra, of which the solid line in Fig. 3 is an example, can be reproduced by considering population by *recombination only* in a low temperature, high density plasma. *There is no set of plasma conditions for which ground state excitation can yield such a spectrum.* The 'excitation' spectrum, however, can be reproduced almost entirely by electron impact excitation from ground state D^0. After measurement of many 'recombining' spectra and from the characteristic difference in the population distributions resulting from recombination and excitation, we conclude that volume recombination occurs routinely in the Alcator C-Mod divertor and in MARFEs when they form in the Alcator C-Mod SOL.

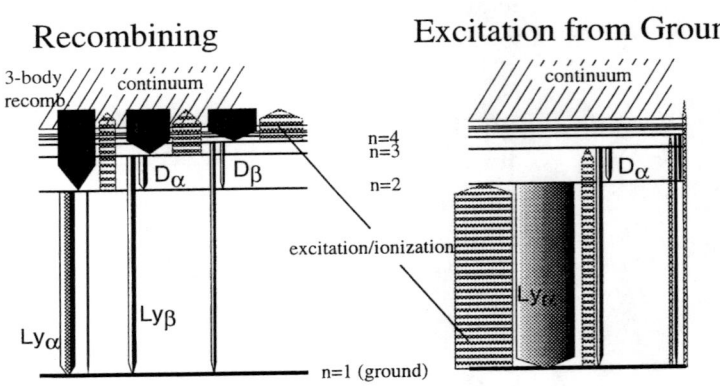

FIGURE 4. A schematic diagram showing the different 'electron flows' associated with D^0 level population by recombination (solid arrows) and by excitation (striped arrows). Photon emission is indicated by the grayish arrows. The emission spectra are distinctly different.

In the next sections we set up the formalism to quantify the level of recombination, to describe the measurements which are needed to apply the formalism, and finally to determine the amount of recombination for some representative C-Mod discharges.

RECOMBINATIONS PER PHOTON AND OPACITY

Having found experimentally those situations when the upper level populations are determined by 3-body (and radiative) recombination, we use this fact to determine the recombination rate from those pathways. The rate is a strong function of temperature at low temperature ($T_e \lesssim 3$ eV) and a somewhat weaker function of density. However, because the upper levels of measured lines are populated by this recombination process, the number of emitted photons is directly related to the number of recombinations. *The photon emission acts as 'monitor' of the recombination rate with a much weaker temperature and density dependence than the recombination rate itself.* This is the essence of the concept of 'recombinations per photon', defined formally for the photon from level n to level j as

$$\Theta_{n \to j} \equiv \frac{n_e n_i <\sigma v_e>_{recombination}}{N_n A_{n \to j}}. \tag{1}$$

The curve showing the number of recombinations per D_α photon in the case for which all D^0 lines are optically thin is shown as the top curve in Fig. 5. (A similar set of 'recombinations per D_γ photon' curves can be found in Ref. [5]). To calculate such a curve a CR model (in this case the Collisional Radiative Atomic Molecular Data (CRAMD)[9] code) is used to determine self-consistently both the line emission and the total 3-body and radiative recombination rate for cases where recombination determines the upper level population. The number of recombinations per 'monitor' photon, although weakly dependent upon temperature (above $\sim$0.8 eV) and density, *is* dependent upon the opacity of the Lyman series lines, particularly Ly_α and Ly_β. Essentially the dependence is a result of the fact that emission of Lyman series photons is typically the last stage in the process of 'complete' recombination which began by recombination into excited states. Since most of the 'complete' 3-body recombinations occur by emission of Ly_α ($\sim$60%) or Ly_β ($\sim$10%), any reduction in these channels due to photon loss reduces the net recombination rate and increases the net ionization rate. After emission, photons can be absorbed by neutral atoms in their paths. If an atom absorbs, say, a Ly_α photon, it can either re-radiate or suffer ionizing collisions (as well as other collisions). If it re-emits the photon, that Ly_α photon continues to diffuse through the absorbers. However, if the absorbing atom is subsequently ionized, the Ly_α photon has been lost and no net recombination has occurred, since the 'complete' recombination of the original atom has resulted in an ionization of an absorbing atom. The term 'trapped' is used here to denote those circumstances where there is a net photon loss due to the combined influence of absorption and collisional processes. The importance of photon trapping has been considered in Refs. [14,16,8,17]. The primary effect on the ionization/recombination balance is to lower the temperature at which ionization and recombination are equal (for fixed n_e and N_0). Radiation transfer is a complex problem both computationally and experimentally. In an initial attempt to compute the effect of trapping on the number of recombinations per photon, CRAMD has been modified to treat the spatial diffusion of Lyman series photons through homogeneous plasmas with differing neutral densities. Diffusion in wavelength has not been considered in this model. The results are shown by the

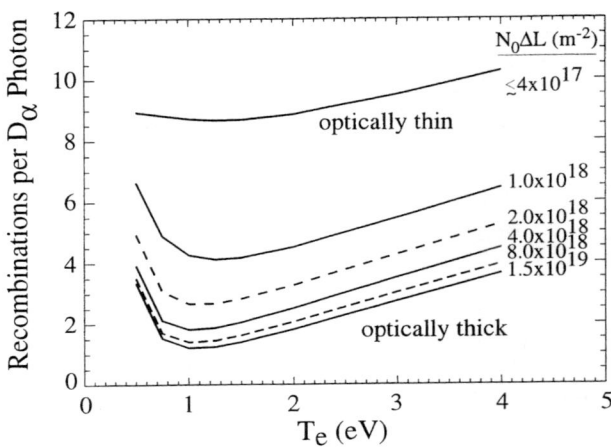

FIGURE 5. The number of recombinations per D_α photon calculated for a recombining plasma with different magnitudes of Lyman line trapping. The T_e dependence is relatively weak for D_α, especially in the optically thin case, applicable for $N_0 \Delta L \lesssim 4 \times 10^{17}$ m^{-2}, shown as the top solid line. The other curves show the effects of trapping on the recombination rates. The trapping depends on the number of absorbers along the line of sight, $N_0 \Delta L$. Each curve is labeled by this quantity (1.0(18) means 1.0×10^{18}).

lower curves in Fig. 5. ΔL is the viewing path length through the region of neutral density N_0. As expected, the trapping of $Ly_{\alpha,\beta}$ decreases the recombination rate, which in turn decreases the recombinations per photon.

DETERMINATIONS OF OPACITIES, T_e, and n_e

Opacity Determination

For reasons just discussed, the strongest dependence of the number of recombinations per 'monitor' D_α photon is on the opacity of emitting plasma to Ly_α and Ly_β. One method by which the magnitude of the trapping can be investigated experimentally is through examination of the line shapes. This approach calls for extremely high instrumental resolution, in fact significantly higher than that is achievable with the VUV spectrograph used for this work. Therefore an alternative experimental approach has been used here, where Ly_β and D_α are measured simultaneously along the same view, the "scannable view" in Fig. 1. Since these lines share the same upper state, their source emissivity ratio is fixed by their spontaneous emission coefficients. Any change in the observed brightness ratio from this determined ratio is a measure of the magnitude of the Ly_β trapping. (D_α is not trapped is these plasmas, since its absorption mean-free-path is ~50x that for Ly_β.) Quantitatively we describe the fraction of Ly_β trapping by defining the Ly_β 'transmission', T_{Ly_β}, to be the magnitude of the measured Ly_β/D_α relative to the optically thin ratio, where $T_{Ly_\beta} \equiv 1$. In Fig. 6 this quantity is plotted vs the D_α brightness for a selected set of shots for which the view was essentially the same. The D_α brightness is a rough measure of the number of absorbers along the line-of sight and is therefore roughly $\propto N_0 \Delta L$. As might be expected, the transmission curve plotted this way is somewhat dependent upon the view. The trend of decreasing Ly_β transmission with increasing D_α brightness is clear and a result of trapping. The diffusive photon transport modelling relates the Ly_β transmission to the trapping of Ly_α. For example, when T_{Ly_β} is ~0.5, over 95% of the Ly_α photons are trapped. And although there is no exact analog of the Ly_β/D_α ratio measurement for Ly_α, we measure a stronger decrease in the Ly_α/D_α ratio with increasing D_α brightness. Both simple estimates and CRAMD modelling indicate that for Ly_β transmission ~0.5, the typical neutral density, N_0, in the recombination region is $\sim 2 \times 10^{20}$ m^{-3} and the curve in Fig. 5 corresponding to $N_0 \Delta L = 4 \times 10^{18}$ m^{-2} is appropriate for this transmission.

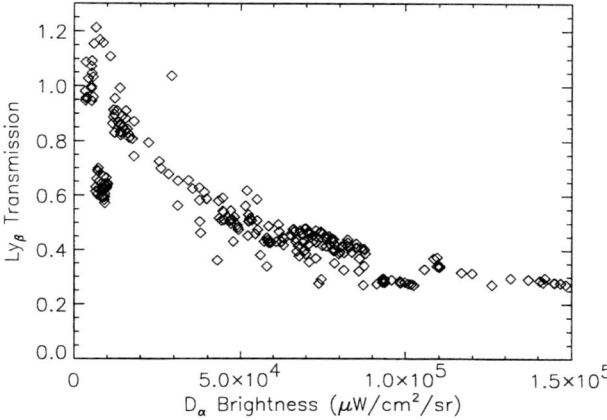

FIGURE 6. The scaling of the transmitted fraction of Ly_β vs. the D_α brightness.

It is practically impossible to measure the Lyman trapping along all lines-of-sight viewing the complicated geometry of the recombination regions in the C-Mod divertor. Nonetheless we can use the scaling shown in Fig. 6 and assume that the transmission along any line-of-sight (and therefore the appropriate recombinations per photon curve) is only a function of the D_α brightness and is given by Fig. 6. Obviously this assumption adds uncertainty to the determination of the recombination rate and deserves further study.

T_e Determination

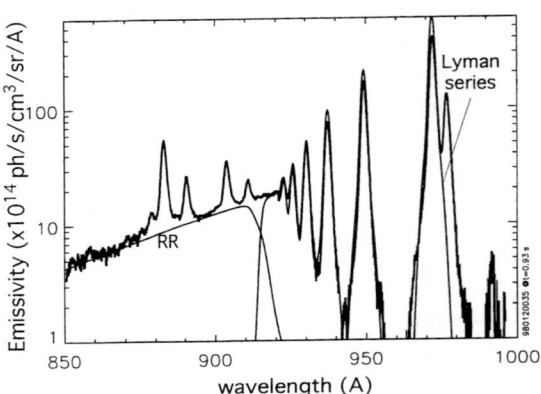

FIGURE 7. Measured spectrum showing the radiative recombination continuum and the Lyman series lines ($n=9, 8, 7,...4 \rightarrow 1$). Also shown are the line and continuum spectra predicted for a plasma of $T_e=0.7$ eV with $n_e=n_i = 2.25 \times 10^{21}$ m^{-3}.

The recombination per photon curves are also a function of T_e, especially when trapping is occurring and $T_e \lesssim 0.8$ eV. The temperatures in the emission regions have also been evaluated spectroscopically using well known techniques[18]. Shown in Fig. 7 are a partial Lyman series spectrum and a spectrum of the continuum emission resulting from radiative recombination into the $n=1$ ground state. The slope of this continuum is a strong function of T_e at wavelengths below the $n=1$ recombination edge (occurring at $\lambda_{edge}=911$ Å), since T_e is low ($\lesssim 1.5$ eV) and the intensity is approximately proportional to $e^{-(h\nu/T_e)}$. The continuum is extremely well fit by that predicted for a $T_e=0.7$ eV plasma. The prediction for a 0.7 eV radiative recombination spectrum is shown in Fig. 7 as the thin line labeled, 'RR'. Along lines-of-sight accessible to the VUV spectrograph, a precise T_e determination is made using the fit to this continuum. Also shown by the thin line labeled 'Lyman series' is the spectrum predicted by the CR model for a 0.7 eV recombining plasma, where the lines have been broadened to the instrumental line shape. Note that the fit of the line spectrum is also good for $n \geq 6$, but predicts line brightnesses for $n=4$

and 5 which are larger than measured. We attribute this effect to opacity which progressively reduces the intensities of lower-n series lines.

The extension of the radiative recombination continuum to wavelengths greater than λ_{edge}, is a result of the high electron density in the atoms' environment. The high n_e produces statistical plasma microfields which move very high-n, otherwise-bound states into the continuum. It also causes Stark broadening. These effects have been incorporated[19,20] into CRAMD and used for these predictions.

Another method for the T_e measurement uses the intensity scaling of Balmer/Lyman series lines. In the recombining regions the populations of higher-$n(\sim 5\text{-}12)$ levels follow a Saha-Boltzmann distribution, and the line intensities, $B_{n\rightarrow i}$ scale as

$$B_{n\rightarrow i} \propto A_{n\rightarrow i} \times \frac{n^2 e^{(IP_n/T_e)}}{T_e^{3/2}}, \qquad (2)$$

where IP_n is the ionization potential from the n-th level, and $A_{n\rightarrow i}$ is the appropriate spontaneous emission coefficient. The dependence is implicit in the predicted 'Lyman series' spectrum shown in Fig. 7, although we have used this technique more with Balmer series spectra, which are available with a more extensive spatial coverage.

n_e Determination

Although it is not shown in Fig. 5, there is a slight density dependence to the recombinations per photon curves, and it is desirable to know the local electron density in the recombining region if the recombination rate is to be determined from the line intensities. (Of course, there is a strong density dependence to the actual recombination rate.) Roughly the recombinations per photon scale as $\sqrt{n_e}$, i.e. if there are 2 recombinations per photon at $n_e = 1\times 10^{21} \mathrm{m}^{-3}$, there are about 3 recombinations per photon at $n_e = 2\times 10^{21} \mathrm{m}^{-3}$. The recombining region is cold and dense enough that the Balmer series line-widths are dominated by Stark broadening. Typical Stark widths (FWHM) for $n=7\rightarrow 2$ are 0.5 nm, large compared to the instrumental resolution of the visible spectrograph ($\Delta\lambda \sim 0.2$ nm), the Zeeman splitting, and the Doppler width. Using the parameterizaton of Ref. [21] in which $n_e \propto (\Delta\lambda_{Stark})^{3/2}$, we find that the densities inferred from the widths of the $n=10$, 9, 8, and $7 \rightarrow 2$ Balmer lines are consistent to within about $\pm 10\%$. Densities as low as $\sim 5\times 10^{19} \mathrm{m}^{-3}$ can be measured, and densities as high as $\sim 3\times 10^{21} \mathrm{m}^{-3}$ have been measured.

EVALUATION OF THE RECOMBINATION RATE

Having discussed the formalism for evaluating the recombination rates from the measured spectra and having detailed the methods by which the quantities needed for such an evaluation were obtained, we now apply these techniques to the experimental spectra.

The first case we will analyse, because of its relative geometric simplicity, is the recombination rate within what is known as a MARFE[7]. As stated earlier, a MARFE is a radially-thin, toroidal belt of cold, high density plasma which sometimes forms around the central column in a tokamak, typically near the plasma midplane. The image of the MARFE in D_α light is shown in Fig. 2. Its Lyman spectrum (shown in Fig. 7) and Balmer spectrum are measured along major radial views from the outside. The 'scannable view' (raised to horizontal) is used. We measure its radial thickness, ΔL, to be 0.025 m by examining the D_α-filtered TV image (Fig. 2) where some of the chords look through the emission region tangentially. Its vertical extent is ≈ 0.09m (the FWHM of the emission) as measured both from the TV image and a vertically resolving array of D_α-filtered detectors[12]. The 'transmission' for Ly_β is measured to be $\approx 40\%$, indicating that this cold plasma is opaque to Ly_α. The temperature from the radiative recombination spectrum is 0.75 eV, and the same spectrum yields an n_e of $2.0 \pm 0.2 \times 10^{21} \mathrm{m}^{-3}$. (For this density measurement knowledge of the thickness of the emitting region was used.) The Stark broadening of the Balmer lines yields a density of $2.45 \pm 0.3 \times 10^{21} \mathrm{m}^{-3}$. The total recombination rate is given by

$$RR(s^{-1}) = 8\pi R_o \int \Theta(\tau_{opacity}, T_e, n_e) B(z) dz, \qquad (3)$$

where $\Theta(\tau_{opacity}, T_e, n_e) \approx 2.6$ is the number of recombinations per D_α photon in the trapped case at 0.75 eV and $n_e = 2 \times 10^{21} \mathrm{m}^{-3}$, B(z) is the D_α brightness (in ph/s/m²/ster), and $R_o=.46$ m is the major radius of the MARFE emission region. Evaluating RR, we find it reaches a steady state value of $\approx 6 \times 10^{22}$ s^{-1} for this discharge. The significance of this value can be assigned by comparing it to the magnitude of the ion current to the divertor plates in the absence of the MARFE. (In the presence of a MARFE of this size and brightness, the ion current is typically less than 5×10^{21} s^{-1}, measured by arrays of probes mounted in both the inner and outer plates[3].) In similar plasmas without MARFEs (and when the divertor is not "detached"), the ion current to the plates is as high as $\sim 7 \times 10^{22}$ s^{-1}. Thus the implication is that essentially all of the ion sink is volume recombination within the cold, dense region of the MARFE.

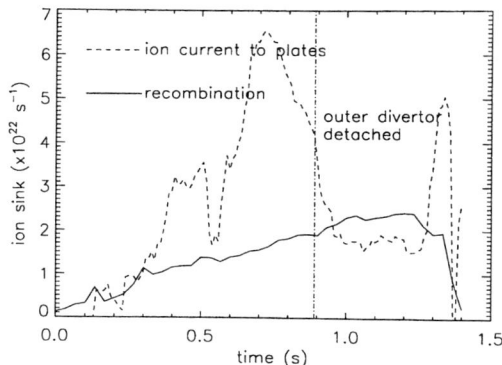

FIGURE 8. Time histories of the two ions sinks: the total ion current to the plates (dashed), and the volume recombination sink (solid). After ~0.89 s the plasma detaches from the outer divertor plate.

As a second example, we have examined the recombination rate in the divertor region as the plasma detaches from the plates. (See Ref. [22] and references therein for details of what "detachment from the divertor plates" means for Alcator C-Mod discharges.) In this case the "monitor" photons were D_γ photons detected by a filtered TV camera which views the divertor regions nearly tangentially with a spatial resolution of a few mm between chords. The images were inverted[23], assuming toroidal symmetry, to yield 2D emissivity profiles of D_γ like the one shown in Fig. 1. The resolution of the emissivity profile is degraded to 1 cm due to large sizes of the matrices involved in the inversion algorithm. Before detachment the emission is concentrated near the outer strike point and in the region above the inner strike point. After the detachment the emission increases strongly in the region between the outer strike point and the X-point, as shown in Fig. 1. The 2D emissivity profile of D_γ, the temperatures, densities and opacities of the emission regions were measured or estimated. There are chordal measurements of the Balmer spectra from the 'top divertor views', shown in Fig. 1. These yield emission-weighted T_e's of 0.5-0.8 eV and n_e's from ~ 0.6 to 1.4 $\times 10^{21}$ m^{-3} depending on the view. The Ly$_\beta$ opacity was measured along the chord shown in Fig. 1, and this measurement indicates that the opaque 'recombinations per D_γ photon' curve was appropriate. For ease of computation we took the recombinations per D_γ photon to be constant in both space and time and equal to 50 (corresponding to 0.7 eV and 1×10^{21} m^{-3}). The resulting time history of the recombination rate is shown as the solid line in Fig. 8. Also shown is the total ion current to the divertor plates (the dashed line). The outer divertor detaches at 0.89 s. It is evident that, although the ion current drops sharply at detachment, there is at most a small change in slope on the recombination rate. Thus it appears that the decrease in ion current is not directly compensated by an increase in recombination. After detachment is it seen that the recombination sink is somewhat greater than the ion loss to the plates. The implication from this analysis is that the ionization source did not remain constant throughout detachment process.

SUMMARY

Volume recombination is diagnosed in the edge regions of Alcator C-Mod plasmas. Clear signatures of the recombination are seen in the D^0 emission spectra. The formalism to quantify the magnitude of the 3-body and radiative recombination is developed, using the concept of 'recombinations per D^0 line photon'. The number of 'recombinations per photon' is a much weaker function of temperature and density than the recombination rate itself. However, both the recombination rate and 'recombinations per photon' depend upon the opacity of the plasma to Lyman series photons. Descriptions of the experimental measurement of the opacity of Ly_β along some lines-of-sight are given, and the quantitative implications of the measured opacity upon the number of recombinations per photon are calculated using a model in which the Lyman series photons diffuse spatially through the plasma. The methods by which the temperature and density are measured are described as well. Temperatures and densities in Alcator C-Mod recombining regions are measured to be $\lesssim 1$ eV and $\sim 1-3 \times 10^{21}$ m^{-3}. Under these conditions volume recombination is a significant sink for ions. The magnitude of the recombination rate in an inboard, edge plasma MARFE is found to be large. The recombination rate is also evaluated before, during, and after plasma detachment from the divertor plates. We find that before detachment the recombination sink is $\sim$0.2-0.3 of the ion current sink. After detachment the recombination is $\sim$50% greater than the ion loss to the plate. The recombination does not directly compensate the change in ion current at detachment.

ACKNOWLEDGEMENTS

This work is supported by the U.S. Department of Energy under the contract # DE-AC02-78ET51013 and under the grant #DE-FG02-910ER-54109. The authors acknowledge help from the entire Alcator C-Mod physics staff and the excellent operational support by the engineering and technical staff.

REFERENCES

[1] G. Matthews, *J. Nucl. Mater.*, **220-222**, 104 (1995).
[2] A. Loarte, *J. Nucl. Mater.*, **241-243**, 118 (1997).
[3] B. LaBombard et al., *Phys. Plasmas*, **2**, 2242 (1995).
[4] S. L. Allen et al., *J. Nucl. Mater.*, **241-243**, 595 (1997).
[5] J. L. Terry et al., *Phys. Plasmas*, **5** (1998).
[6] B. Lipschultz et al., submitted to Phys. Rev. Lett.
[7] B. Meerson, *Rev. of Mod. Phys.*, **68**, 215 (1996).
[8] D. Post, *J. Nucl. Mater.*, **220-222**, 143 (1995).
[9] A. Pigarov and S. Krashenninikov, *Phys. Lett. A*, **222**, 251 (1996).
[10] I. H. Hutchinson et al., *Phys. Plasmas*, **1**, 1511 (1994).
[11] H. Ohkawa, *Determination of the Spectral Sensitivity of a VUV, Grazing Incidence Spectrograph*, M.I.T. M.S. Thesis, PSFC Report PSFC/RR-97-11 (1997).
[12] J. L. Terry, J. A. Snipes, and C. Kurz, *Rev. Sci. Instrum.*, **66**, 555 (1995).
[13] This way of illustrating "electron flow" through the states of D^0 is from the papers of T. Fujimoto and K. Sawada.
[14] L. C. Johnson and E. Hinnov, *J. Quant. Spectrosc. Radiat. Transfer*, **13**, 333 (1973).
[15] T. Fujimoto, S. Miyachi, and K. Sawada, *Nucl. Fus.*, **28**, 1255 (1988).
[16] S. Krasheninnikov and A. Y. Pigarov, *Contrib. Plasma Phys.*, **28**, 443 (1988).
[17] K. Behringer, private communication (1997).
[18] H. Griem, *Spectral Line Broadening by Plasmas*, Academic Press, New York and London, 1974.
[19] A. Y. Pigarov, J. Terry, and B. Lipschultz, *Proc. of the 24th European Phys. Soc. Conf. on Contr. Fus. and Plasma Phys. - Vol 21A - PartII*, Page 577, European Physical Society, Petit-Lancy, 1997.
[20] A. Pigarov, J. Terry, and B. Lipschultz, to be published in Plasma Phys. and Controlled Fusion (1998).
[21] R. Bengtson, J. Tannich, and P. Kepple, *Phys. Rev. A*, **1**, 532 (1970).
[22] B. Lipschultz et al., *J. Nucl. Mater.*, **241-243**, 771 (1997).
[23] A. Allen, *Capture, Storage, and Analysis of Video Images on the Alcator C-Mod Tokamak*, M.I.T. M.S. Thesis, PSFC Report PSFC/RR-97-1 (1997).

# Monte Carlo Impurity Transport Modeling in the DIII–D Tokamak

T.E. Evans* and D.F. Finkenthal[†]

*General Atomics, PO Box 85608, San Diego, CA 92186
[†]Palomar College, 1140 West Mission Road, San Marcos, CA 92069

Abstract. A description of the carbon transport and sputtering physics contained in the Monte Carlo Impurity (MCI) transport code is given. Examples of statistically significant carbon transport pathways are examined using MCI's unique tracking visualizer and a mechanism for enhanced carbon accumulation on the high field side of the divertor chamber is discussed. Comparisons between carbon emissions calculated with MCI and those measured in the DIII–D tokamak are described. Good qualitative agreement is found between 2D carbon emission patterns calculated with MCI and experimentally measured carbon patterns. While uncertainties in the sputtering physics, atomic data, and transport models have made quantitative comparisons with experiments more difficult, recent results using a physics based model for physical and chemical sputtering has yielded simulations with about 50% of the total carbon radiation measured in the divertor. These results and plans for future improvement in the physics models and atomic data are discussed.

I. INTRODUCTION

The control of non-hydrogenic core and boundary layer impurities is a critical issue for fusion energy production in magnetic confinement devices. In large magnetic fusion devices such as ITER small amounts of non-hydrogenic impurities significantly degrade the performance of the device both by diluting the DT fuel and by radiating energy from the core where high plasma temperatures are required for the efficient production of fusion power. Calculations based on the ITER design indicate that the effective charge (Z_{eff}) of the core plasma must be kept below 1.6 in order to maintain acceptable operating conditions (1). In addition, heat conduction from the core plasma into the edge requires that the radiated power exceed 80% of the total loss power to prevent target plate damage. Experiments in existing tokamaks have demonstrated that these levels of radiation can, in fact, be obtained without significantly degrading the core confinement but in some cases the core impurity concentration exceeds Z_{eff} = 1.6. DIII–D has achieved good performance using D_2 puffing to increase the radiation from D_α and intrinsic carbon impurities. Lower wall heat fluxes were achieved at Z_{eff} < 1.8. In JET it was found that by combining D_2 and N_2 gas puffs a high radiated power fraction (~85%) could be obtained in the divertor region with good core confinement. Unfortunately, the core impurity concentration increased steadily throughout these shots exceeding Z_{eff} = 2.5 over the entire high confinement phase and resulting in a steady decline in the neutron production rate (2). Thus, it is not yet clear whether radiative divertors sustained by injecting either pure D_2 or a mixture of D_2 and a radiating impurity, are capable of providing a complete solution for particle and heat flux control. Consequently, comprehensive numerical models of the edge plasma and its interactions with plasma facing material surfaces are essential for developing a better understanding of the complex processes involved in optimizing the radiative divertor approach. The DIII–D Monte Carlo Impurity (MCI) model discussed here is one part of an array of numerical models being developed for this purpose.

II. DIVERTOR MODELING BACKGROUND

As originally envisioned, the roll of a perfect axisymmetric poloidal divertor is to move impurity sources as far as possible from the good confinement region by creating a layer where all of the magnetic field lines terminate on material surfaces buried deep inside chambers which are well isolated from the main plasma. Conceptually, the idea is that these open field lines form a so-called scrape-off layer (SOL) around the hot plasma where the flow of background deuterium ions (D^+), which have escaped the main plasma, drag impurities away from the well confined region into the isolation chamber. In the divertor chamber, D^+ and impurity ions are removed from the system with cryogenic pumps. Since impurities are relatively good radiators, the energy conducted and/or convected into the divertor is converted to line radiation and uniformly distributed on the divertor walls where it can be removed.

Realistic poloidal divertor geometries have taken several forms over the last fifteen years. Initially, closed divertor designs with both upper and lower chambers, such as in the ASDEX tokamak, were used. High confinement H–modes were originally obtained with closed divertors in ASDEX (3). One serious drawback of the closed divertor design is a severely reduced ability to shape the well confined plasma volume and to control the SOL geometry. A second generation of poloidal divertors used an open chamber design which allowed much more shaping flexibility. Tokamaks such as DIII–D and JET have successfully used the open divertor approach to produce record core plasma pressure and confinement levels. Recently, with the addition of cryogenic pumping systems, baffles, contoured graphite divertor surfaces and well controlled distributed gas puffing systems, open divertors have substantially exceeded the performance levels attained with closed divertors.

The basic elements of an open axisymmetric poloidal divertor are shown in Fig. 1. The white region in the figure shows where the well confined main plasma resides in a typical H–mode discharge. The shaded region is located outside the 98% flux surface. It covers the boundary layer, the separatrix and the SOL out to the vacuum vessel walls. The separatrix nominally divides the inner, closed field line region, from the outer open field line region. The detailed wall geometry, as well as the material used for plasma facing surfaces, vary from one tokamak to another. Figure 1 shows the

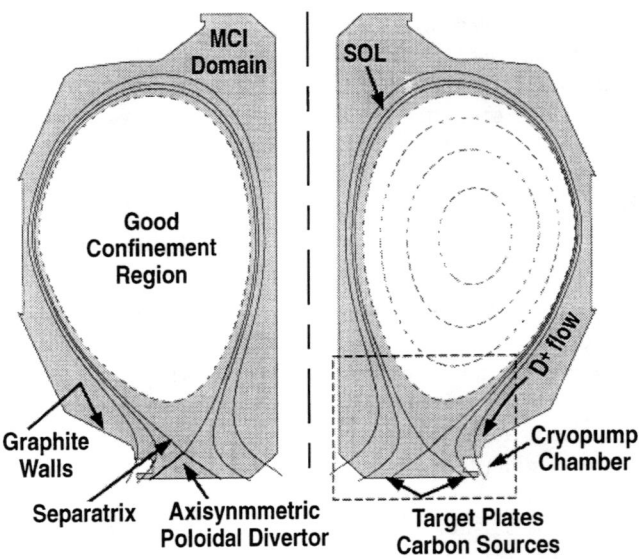

FIGURE 1. Poloidal cross section of a lower single-null DIII–D configuration. The principal components needed to understand the operation of axisymmetric divertors are illustrated. The shaded area indicates the typical edge transport modeling domain with the MCI code.

wall geometry for DIII–D in a lower single null divertor configuration. With the exception of a few specialized components such as rf antennas, all of the DIII–D plasma facing surfaces are covered with graphite.

The primary goal of the divertor is to compress impurity ions and helium produced during DT fusion reactions into a region as close as possible to the target plates. Ideally, drag from the D^+ flow along with parallel electric fields in the SOL provide the compressive forces for this process. These forces are nominally directed toward the target plates while thermal gradient forces acting on impurity ions typically oppose the compressive forces and act to pull impurities back up the SOL field lines toward the mid-plane region where they can easily be transported into the main plasma. Unfortunately, the nominal picture of the compressive forces is not always applicable to fusion relevant divertor operations and the dynamics of scrape-off layer flows are more complex than the simple picture envisioned above. In fact, it is now known that in tokamaks sometimes conditions in isolated areas of the SOL are such that the D^+ flows reverse direction and actually assists the thermal gradient forces in pulling impurities out of the divertor region. The physics involved in the onset of these flow reversal regions and the detailed 2D structure of the flow patterns are known to be sensitive to impurity cooling rates and thus to the production and transport of impurities ions from each of the plasma facing material surfaces.

Recently, new diagnostics which measure the parallel D^+ and carbon impurity ion flows in the DIII–D divertor have acquired data which may give us a better understanding of how to modify or control their properties. Techniques such as heavy deuterium gas puffing in the SOL to increase the D^+ flow have been effective in substantially reducing the electron temperature (T_e) and increasing the density (n_e) in the divertor while increasing the radiated power near the target plates. Substantial differences in the flow patterns are observed after injecting large deuterium gas puffs into the SOL. Impurity gas puffs, using species such as neon, argon and nitrogen, have also been used to increase the divertor radiation. This approach sometimes results in higher core plasma Z_{eff} values but may, if properly designed and controlled, ultimately lead to better control of divertor flows and atomic processes which influence other aspects of the divertor performance. These so-called radiative divertor approaches using either D_2 puffing alone or combination of D_2 and impurity puffs are currently an active area of research in diverted tokamaks.

In principle, a pure D_2 radiative divertor is particularly promising if T_e near the target plates can be driven low enough (~1 eV) to obtain a fully recombining divertor plasma without disrupting the main plasma. This scenario is especially attractive because it avoids difficulties with non-intrinsic impurity puffs which can result in a core plasma with Z_{eff} above 1.6. A fully recombining D_2 radiative divertor also reduces intrinsic carbon impurities from physical sputtering and theoretically leads to a reduction in the thermally activated contribution of the chemical sputtering at the target plates. On the other hand, the recombining divertor approach is more difficult with advanced tokamak confinement scenarios where current and pressure profile control, critical for maintaining stability, requires low density/high temperature operation.

Thus, the next major step in the development of tokamaks and generally in the advancement of fusion energy research hinges on our ability to control the boundary layer, SOL, and divertor without impinging on the shaping flexibility needed to maintain high core confinement levels with good stability properties. This is a sizable challenge requiring creative new ideas, advanced hardware concepts, and a substantial theoretical and modeling commitment which must be benchmarked against a continuing stream of experimental data from our existing machines. Since atomic and molecular physics is a keystone in the dynamics of a fusion plasma's boundary, increasing demands are being put on the quality, accuracy, and extent of the atomic data used as well as our ability to produce and benchmark this data. The objective here is to highlight the role of atomic physics in divertor modeling and point out what types of data are needed for developing a better understanding of the complex processes found at the edge of a fusion plasma.

III. MCI MODELING APPROACH

The basic transport physics used in edge plasma fluid and Monte Carlo codes is essentially identical. Single and multi-species fluid codes are primarily used in collisionally dominated regions of the edge plasma while Monte Carlo codes are required in regions with large gradients or when the mean free path of a particular ion species is long compared to the local geometric scale, for example near plasma facing surfaces or in regions of low collisionality. A general review of edge plasma fluid and Monte Carlo codes can be found in a paper by Stangeby (4).

The current generation of Monte Carlo impurity codes are quasi-kinetic in the sense that they do not fully account for drift kinetic or velocity space diffusion effects such as is done in a Fokker-Planck code. A full Fokker-Planck treatment of the impurities would be computationally prohibitive with today's CPUs. The general approach used in Monte Carlo impurity codes is to follow an individual "particle" through a computational domain in which the background plasma parameters change as a function of space but are fixed in time. The computational domain for the DIII–D MCI code is depicted by the shaded region in Fig. 1. It extends from the vessel walls into the edge of the well confined plasma just inside the separatrix. A simulation "particle," which starts as neutral atom or molecule from a plasma facing surface, moves from one cell on a grid to a neighboring cell and interacts with the background plasma in each cell on the grid.

The trajectory of a "particle" is determined from the three-vector force components acting on it at each position in the cell. In addition, the properties of the "particle," such as its charge state, temperature, and flow velocity are tracked and stored for additional statistical analyses. As the simulation evolves, spatial and temporal histories of the "particles" are used to calculate spatial density distributions and average residence times in the plasma. In order to obtain reasonable simulation times for individual code runs, each "particle" actually represents a specified number of impurity atoms or molecules. This approach provides an accurate description of the impurity distribution as long as a statistically significant number of "particle" histories are tracked for each simulation.

A. Transport Physics

In the MCI code, impurity ion transport is driven by both kinetic and collective effects such as: collisional diffusion, electric field drifts, drag from flows in the background ions, and thermal gradient forces due to the background ions and electrons. Kinematically, impurity ions undergo a classical guiding center motion along the magnetic lines of force i.e., parallel motion and an anomalous motion normal to the magnetic field due to turbulent transport processes i.e., perpendicular motion.

The first step in the transport process is to calculate a characteristic time interval, $\Delta t'$, in each cell. $\Delta t'$ is determined by the local electron-ion collision time. An internal transport time interval Δt is then chosen such that $\Delta t = \min(\Delta t_b, 0.02\Delta t')$. $\Delta t_b = S_{pol}|v_{\parallel Tz}|^{-1}$ is the time required to reach either a cell boundary or material surface, where $|v_{\parallel Tz}| = (kT_z/m_z)^{1/2}$ is magnitude of the impurity ion thermal velocity. $S_{pol} = S_{\parallel}\sin\zeta$ is the poloidally projected distance along the field line, k is the Boltzmann constant, m_z is the mass of the impurity ion, T_z is the temperature of the impurity ion, ζ is the angle the magnetic field line makes with a unit vector in the toroidal direction and $S_{\parallel}$ is the parallel distance to the cell boundary or a material surface.

Once the internal time step has been established a new spatial position along the field line ($s_{\parallel,i+1}$) is calculated based on the current position ($s_{\parallel,i}$) and the current parallel flow velocity ($v_{\parallel,i}$) of the impurity ions using the following expression:

$$S_{\parallel,i+1} = S_{\parallel,i} + v_{\parallel,i}\Delta t + \frac{1}{2}C_i(\Delta t)^2 + \gamma_{rs}\sqrt{2D_{\parallel,i}\Delta t} \tag{1}$$

The acceleration coefficient C_i is given by:

$$C_i = \frac{v_b - v_{\parallel,i}}{\tau_{sl}} + \frac{1}{m_z}\left[0.71 Z_z^2 \frac{\partial T_e}{\partial s_\parallel} + \beta_z \frac{\partial T_b}{\partial s_\parallel}\right] + \frac{Z_z eE}{m_z} \tag{2}$$

In Eq. (2) v_b and T_b are the background deuterium ion D^+ parallel flow velocity and temperature respectively, Z_z is the charge of the impurity ion, E is the parallel electric field, e is charge of an electron, τ_{sl} is the impurity ion collision time with the background ions given as:

$$\tau_{sl} = \frac{m_z T_b \sqrt{(T_b/m_b)}}{6.8 \times 10^4 Z_z^2 Z_b^2 n_b \ln(\Lambda)\left[1 + \frac{m_b}{m_z}\right]} \tag{3}$$

and β_z is the ion thermal gradient coefficient (5) given by:

$$\beta_z = \frac{3\left[\mu + 7.07 Z_z^2 \left(1.1\mu^{5/2} - 0.35\mu^{3/2}\right) - 1\right]}{5.4\mu^2 - 2\mu + 2.6}; \mu = \frac{m_z}{m_b + m_z} \tag{4}$$

In Eq. (3) $\ln(\Lambda)$ is the Coulomb logarithm which typically ranges between 10 and 20 for cases of interest while n_b, Z_b, and m_b are the density, charge, and mass of the background ions respectively. The fourth term on the right hand side of Eq. (1) accounts for the classical diffusion of impurity ions along the magnetic field lines due to collisions between the impurity ions and the background ions. The coefficient γ_{rs} is a random sign operator which, on average, switches only once between positive and negative during each ion-ion collisional time i.e., approximately every 50 $\Delta t'$ internal time steps based on the constraint $\Delta t = \min(\Delta t_b, 0.02\Delta t')$. The classical parallel diffusion coefficient $D_{\parallel,i}$ in Eq. (1) is calculated with the current impurity ion temperature $T_{z,i}$ and the parallel impurity ion collision time $\tau_{\parallel,i}$ using the following expression:

$$D_{\parallel,i} = 1.22 \times 10^8 \left(\frac{\tau_{\parallel,i} T_{z,i}}{m_z}\right) \tag{5}$$

where $\tau_{\|,i}$ is given by:

$$\tau_{\|,i} = \frac{m_z T_{z,i}\sqrt{(T_b/m_b)}}{6.8\times10^4 Z_z^2 Z_b^2 n_b \ln(\Lambda)} \quad (6)$$

In addition to calculating the new parallel spatial position $s_{\|,i+1}$, a new impurity ion temperature $T_{z,i+1}$:

$$T_{z,i+1} = T_{z,i} + (T_b - T_{z,i})\frac{\Delta t}{\tau_{se}} \quad (7)$$

and flow velocity $v_{\|,i+1}$:

$$v_{\|,i+1} = v_{\|,i} + C_i \Delta t \quad (8)$$

are obtained after each internal time step. In Eq. (7) τ_{se}, the collisional Spitzer energy transfer time, is given by:

$$\tau_{se} = \frac{m_z T_b\sqrt{(T_b/m_b)}}{1.4\times10^5 Z_z^2 Z_b^2 n_b \ln(\Lambda)} \quad (9)$$

and C_i is the acceleration coefficient given by Eq. (2). The spatial-diffusion approach used to simulate impurity pressure gradient driven motion parallel to the field lines, as specified by $D_{\|,i}$ in Eqn. (5), is not satisfactory in all situations encountered. Thus, a velocity-diffusion model is currently under development which can be compared with the approach outlined above.

Perpendicular transport is modeled with an anomalous cross-field diffusion coefficient, $D_\perp$, using a random Monte Carlo algorithm similar to that of parallel diffusion. In terms of the cross-field variable x_i the new perpendicular position x_{i+1} of an impurity ion after each internal time step is given by:

$$x_{i+1} = x_i + \gamma_{rs}^*\sqrt{2D_\perp \Delta t} + v_{pinch} \quad (10)$$

Since we lack an experimentally verifiable theory of anomalous cross-field diffusion in the SOL and divertor, MCI uses a semiempirical value for $D_\perp$ along with a constant pinch velocity v_{pinch}. The values of $D_\perp$ used matches those obtained in edge transport codes when reproducing experimentally measured radial density profiles. This value typically ranges between $0.1 \leq D_\perp \leq 1.0$ m²/s for most DIII–D plasmas.

B. Intrinsic Versus Non-Intrinsic Impurities and Nonlinear Effects

MCI is used to simulate both intrinsic and non-intrinsic impurities. In DIII–D and other tokamaks non-intrinsic gas puffs, with impurities such as neon, argon, and nitrogen, are frequently used either for diagnostic purposes or for radiative divertor experiments. Thus, it is of considerable interest to model the transport of these impurities along with the intrinsic impurities. On the other hand, prior to performing simulation of non-intrinsic impurities, it is necessary to understand the sputtering, transport and radiation physics of the intrinsic impurities in the discharge of interest. This is particularly important because, in many cases, radiative cooling by the intrinsic impurities is known to produce substantive changes in the properties of the background plasma which then affect the transport of the non-intrinsic impurities. The question of how to feed back radiative cooling effects on the background plasma is a central issue for divertor modeling codes. This is presently beyond the scope of all the existing Monte Carlo models due to the intrinsically nonlinear, time dependent, nature of the problem. Nevertheless, a concerted effort is being made to address this issue since virtually all of the important mechanisms involved in establishing how the divertor performs when approaching fusion relevant conditions, such as detachment and flow reversal, depend on the nature of cooling processes.

Since essentially all of the DIII–D plasma facing material is graphite, the primary intrinsic impurity tracked in MCI is carbon. In this paper we limit our discussion to the linear modeling of intrinsic carbon in DIII–D.

C. Neutral Sources, Atomic Data, and Computational Grids

Neutral carbon "particles" are sputtered from plasma facing surfaces with a Thompson energy distribution (6) which depends on the energy of the incident ion or neutral. The sputtered neutral atoms and molecules follow straight line trajectories, based on their launch angle and energy, until they are ionized by collisions with the background plasma. Ionization mean free paths of carbon atoms depend on their launch angle, energy, and the plasma density and temperature at the location of the sputtering site.

In MCI, the entire plasma facing surface is divided into small segments where the surface properties are individually specified according to known or experimentally measured conditions and sputtering yields are calculated with a variety of semiempirical and physics based models. These models are driven by background plasma ion and neutral fluxes impinging on each segment and by carbon ions and neutrals which have been transport across the plasma after being produced at the same or other wall segments (*i.e.*, self-sputtering events). Two basic types of sputtering models are used in MCI, physical and chemical sputtering. Currently, MCI has four options for physical sputtering and one for chemical sputtering. In addition, the chemical sputtering model is driven by a physical sputtering process so any of the four physical sputtering processes can be used in chemical sputtering model. Each of the physical sputtering models account for electrostatic sheath effects, the Maxwell-

Boltzmann properties of the incident flux, and threshold energy effects while the chemical sputtering model has an explicit dependence on the local surface temperature which can, in some cases, be experimentally measured with an infrared camera. Thus, MCI has a complete set of erosion and redeposition models which are coupled to a realistic divertor and core plasma. As comparisons are made with experimentally measured erosion and redeposition profiles we are learning which of the various models best fit a particular set of experimental conditions and are able to identify additional physics constraints which need to be included in the models.

The current version of the MCI chemical sputtering model is based on the Roth García-Rosales (7) approach. This model only describes the production of methane (CD_4) and does not address processes associated with the release of heavier hydrocarbons in the C_2D_x and C_3D_x complex. There is experimental evidence indicating these larger hydrocarbon chains (8,9) may also be important in low temperature divertors but as yet we do not have access to good models for these sputtering processes.

During the transport process, collisions between the background plasma particles and impurity ions, neutral atoms or neutral molecules can result in a change in the impurity's charge state. As seen from the transport equations, the parallel forces on the impurity ions have a quadratic dependence on the charge state. After each simulation step Δt a series of Monte Carlo tests are made using tabulated atomic rates from the ADPAK database. During these tests, MCI calculates the probability of a change in the charge state due to an ionization, recombination or neutral charge exchange event and adjusts the transport accordingly. Thus, the accuracy of the atomic data can have a substantial impact on the simulation results. The ADPAK database uses a compilation of experimental and theoretical rates. Polynomial fits to the data are used to estimate ionization, recombination, charge exchange and radiation rates as needed for various background plasma condition. The ADPAK database is now about 15 years old.

The local electron temperature is used when calculating atomic rates with ADPAK data. In MCI, this temperature is obtained, along with the other background plasma properties such as the ion density and temperature, from the UEDGE (10) fluid code. Relatively small changes in the electron temperature can have a measurable impact on the impurity ion transport through the strong charge dependence. The electron temperature also plays a key role in the line emission rates which are used to calculate 2D radiation patterns that are compared to experimental measurements and to make quantitative predictions of the total radiated power and cooling rates. Thus, the accuracy of the electron temperature in each cell on the simulation grid is critical for the success of the simulation results. If the cells on the simulation grid are too large compared to the gradient scale lengths established by the plasma, then the discreteness of the background plasma temperature artificially skews the transport and line emission modeling in the MCI code.

MCI is unique compared to other Monte Carlo impurity codes such as DIVIMP (11) in that it has its own internal grid generator. The grid is constructed from an unstructured mesh making it possible to simulate practically any geometric boundary shape. The MCI grid generator is also used to identically reproduce the UEDGE grid and extend it out to the vessel walls. Figure 2 shows the lower divertor region, as references by the dashed rectangular outline in Fig. 1, for two of the grids available in MCI. An unextended UEDGE grid is shown on the left and the corresponding standard MCI grid for the same plasma equilibrium is shown on the right. MCI simulations may be run on either grid or on a nonorthogonal UEDGE grid extended to the walls with an orthogonal MCI grid. Note that the resolution of standard MCI grid is higher than that of the standard UEDGE grid and may be increased of decreased to better match the experimental measurements. Computational time is only moderately increased with grid density in MCI.

IV. DISCUSSION AND MCI RESULTS

MCI is presently being used primarily as an interpretive tool for better understanding the dominant physical processes controlling the divertor performance. Over the course of its development the code has been tested and benchmarked against both fluid and Monte Carlo codes as well as against a variety of experimental data. These benchmarks have been used to identify weak points in the MCI modeling approach and to improve the underlying physics models used by the code. In the following discussion we will highlight some of the results from these benchmarks and summarize the conclusions drawn from them.

A. Comparisons with Other Codes

The standard simulation approach used in MCI involves detailed source modeling and realistic wall conditions as outlined in the previous section. It is also possible to reproduce the boundary conditions used in the UEDGE code and to compare the carbon transport results obtained with each of these codes. This comparison was done

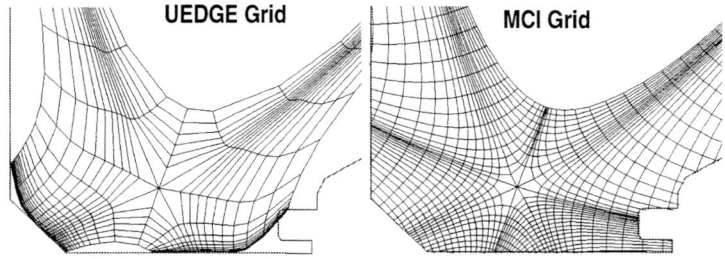

FIGURE 2. Comparison of two grid options which are possible with the MCI grid generator. The low resolution grid on the left is the standard UEDGE grid and the high resolution grid on the right is the standard MCI grid. The standard UEDGE grid can also be extended out to the walls with the MCI grid generator.

for a case in which the divertor plasma was attached to the outer target plate and detached from the inner target plate. These plasmas are typically more straightforward to model because they are less likely to be dominated by nonlinear processes. Several of the changes which had to be implemented in order to reproduce the UEDGE boundary conditions were: the use of a constant sputtering yield, $Y = 10^{-3}$, on each plasma facing surface segment and a modification of the way in which the carbon neutrals are launched from the surface segments. A 2D cosine launch distribution was used and the launch energy was set to match that of the deuterium ions in the cells intersecting the wall segment where a neutral is launched. The edge of the standard UEDGE grid, as shown on the left in Fig. 2, became the plasma facing surface rather then the vessel walls. The sputtering probability along the bottom of the UEDGE grid for this case is shown in Fig. 3.

The integrated source rate and distribution for the two codes, as shown in the figure, is 31.6 Amps of neutral carbon current. UEDGE simulation resulted in a total carbon content, integrated over all charge states, of 1.11×10^{18} particles. This is all the carbon accumulated in the computational domain once a steady state has been reached. It does not include carbon in the well confined plasma volume. The MCI simulation resulted in less carbon by about a factor of 2.6 $i.e.$, a total content of 3.04×10^{17} carbon particles. The largest discrepancy between the two codes is in the total number of C^{2+} ions. The UEDGE simulation yields about 12 times more C^{2+} than MCI. On the other hand, UEDGE and MCI agree relatively well in their C^{5+} and C^{6+} content. MCI and DIVIMP have also been compared using the same data as for the UEDGE comparison. The two Monte Carlo codes agree in total carbon content to within a few percent (12). Based on a series of sensitivity studies done with MCI, as described in the next section, some of the differences between the two codes appear to be related to small variations in the atomic data used. UEDGE and MCI utilized slightly different versions (different internal database switch settings) of ADPAK data for these comparisons. Direct cell-to-cell comparisons of the atomic data are underway to resolve these differences. UEDGE also includes elastic carbon-deuterium neutral collisions in its calculation of the neutral carbon transport. This option is currently being implemented in MCI.

B. The Effects of Atomic Data on Carbon Transport

MCI is used to assess how uncertainties in atomic data affect impurity ion transport and radiation physics. Uncertainties in atomic data can have a relatively large impact on simulation results when parts of the divertor plasma are entering a nonlinear regime. This is typically the case during strongly driven target plate detachment processes or during the formation of a high density, highly radiating, cold plasma (MARFE) (13) near the divertor X–point. We see from Eqs. (2) and (3) that the background ion temperature plays an important role in both the drag term, with a

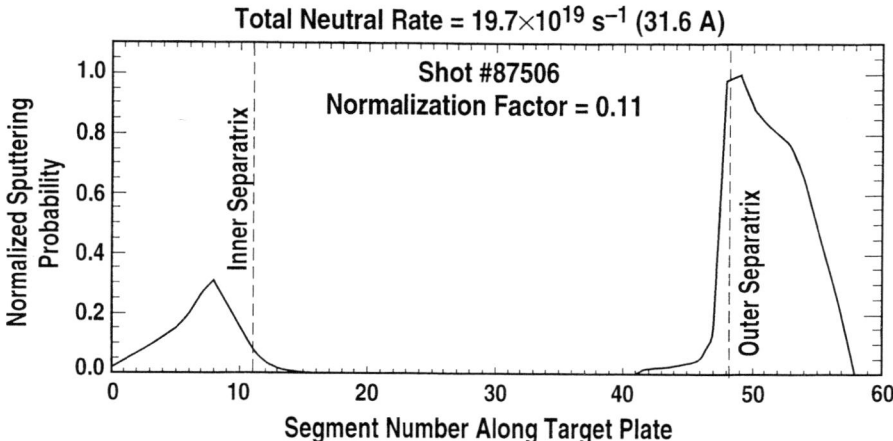

FIGURE 3. Normalized carbon sputtering probability along the lower boundary of the UEDGE grid shown in Fig. 2 for a constant $Y = 10^{-3}$ sputtering yield.

$T_b^{-3/2}$ dependence, and in the ion thermal gradient term when regions of high radiation are present in the divertor. This changes the impurity transport and thus the availability of impurities for assisting in the local cooling of the background plasma. In this section we briefly describe the results of a series of test which were done on the C^0, C^+, D^0, and D^+ charge exchange rate.

For this comparison we increased the charge exchange rate of $C^+ + D^0 \rightarrow D^+ + C^0$ from essentially zero in each cell to a rate equivalent to that for the $C^0 + D^+ \rightarrow D^0 + C^+$ charge exchange process. With the same parameters as used for the comparison discussed in the previous section, this corresponded to a $<\sigma v>_{cx}$ of 2.45×10^{-9} cm^{3}s^{-1} in the cells just above wall segments 4 → 12, as shown in Fig. 3, where $T_e = 1.5$ eV. With both change exchange processes, the integrated carbon sputtering source rate increased from 31.6 to 38.6 Amps while the total UEDGE carbon inventory decreased from 1.11×10^{18} to 6.23×10^{17} particles and MCI's total carbon inventory decreased from 3.04×10^{17} to 2.24×10^{17} particles. In addition, the average lifetime of the carbon particles decreases from 1.55 to 0.93 ms in the MCI simulations. It is interesting to note that in the case with only the energetically allowed $C^0 + D^+ \rightarrow D^0 + C^+$ charge exchange process the total carbon content increased even though there was a smaller sputtering source rate. In this case, MCI estimates of the total power radiated by the carbon in the divertor increased to approximately 46.7 kW whereas with both charge exchange processes the carbon radiation in the divertor was 37 kW. While this radiated power level is still considerably lower than the 382 kW estimated from bolometer measurements it does produce a modest improvement in the radiated power by increasing the total carbon inventory.

An impurity ion tracking algorithm was developed for the MCI code to facilitate studies of how the various terms in the transport equations given above affect the trajectories of impurities as they move through the computational domain. The histories of individual impurity ions are recorded during a simulation and played back in slow motion. As the impurity moves from one time step to another and from cell to cell, the current charge state is displayed and statistics are gathered about the probability of an impurity in a particular charge states escaping from the divertor along with details about the motion of the impurities in various regions of the plasma. In a plasma with a attached outer separatrix, a majority of the carbon impurities are generated near the region where the outer separatrix intersects the divertor target plate *i.e.*, the outer strike point as shown in Fig 3. Thus we are primarily interested in the dynamics of these particles.

Studies of the most statistically significant transport pathways for carbon originating from the outer strike point indicate that C^+ ions created very near the outer strike point target plate have a relatively high probability of escaping the divertor compared to the other transport pathways available. Since the recombination rate in this region is orders of magnitude lower than the ionization rate to the C^{2+} charge state, we find a high probability for $C^+ \rightarrow C^{2+}$ in the first few cells above the target plate. In both charge states, the dynamical balance between the drag due to the deuterium ion flow and the ion thermal gradient force is relatively well maintained. On the other hand, the C^+ ions are relatively cold compared to the background ions in these cells *i.e.*, their thermalization time is long compared to their residence time. Once they become C^{2+} ions their thermalization time is reduced by a factor of four and their parallel diffusion rate increases accordingly. The tracking results clearly show that the escape probability increases with the charge of the impurity ion and reaches approximately 50% in the C^{3+} charge state.

Thus, we see from the discussions of the forces on the C^+ ions given above that the artificial inclusion of an enhanced $C^+ + D^0 \rightarrow C^0 + D^+$ charge exchange rate can be expected to result in a reduction of the carbon inventory in the plasma because it removes C^+ ions from the cells near the targets. This effectively lowers the escape probability of the carbon from the divertor. Thus, as noted above the carbon inventory was reduced when the $C^+ + D^0 \rightarrow C^0 + D^+$ process was included.

Once a carbon ion escapes the divertor region it either goes into the well confined plasma or travels over the top of the discharge, along the inner wall, and down to the inner strike point region. Since the inner strike point is typically detached and the plasma is relatively cold in this region the recombination rate becomes significant compared to the ionization rate and most of the carbon ions are neutralized. The carbon neutrals tend to build up near the inner wall but can also re-ionize and move back up an outer SOL flux tube. A majority of these second generation ions get transported into flux tubes where the flow toward the inner strike plate once again dominates the force balance and eventual re-enter the high recombination zone. Thus, the dynamics in the inner SOL region form a kind of closed loop which results in a build up of the local carbon density. This build up is typically seen in the C^{2+} emissions from the region as shown in the next section.

C. Benchmarks on Experimentally Measured CIII Images

An example of how MCI simulations are benchmarked against experimental data is shown in Fig. 4. An experimentally measured 2D distribution of the C^{2+}, $\lambda = 465$ nm, line emission is shown in the upper part of the figure and the calculated emission pattern, using the carbon source from Fig. 3, is given in the lower part of the figure. Reasonably good qualitative agreements are found in regions with high emission rates. As mentioned in Section B above, an estimate of the total power radiated by the carbon for this case is about a factor of ten below that implied by bolometer measurements of the region suggesting that the simulated carbon content is too low.

D. Comparison of Physical and Chemical Sputtering

Detailed comparisons between the case discussed above *i.e*, $Y = 10^{-3}$ physical sputtering only and simulations with both physical and chemical sputtering have been described in a previous paper (14). Here we briefly summarize the key results. The

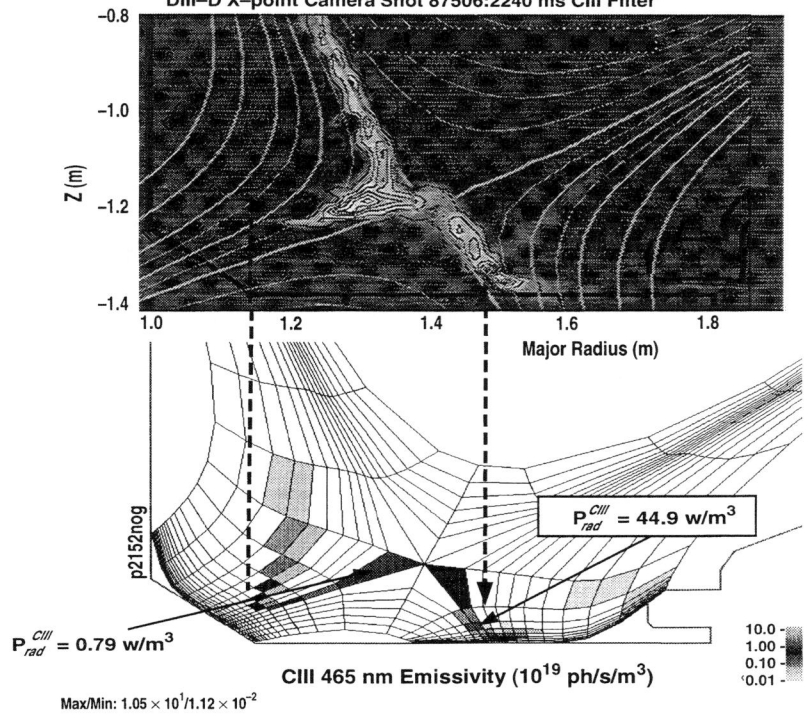

FIGURE 4. Comparison of the measured C^{2+} emissions (above) and those calculated with MCI (below) for shot 87506.

source distribution for the combined sputtering case was similar to that shown in Fig. 3 but the integrated neutral carbon source rate increased from 1.97×10^{20} s^{-1} without chemical sputtering to 1.70×10^{21} s^{-1} with chemical sputtering. While the total carbon inventor increased to 6.11×10^{17} particles with chemical sputtering there was essentially no increase in the neutral carbon inventory. In fact, the C$^+$ content doubled while the C^{2+} and C^{3+} content went up by a factor 4 to 5. As a result, the power radiated from the carbon in the divertor, as estimated from the MCI simulation, increased to approximately 165 kW or roughly half that measured with the bolometer.

V. SUMMARY AND CONCLUSIONS

An overview of the MCI model has been presented along with a representative summary of recent results. MCI simulations are in reasonably good agreement with other fluid and Monte Carlo impurity codes and with experimental measurements in DIII–D. Simulated 2D carbon line radiation patterns agree qualitatively with line filtered images of the C^{2+} emission patterns observed in the DIII–D divertor and total radiated power measurements agree to within a factor of two with those calculated in MCI. The next step required to improve MCI's results is to address the role of nonlinear effects in the simulations. Improvements in the accuracy of the atomic data are essential for simulating nonlinear effects.

ACKNOWLEDGMENT

Work supported by U.S. Department of Energy under Contract No. DE-AC03-89ER51114. Valuable comments by G.D. Porter and W.P. West on comparisons between MCI and UEDGE are gratefully acknowledged. We would also like to thank P. C. Stangeby for suggestions on comments on comparisons with DIVIMP.

REFERENCES

1. Janeschitz, G., Plasma Phys. Contrl. Fusion **37** (1995) A19.
2. Keilhacker, M., *et al.*, Plasma Phys. Contrl. Fusion **37** (1995) A3.
3. Wagner, F., *et al.*, Phys. Rev. Lett. **49** (1982) 1408.
4. Stangeby, P.C., Contrib. Plasma Phys. **28** (1988) 507.
5. Neuhauser, J., et al., Nucl. Fusion **24** (1984) 39.
6. Thompson, M.W., Philos. Mag., **18** (1968) 377.
7. Roth, J., and García-Rosales, C., Nucl. Fusion **36** (1996) 1647, and Corrigendum, Nucl. Fusion **37** (1997) 897.
8. Haasz, A.A., Davis, J.W., J. Nucl. Mater. **175** (1990) 84.
9. Yamada, R.J., J. Nucl. Mater. **174** (1990) 118.
10. Rognlien, T.D. *et al.*, J. Nucl. Mater. **196-198** (1992) 347.
11. Stangeby, P.C., and Elder, J.D., J. Nucl. Mater. **196-198** (1992) 258.
12. Stangeby, P.C., personal communication (1997).
13. Petrie, T.W., *et al.*, Nucl. Fusion **37** (1997) 643.
14. Evans, T.E., *et al.*, Contrib. Plasma Phys. **38** (1998) 260.

Calculated Radiative Power Losses from Mid- and High-Z Impurities in Tokamak Plasmas

Kevin B. Fournier*, M. J. May[†], D. Pacella[¶], B. C. Gregory[¶,1], J. E. Rice[§], J. L. Terry[§], M. Finkenthal[†,2] and W. H. Goldstein*

Lawrence Livermore National Laboratory, Livermore, CA 94551 USA
[†]*Plasma Spectroscopy Group, The Johns Hopkins University, Baltimore, MD 21218*
[¶]*Associazione EURATOM-ENEA sulla Fusione, Centro Riccerche Frascati C.P. 65 – 00044 Frascati, Rome, Italy*
[§]*Plasma Fusion Center, Massachusetts Institute of Technology, Cambridge, MA, 02139*

Abstract. This paper summarizes recent calculations of the radiative cooling coefficient for molybdenum (Z=42), krypton (Z=36) and argon (Z=18). The radiative processes considered are collisional-radiative line emission, dielectronic recombination line emission, and radiative recombination and bremsstrahlung continuum emission. Collisional-radiative line emission dominates the power loss channels for a given impurity at all but the highest plasma electron temperatures. The atomic data for the line emission are computed *ab initio* with the HULLAC atomic physics suite of codes. Relativistic, *ab initio* atomic physics data are used to compute ionization and recombination rate coefficients; the resulting charge state distribution and recombination rates are used to estimate the radiative power from recombination processes. The calculations in the present work are benchmarked against absolute measurements of ion brightness profiles in the Frascati Tokamak Upgrade plasma. Integrated measurements from tokamak plasmas such as bolometery are then simulated. The atomic physics data used to predict the emissivity of individual ions is validated; the calculated cooling coefficients agree well with bolometric measurements.

INTRODUCTION

Computing the radiative cooling coefficient for multi-electron elements involves detailed calculation of an enormous amount of data. The rates of excitation at different temperatures for all transitions in each charge state of a given impurity are required. The beauty of older calculations such as the ADPAK database [1] is that they are generated using highly approximate models which can be implemented in a straight forward manner in computer simulations [2]. Advances in

---

[1)] INRS et Centre Canadien de Fusion Magnétique, Varennes, Quebéc, Canada
[2)] Racah Institute of Physics, Hebrew University, Jerusalem, Israel

computer power in the last two decades and the development of robust atomic physics codes mean that the actual radiative cooling coefficient can be computed from all the necessary, detailed atomic data [3]. In order to produce a single, final number, in this case the radiative power losses at a given temperature, one must sum over many independent quantities (all transitions in all populated ions, rates of recombination, etc.). With this in mind, a comparison of that single, final number with experimental data such as bolometery, which represents the total power radiated from the plasma does not benchmark the calculated cooling coefficient. If the calculated cooling coefficient agrees with the measured bolometric losses, then one knows that the calculated cooling coefficient is *not wrong*. However, in order to truly benchmark the cooling coefficient, the contribution to the total cooling rate from the independent components must be validated. This is accomplished in the present work by taking radial scans of the strong lines of impurity ions. By successfully modeling the emissivity profile of the lines of individual ions, we validate the calculated emissivity coefficient for each ion as well as the calculated charge state distribution of the given impurity. When the components of the calculated radiative cooling coefficient are thus validated, then the total cooling coefficient can be said to be well benchmarked.

The success of current and future magnetically confined fusion plasma experiments will depend on exhausting adequate power from the plasma edge and divertor using impurity radiation [4]. Due to the low sputtering yields and refractory nature of heavy elements, a fusion reactor (with a sustained, ignited plasma) will need to have divertor and plasma facing components made of such materials (e.g. molybdenum or tungsten) [5,6]. For example, tungsten ($Z=74$) plasma facing components are estimated to have a lifetime nearly 300 times longer than beryllium ($Z=4$) or carbon ($Z=5$) plasma facing components in a fusion reactor [7]. However, mid- and high-Z elements have historically been regarded as deleterious to magnetically confined fusion experiments [8]. This is because multi-electron impurity species are incompletely stripped in the high temperature plasma core and thus, they radiate energy from the reaction and compromise energy and particle confinement [9,10]. By coating high-Z plasma facing components with a low-Z material [11,12] (such as boron or graphite), core concentration of strongly radiating impurities can be further reduced. Low-Z materials, on the other hand, suffer from large erosion losses. The controlled introduction of a strongly radiating impurity in the outer part of the plasma can ameliorate erosion of both high- and low-Z materials [4,13]. With this in mind, work was performed during 1995–1996 that generated a database of rates (excitation, ionization and recombination) for highly stripped molybdenum ($Z=42$) ions [14]. The calculations were benchmarked against absolute measurements of x-ray and VUV brightness profiles in the Alcator C-Mod plasma [15]. This database was then used with other, extensive calculations of molybdenum collisional-radiative line emission and recombination line and continuum emission to compute the total radiative cooling coefficient for molybdenum in a low density plasma [16]. This calculation was tested against bolometric measurements of the radiative losses profile at the Frascati Tokamak Upgrade (FTU) and at Alcator

C-Mod [17].

The work has been extended with an emphasis on noble gases, Ne, Ar, Kr and Xe. It is an open question as to which of these elements would be the best coolant in a sub-hundred eV (divertor or scrape-off layer) plasma. Along these lines, a large database of quantities necessary for computing the radiative cooling coefficient has been produced for two inert gases, Ar (Z=18) [18,19] and Kr (Z=36). The cooling coefficient for each element has been computed [20]. In autumn 1997, injections of Ar and Kr into the FTU plasma [21] were made to measure plasma transport properties and the radiative patterns of these impurities. Absolutely calibrated x-ray, XUV and VUV spectra have been recorded at different radial positions and at different times in the plasma. The database of emissivity, ionization and recombination rates for each ion of the above impurities can now be completely validated.

COOLING CURVE CALCULATION

The method used to calculate the radiative cooling coefficients in the present work is described in our previous work [16] for molybdenum. In that work, radiative losses from collisional-radiative (CR) line emission and line emission from dielectronic recombination (DR) and continuum radiation from radiative recombination (RR) were considered. Subsequently [20], the calculations were extended to include bremsstrahlung [22] losses. We briefly remind the reader of some of the quantities computed: the power loss coefficient for line radiation from an ion (with charge q+) at a fixed temperature is given by

$$L_q^{CR}(T_e) = \sum_{j,f} \frac{\hbar c}{\lambda_{j,f}} \times J_{j,f}(T_e), \qquad (1)$$

where $J_{j,f}(T_e)$ (in units of photon cm^3/sec) is found from the collisional-radiative model, $\lambda_{j,f}$ is the transition wavelength, and the sum is taken over all possible radiative transitions (including "forbidden" decays). The emissivity coefficient for transition $j \to f$ from the CR model, $J_{j,f}(T_e)$, has had the dependence on electron density divided out

$$J_{j,f}(T_e) = \frac{I_{j,f}(n_e, T_e)}{n_e n_q} = \frac{n_j^q(n_e, T_e) A_{j,f}^{rad}}{n_e \sum_i n_i^q(n_e, T_e)} \qquad (2)$$

where $I_{j,f}$ is the computed transition intensity, n_q is the total density of ion q+ found by summing over the population in all levels of ion q+, n_j^q is the population in the upper state of the transition in question, and n_e and T_e are the electron density and temperature at which the transition intensity is computed. Thus, the contributions to the total radiative emissivity for a given ion from forbidden decays and from transitions fed by metastable levels other than the ground level are included in $J_{j,f}(T_e)$. $J_{j,f}$ is found to be linearly dependent on electron density (to a few percent) up to $n_e \lesssim 10^{15}$ cm^{-3}. The amount of power radiated per unit volume by collisional-radiative line emission from an impurity atom with atomic number Z is given by

$$P_Z^{CR}(T_e) = n_e \sum_q n_q L_q^{CR}(T_e), \qquad (3)$$

where the sum is taken over all charge states of the atom. The sum in equation 3 can be expressed as

$$P_Z^{CR}(T_e) = n_e n_Z \sum_q f_q^{CE}(T_e) L_q^{CR}(T_e), \qquad (4)$$

where n_Z is the total number density of atom Z and $f_q^{CE}(T_e)$, the coronal equilibrium fractional abundance of ion $q+$. Thus, the power lost per unit volume by CR line radiation can be written

$$P_Z^{CR}(T_e) = n_e n_Z L_Z^{CR}(T_e), \qquad (5)$$

where $L_Z^{CR}(T_e)$ is the CR *radiative cooling coefficient* for element Z.

Great care has been taken to match the grid of temperatures chosen for the CR models for each ion to the range of temperatures where that ion is predicted to exist. In Ref. [16], a grid of temperatures equal to 20, 40, 60, $\cdots$, 120% of the ionization potential of each ion was used. Later it was found that this grid provides inadequate resolution for some ions that exist only at temperatures far below their ionization energy [20,23]. Thus, grids of 5, 10, 15, $\cdots$, 30%, and 10, 20, 30, $\cdots$, 60% of each ion's ionization energy were adopted where appropriate for the argon [20] and krypton cooling coefficients. The budget for power emitted from each configuration in the CR model used for Ar XI is given in Table 1. In this case, dipole allowed transitions from the levels of just a few configurations dominate the CR radiative cooling coefficient.

An integral part of the calculation of the radiative cooling coefficient for some element is the (coronal) charge state distribution (CSD). The relative populations of two adjacent ionization states are given (in the coronal limit) by

$$\frac{n_{q+1}}{n_q} = \frac{S^q(T_e)}{\alpha^{q+1}(T_e)}, \qquad (6)$$

where $S^q(T_e)$ is the total rate of ionization out of the ground level of ion $q+$ and $\alpha^{q+1}(T_e)$ is the total rate of recombination out of the ground level of ion $(q+1)+$. The total rate of ionization includes the contribution from direct electron impact ionization (DI) and collisional excitation followed by autoionization (EA). The total rate of recombination includes contributions from radiative (RR) and dielectronic recombination (DR). DI rates for all molybdenum ions are generated using relativistic calculations [24] of shell binding energies and the formulas of Sampson, Moores and Golden [25–27], krypton and argon DI rates are generated using the

Lotz formula [28,29]. RR rates for all molybdenum, krypton and argon ions are generated from the Hartree-Slater photo-ionization cross sections of Saloman et al. [30]. EA and DR data for the elements in the present work have been generated with the HULLAC suite of codes [24,31,32], details are given elsewhere [14,18,19]. By requiring that

$$\sum_q n_q = 1$$

for each temperature, where the sum is over all ions, the fractional population of each ionization stage

$$f_q^{CE}(T_e) = \frac{n_q}{n_Z} \qquad (7)$$

is found. The CR radiative cooling coefficient for molybdenum, krypton and argon is shown in Figs. 1, 2 and 3, respectively. Also shown are the other cooling channels considered in the present work. The calculated charge state distribution for molybdenum, krypton and argon is shown in the bottom frame of Figs. 1, 2 and 3, respectively.

At all temperatures, CR line emission dominates the *total* radiative loss coefficients (solid line) in Figs. 1 to 3; the only place the CR line cooling appreciably

TABLE 1. Budget for CR line power from each configuration considered in O-like Ar XI at three temperatures and one density. The temperatures shown are ~ 20, 30 and 40% of the ion's ionization energy.

Configuration	Number of Levels	Emission at $n_e = 1 \times 10^{12} cm^{-3}, T_e =$					
		110 eV	%	160 eV	%	220 eV	%
$2s^2 2p^4$	5	1.93[−09]	0.8	2.16[−09]	0.5	2.04[−09]	0.4
$2s^1 2p^5$	4	1.49[−07]	60.9	1.61[−07]	40.7	1.68[−07]	29.8
$2s^2 2p^3 3s^1$	10	1.83[−08]	7.5	4.16[−08]	10.5	6.53[−08]	11.6
$2s^2 2p^3 3p^1$	28	1.00[−08]	4.1	2.26[−08]	5.7	3.53[−08]	6.3
$2s^2 2p^3 3d^1$	38	4.40[−08]	18.0	1.07[−07]	27.0	1.79[−07]	31.8
$2s^2 2p^3 4s^1$	10	2.54[−10]	0.1	7.79[−10]	0.2	1.45[−09]	0.3
$2s^2 2p^3 4p^1$	28	1.23[−09]	0.5	3.51[−09]	0.9	6.26[−09]	1.1
$2s^2 2p^3 4d^1$	38	5.60[−09]	2.3	1.71[−08]	4.3	3.23[−08]	5.7
$2s^2 2p^3 4f^1$	40	3.67[−10]	0.1	1.08[−09]	0.3	1.96[−09]	0.3
$2s^2 2p^3 5s^1$	10	2.68[−11]	0.0	1.01[−10]	0.0	1.94[−10]	0.0
$2s^2 2p^3 5p^1$	28	2.43[−10]	0.1	7.75[−10]	0.2	1.48[−09]	0.3
$2s^2 2p^3 5d^1$	38	1.64[−09]	0.7	5.92[−09]	1.5	1.12[−08]	2.0
$2s^2 2p^3 5f^1$	40	6.30[−11]	0.0	2.15[−10]	0.1	4.12[−10]	0.1
$2s^1 2p^4 3s^1$	16	4.80[−09]	2.0	1.19[−08]	3.0	1.97[−08]	3.5
$2s^1 2p^4 3p^1$	42	2.86[−09]	1.2	8.07[−09]	2.0	1.48[−08]	2.6
$2s^1 2p^4 3d^1$	56	3.50[−09]	1.4	9.87[−09]	2.5	1.78[−08]	3.2
$2s^1 2p^4 4s^1$	16	1.75[−10]	0.1	5.80[−10]	0.1	1.14[−09]	0.2
$2s^1 2p^4 4p^1$	42	1.95[−10]	0.1	7.06[−10]	0.2	1.52[−09]	0.3
$2s^1 2p^4 4d^1$	56	3.92[−10]	0.2	1.39[−09]	0.4	2.87[−09]	0.5
$2s^1 2p^4 4f^1$	60	3.86[−11]	0.0	1.42[−10]	0.0	2.82[−10]	0.1
Totals	604	2.45[−07]	100	3.97[−07]	100	5.62[−07]	100

differs from the total radiative cooling is at the highest temperatures shown, where each ion is nearly fully stripped. The radiative cooling coefficients for DR and RR recombination emission and bremsstrahlung are also shown in Figs. 1 to 3. The radiative power loss coefficient from DR line emission is approximated using the formula in Ref. [1], computed level energies [24] and DR rate coefficients from the literature [14,18,20]. The volumetric power loss from RR continuum emission is also approximated using the formula in Ref. [1], the average charge on an impurity ion at electron temperature T_e as shown in the figures, and the radiative recombination rate coefficients from the literature [30]. Bremsstrahlung continuum emission power losses have been computed using the formula of Ref. [22] and the average charge on an argon ion as shown in Figs. 1 to 3. For comparison,

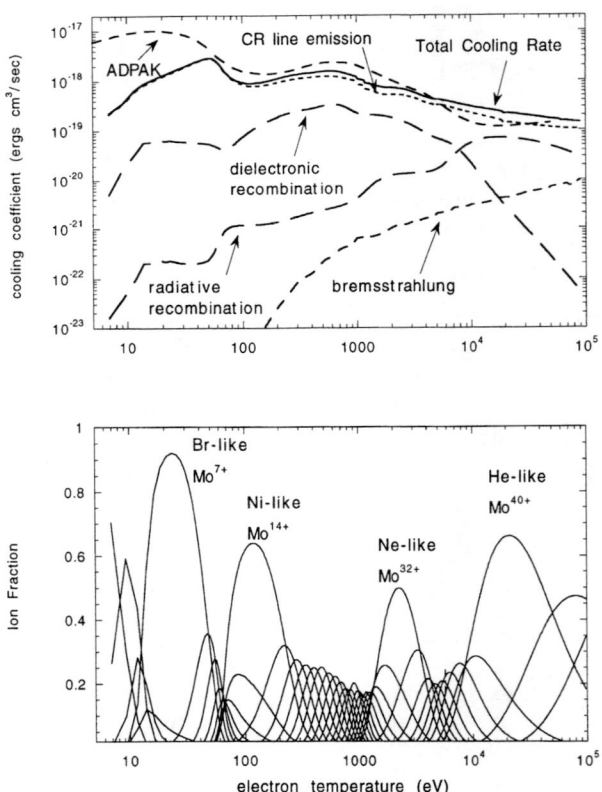

FIGURE 1. Molybdenum radiative cooling channels and the ADPAK radiative cooling coefficient (top) and the coronal molybdenum charge state distribution (bottom).

calculations of the total power loss coefficient by other authors are also shown in Figs. 1 to 3. The total radiative loss coefficient for molybdenum, krypton and argon (including recombination radiative channels) from the ADPAK package [1] is given by the dashes. In Fig. 3, the total radiative cooling coefficient from a recent update of the ADPAK package by Clark, Abdallah and Post [3], which employs detailed collisional-radiative models below 2000 eV, is given by the dash-dot-dot-dot trace. The models generally differ by ∼ a factor of two at higher temperatures, and by more than an order of magnitude at very low temperatures. The factor of two agreement at higher temperatures is remarkable given how different the methods of calculation are between the present work and ADPAK; the lower temperatures, where there is much larger difference between the two models, is the

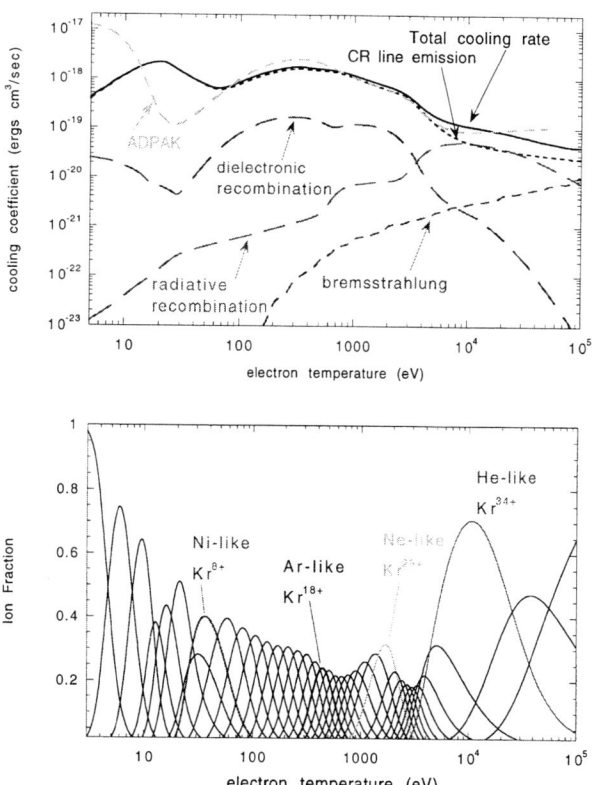

FIGURE 2. Krypton radiative cooling channels and the ADPAK radiative cooling coefficient (top) and the coronal krypton charge state distribution (bottom).

range of temperatures appropriate to divertor and scrape-off layer plasmas. The low temperature calculations must be tested directly in experiments.

MEASURED ION PROFILES

For the present work, absolute measurements of impurity ion radial profiles are used to benchmark the calculated cooling curves. The ion radial profiles are measured in the x-ray (1 to 10 Å), the XUV (20 to 300 Å) and the VUV spectral range (90 to 1200 Å). The observations at FTU in the x-ray range are made with an absolutely calibrated rotating crystal spectrometer and a multi-wire proportional counter detector [33]. The observations in the 20 to 300 Å range are made with a grazing incidence, time resolving grating spectrometer (GRITS) with an electron

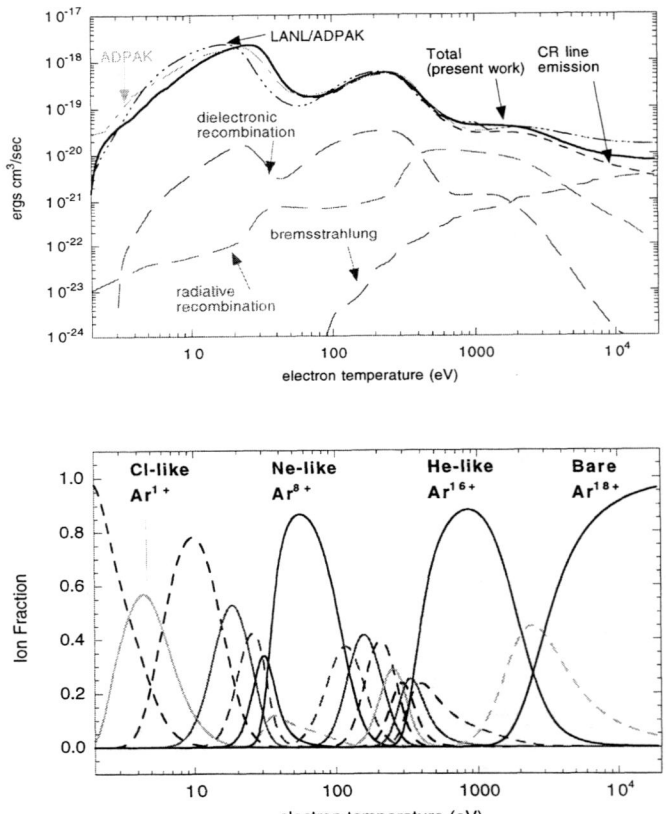

FIGURE 3. Argon radiative cooling channels and the ADPAK radiative cooling coefficient (top) and the coronal argon charge state distribution (bottom).

multichannel plate detector. The spectrometer and detector were photometrically calibrated with synchrotron light at the National Institute of Standards and Technology. A SPRED spectrometer with two gratings provides coverage from 90 to 1200 Å; the grating which covers from 90 to 300 Å has been cross calibrated with the signal of the GRITS. A typical x-ray spectrum from a time before and after the krypton injection is shown in Fig. 4; $\Delta n=1$ lines from ions around the Ne-like Mo^{32+} ion are visible at both times, the $\Delta n=1$ lines from near Ne-like krypton ions are visible after the injection. The signal in the region of the krypton features is nearly zero before the injection, the level of the molybdenum features drops during the injection. Krypton XUV lines as measured with GRITS are seen in Fig. 5; the lines from Cl-like to K-like Kr^{19+} to Kr^{17+} [34–36] sit on a background of $\Delta n=0$ transitions from lower charge states [37], thus making the signal difficult to model [17]. The strong, resonant 3p→3s lines of Na- and Mg-like krypton as seen with the SPRED spectrometer are shown in Fig. 6. The lines of intrinsic molybdenum, iron, chromium and oxygen are also seen in Fig. 6; the level of the intrinsic impurities does not change during the injection.

With coronal ionization-state distributions, detailed excitation rates for observed line transitions and a plasma transport model, the total impurity density for some impurity species can be found for a given shot. The brightnesses of the XUV and x-ray lines are related to the absolute impurity density in the plasma by

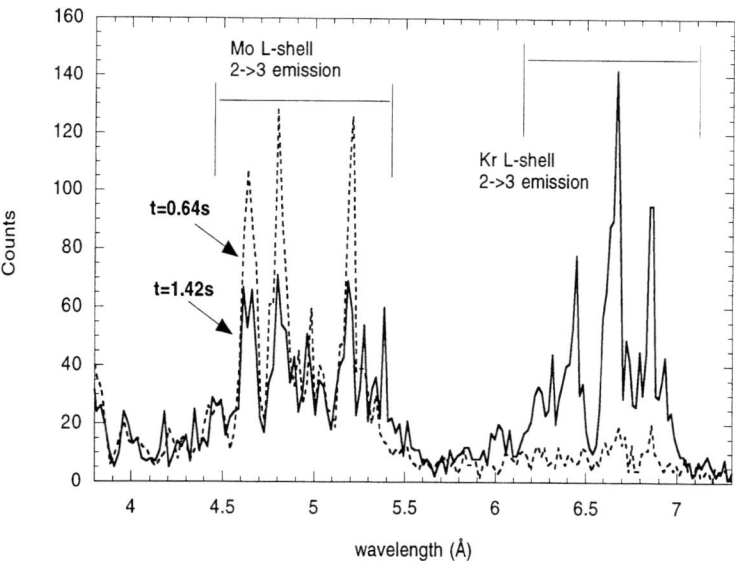

FIGURE 4. The x-ray emission spectra from the FTU plasma at a time before (broken line) and after (solid line) the krypton injection in shot 13719. The L-shell n=2←n=3 x-ray lines of molybdenum and krypton dominate the two spectra.

$$B_{ij}^{\text{obs}} = \frac{\omega}{4\pi} \int_{-L}^{L} n_e(\psi) n_Z f_Z^{q+}(\psi) Q_{ij}^{\text{ex}}(T_e(\psi)) dl \qquad (8)$$

where B_{ij}^{obs} is the measured brightness of a particular line of interest, n_e and T_e are measured electron density and temperature profiles, f_Z^{q+} is the normalized profile of the appropriate charge state, Q_{ij}^{ex} is the calculated excitation rate for the transition, ω is the branching ratio for the transition and ψ is the flux surface coordinate. The use of ψ implies the assumption that the dependent quantity is constant over a flux surface. The integral is performed along the observation line of sight. The impurity charge state distribution is related to the impurity ion density through the radial continuity equation. The continuity equation contains the source and sink terms for each charge state as well as a gradient driven anomalous transport term for the ion flux. Transport coefficients for our simulations are estimated from (at C-Mod) impurity injection experiments [15] and (at FTU) from simulations of well understood spectra [38]; either the transport coefficients or the atomic data for observed transitions must be "known" in order to investigate the other set of data. For ions which exist in the center of the plasma, where temperature and density profiles are flat, the effect of anomalous transport on the ions' distributions is negligible, and the observations test directly the calculated excitation rate coefficient. Impurity ion densities have been measured in this manner on the Alcator C-Mod tokamak [15] and the collisional excitation rates used to compute the molybdenum cooling

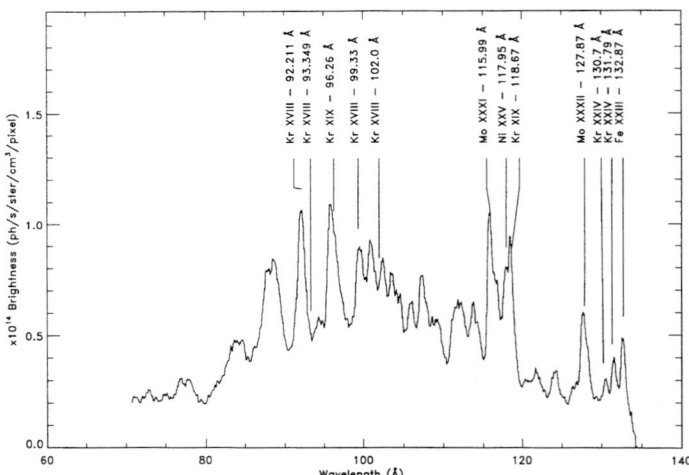

FIGURE 5. The XUV emission spectra from M-shell krypton ions following injection in shot 13728. The lines are from $3p\leftarrow 3d$ transitions in Kr^{17+} to Kr^{19+}; the lines sit on a background of $\Delta n=0$ n=3$\leftarrow$n=3 transitions from lower Kr chargestates.

coefficient have thus been benchmarked. The measured x-ray brightness profiles for Na-, Ne- and F-like molybdenum ions are shown in Fig. 7 [15]; the simultaneous agreement between the data and simulations for the three ions is impossible to achieve using ADPAK atomic physics and varying the plasma transport parameters. Only by including the ionization and recombination data of Ref. [14] in the simulations is the agreement in Fig. 7 achieved. The radial brightness profiles of central and non-central ions of krypton and argon have been measured in the FTU plasma and are currently being analyzed.

COMPARISONS WITH BOLOMETERY

In order to compare the radiative cooling coefficients of argon and krypton with FTU bolometer data, we have derived the total impurity density profile using bremsstrahlung measurements; the impurity density profiles and measured radiative losses are used to derive an impurity radiative cooling coefficient. This work will be presented in more detail by B. Gregory at the European Physical Society meeting this summer [39]. Briefly, the bremsstrahlung signal at a time before the injection of the gas and at a time after the injection are inverted to yield the bremsstrahlung emissivity profile. The profiles before and after the injection are

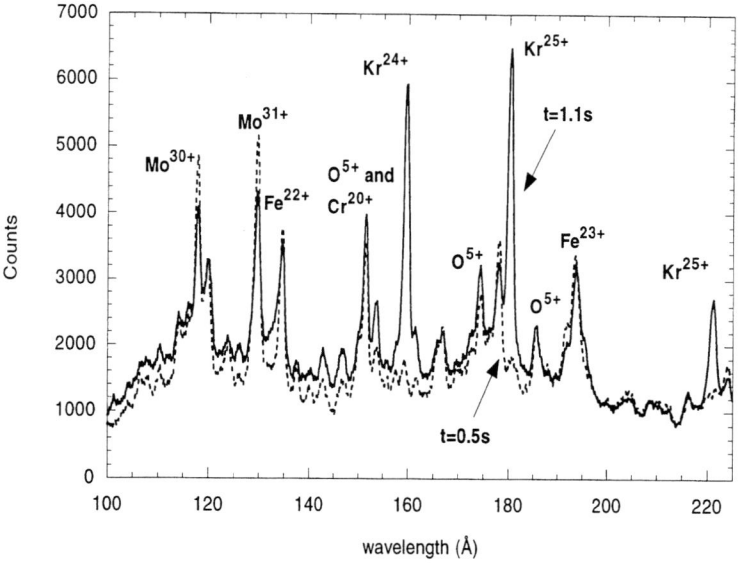

FIGURE 6. The VUV emission spectra from the FTU plasma at a time before (broken line) and after (solid line) the krypton injection in shot 13719. The lines from molybdenum, iron, chromium and oxygen are also labeled in the figure; the level of the intrinsic impurities does not change during the injection.

subtracted and the contribution to the bremsstrahlung exclusively from the impurity is found. Then, using the average charge on an impurity ion at a given temperature [1], we find the density profile of that impurity [22]. After inverting the signal from a 13 channel bolometer array at a time before and after the injection, the radiation emitted from the plasma due to the presence of the impurity is found. This technique relies on the fact that the gas injection does not change the intrinsic impurity levels (see Fig. 6). Dividing the impurity radiation profile by the measured electron and impurity density profiles, the total cooling rate for the impurity in the plasma is found

$$L_Z^{\text{tot}}[T_e(r)] = \frac{1}{n_Z(r)} \left[\frac{P_{\text{rad}}(r)}{n_e(r)} - \frac{P'_{\text{rad}}(r)}{n'_e(r)} \right], \quad (9)$$

where the primed quantities are from a time before the injection. With the measured temperature profile for the plasma, the cooling rate for each impurity is known as a function of temperature. The derived cooling coefficients for krypton and argon using the above procedure are shown in Figs. 8 and 9, respectively. The derived radiative cooling coefficient is shown with a ± 40% error; this is a conservative estimate of the uncertainties from the inversion of the bremsstrahlung

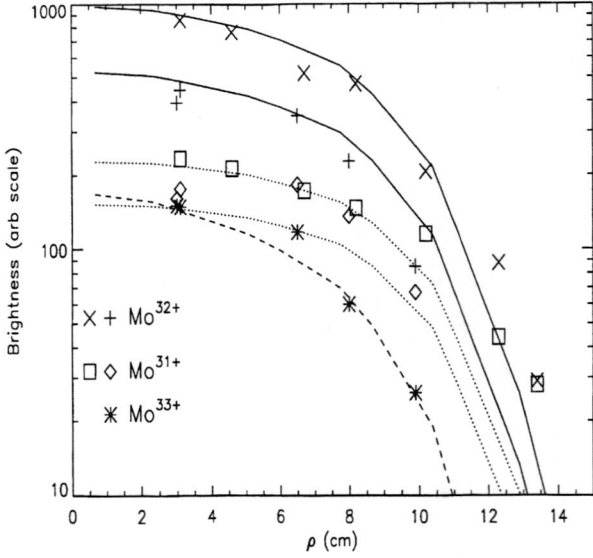

FIGURE 7. Measured x-ray brightness profiles for Na-, Ne- and F-like molybdenum ions. Measurement courtesy of J. Rice, Plasma Science and Fusion Center, Massachusetts Institute of Technology (figure previously in Ref. 15).

and bolometery data and in the measured electron density and temperature profiles. We conclude that the measured krypton radiative cooling coefficient exhibits a much broader radiation barrier, which peaks at a lower ($\sim$ factor of two) value, than that predicted with ADPAK [1]. At low temperatures, the cooling coefficient of the present work matches the experimental values much better than the ADPAK coefficient. The derived argon radiative cooling coefficient agrees well with both ADPAK and the cooling coefficient of the present work for temperatures above 1 keV.

CONCLUSION

This paper reports on work in progress; several encouraging results have been found so far. First, an enormous database of ionic rates for several elements relevant to magnetically confined fusion experiments has been developed and is in the process of being validated. Application of the database to plasma transport models [40] and analysis of plasma behavior in the presence of a strongly radiating impurity [41] is underway. Measurable differences are found between our calculations of radiative losses for high-Z materials and older models [1] across all ranges of temperature. Continued efforts to analyze the radial brightness profiles of the ions of noble gases are underway; progress will be made in the benchmarking of atomic data and in the understanding of anomalous plasma transport. Finally, once the

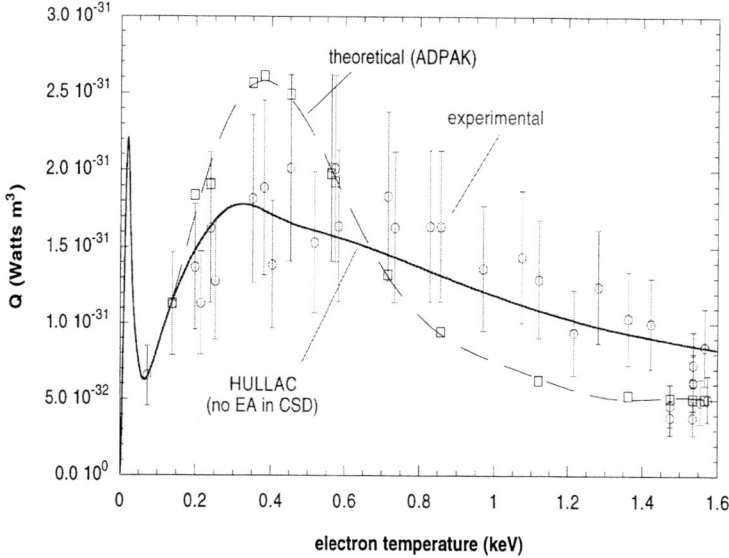

FIGURE 8. Krypton radiative cooling rate as derived from FTU bolometery, and the HULLAC and ADPAK (Ref. 1) radiative cooling coefficient.

emissivity model for each ion of a strongly radiating impurity is benchmarked, then the calculation of the total impurity cooling coefficient for that element can be said to be validated. This work was performed under the auspices of the U. S. Department of Energy at the Lawrence Livermore National Laboratory under contract No. W-7405-ENG-48.

REFERENCES

1. D. E. Post, R. Jensen, C. B. Tarter, W. Grasberger, and W. Lockke, At. Data Nucl. Data Tables **20**, 397 (1977).
2. R. Hulse, Nuclear Tech./Fusion **3**, 259 (1983).
3. R. Clark, J. Abdallah, and D. E. Post, J. of Nuclear Materials **220-222**, 1028 (1995).
4. D. E. Post, J. Abdallah, R. Clark, and N. Putvinskaya, Phys. Plasmas **2**, 2328 (1995).
5. D. E. Post, J. of Nuclear Materials **222**, 143 (1995).
6. N. Noda, V. Philipps, and R. Neu, J. of Nuclear Materials **241–243**, 227 (1997).
7. J. N. Brooks et al., J. of Nuclear Materials **220–222**, 269 (1995).
8. E. Hinnov et al., Nucl. Fusion **18**, 1305 (1978).
9. R. V. Jensen, D. Post, and D. L. Jassby, Nucl. Sci. Eng. **65**, 282 (1978).
10. N. Peacock, R. Barnsley, N. Hawkes, K. Lawson, and M. O'Mullane, Spectroscopy for impurity control in iter, in *Diagnostics for experimental thermonuclear fusion*

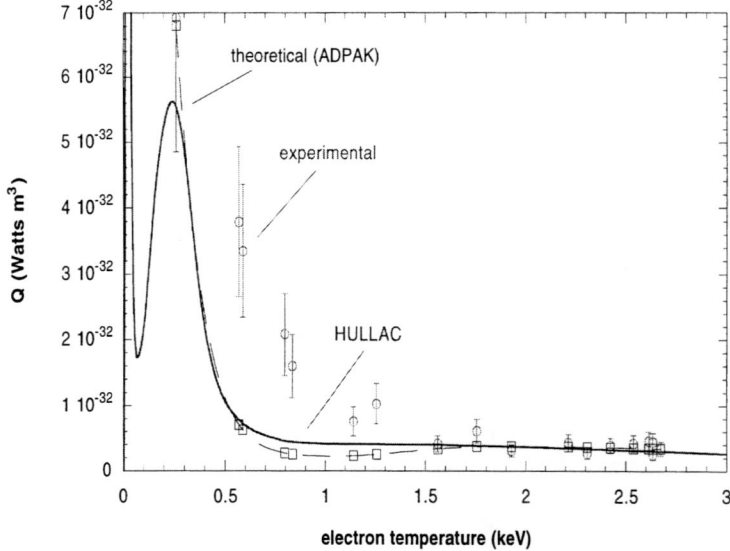

FIGURE 9. Argon radiative cooling rate as derived from FTU bolometery, and the HULLAC and ADPAK (Ref. 1) radiative cooling coefficient.

reactors, Varenna (Italy), edited by P. Stott, G. Giuseppe, and E. Sindoni, pages 291 – 305, New York, 1996, Plenum Press.

11. E. S. Marmar et al., Bull. Am. Phys. Soc. **41**, 7R24 (1996).
12. C. Reddy et al., Bull. Am. Phys. Soc. **41**, 7R23 (1996).
13. S. Allen et al., J. of Nuclear Materials **196**, 804 (1992).
14. K. B. Fournier et al., Phys. Rev. A **54**, 3870 (1996).
15. J. E. Rice et al., J. Phys. B: At. Mol. Phys. **29**, 2191 (1996).
16. K. B. Fournier, D. Pacella, M. J. May, M. Finkenthal, and W. H. Goldstein, Nucl. Fusion **37**, 825 (1997).
17. M. J. May et al., Nucl. Fusion **37**, 881 (1997).
18. K. B. Fournier, M. Cohen, and W. H. Goldstein, Phys. Rev. A **56**, 4715 (1997).
19. M. Cohen, K. B. Fournier, and W. H. Goldstein, Phys. Rev. A **57**, (April, 1998).
20. K. B. Fournier, M. Cohen, M. J. May, and W. H. Goldstein, At. Data Nucl. Data Tables (submitted Jan., 1998).
21. F. Alladio et al., Plasma Phys. Control. Fusion **36**, B253 (1994).
22. K. Kadota, M. Otsuka, and J. Fujita, Nucl. Fusion **20**, 209 (1980).
23. Corrigendum submitted to Nuclear Fusion, February, 1998.
24. M. Klapisch, J. Schwob, B. Fraenkel, and J. Oreg, J. Opt. Soc. Am. **67**, 148 (1977).
25. D. Sampson and L. Golden, Astrophys. J. **170**, 169 (1971).
26. D. Moores, L. Golden, and D. Sampson, J. Phys. B: At. Mol. Phys. **13**, 385 (1980).
27. L. Golden, R. Clark, S. Goett, and D. Sampson, Astrophys. J. Suppl. Ser. **45**, 603 (1981).
28. W. Lotz, Z. Phys. **216**, 241 (1968).
29. W. Lotz, Z. Phys. **232**, 101 (1970).
30. E. B. Saloman, J. H. Hubble, and J. H. Scofield, At. Data Nucl. Data Tables **38**, 1 (1988).
31. A. Bar-Shalom and M. Klapisch, Computer Phys. Comm. **50**, 375 (1988).
32. A. Bar-Shalom, M. Klapisch, and J. Oreg, Phys. Rev. A **38**, 1773 (1988).
33. R. Bartiromo et al., Nuclear Instruments and Methods in Physics Research B **95**, 537 (1995).
34. V. Kaufman, J. Sugar, and W. Rowan, J. Opt. Soc. Am. B **6**, 1444 (1989).
35. Equipe TFR and J. F. Wyart, Phys. Scripta **37**, 66 (1988).
36. V. Kaufman, J. Sugar, and W. Rowan, J. Opt. Soc. Am. B **6**, 142 (1989).
37. R. E. Stewart, D. D. Dietrich, R. J. Fortner, and R. Dukart, J. Opt. Soc. Am. B **4**, 396 (1987).
38. L. Carraro et al., Phys. Scripta **55**, 565 (1997).
39. D. Pacella et al., Proceedings of the 25rd European Conference on Controlled Fusion and Plasma Physics *Praha, Czech Republic* (1998).
40. D. Pacella, L. Gabellieri, G. Mazzitelli, K. B. Fournier, and M. Finkenthal, Plasma Phys. Control. Fusion **39**, 1501 (1996).
41. M. J. May et al., Plasma Phys. Control. Fusion (submitted February, 1998).

ATOMIC PROCESSES IN LASER PLASMAS

Intense and Ultrashort Pulse Laser Interactions with Matter

R.W. Falcone

Department of Physics, University of California, Berkeley, Berkeley CA 94720-7300

High-intensity laser interaction with matter on ultrashort time scales can lead to the generation of dense plasmas with high temperatures, and the emission of x-rays ranging from short-pulse, incoherent light to coherent harmonics. Results from solids, clusters, and isolated atoms will be discussed. At reduced intensity, short pulses will lead to rapid disorder at the surface of crytalline materials, which is detectable by time resolved x-ray diffraction on picosecond timescales. This disordering effect is reversible.

Studies of Ultrafast Laser-Produced X-ray Sources

J.-C. Gauthier*, J.P. Geindre*, P. Audebert*, S. Bastiani*,
Th. Schlegel*, C. Quoix[†], A. Rousse[†], and A. Antonetti[†]

*Laboratoire pour l'Utilisation des Lasers Intenses, UMR 7605 CNRS,
Ecole Polytechnique, 91128 Palaiseau, France
† Laboratoire d'Optique Appliquée, UMR 7639 CNRS,
Rue de la Hunière, 91167 Palaiseau, France

Abstract. Experimental efforts have started for bringing (sub-)picosecond time resolution in pump-probe X-ray diffraction and absorption spectroscopy of transient chemical, biological or physical phenomena. Among the various pulsed X-ray sources available, ultrafast laser-produced plasmas are very promising for the generation of sub-picosecond X-rays. We have significantly progressed toward a better understanding of ultrafast ($\approx 100 fs$) laser-matter interaction at moderate ($\approx 5 \times 10^{16} W/cm^2$) intensities. We have found that fast electron production and non-thermal K_α emission peak for a plasma scale length where resonance absorption is maximized. The plasma scale length can be efficiently controlled by using a prepulse of variable relative intensity and delay. Targets of different substrate and overcoat layer materials have been used to obtain informations on the anisotropy of the electron distribution function and on hot electron energy transport. Experimental results are found in satisfactory good agreement with 1-D hydrodynamic and particle-in-cell simulations.

INTRODUCTION

Many fundamental processes in physics, materials science, chemistry, and biology occur on the ultrafast (picosecond or subpicosecond) time scale. Some of these processes can be initiated by transient optical excitation, and followed in their time evolution by ultrafast infrared, visible and ultraviolet spectroscopy. Pump-probe optical techniques are sensitive to electronic excitations, whereas extension of this measurement method to the subnanometer wavelength range should make possible the direct monitoring of atomic positions.

Accordingly, the great potential of ultrafast x-ray techniques in absorption spectroscopy and time-resolved crystal diffraction has initiated the development of ultrashort pulse x-ray sources and innovative detection systems. The x-ray sources must satisfy particular requirements concerning their pulse duration, spectrum, brightness, and overall photon flux. In addition, it would be particularly helpful to accurately synchronize the x-ray source with other events that can be driven, triggered or stimulated by laser light. Among the newer short pulse x-ray sources available, third generation synchrotron radiation machines, high order harmonic generation in gases and on solid surfaces, and ultrafast laser-produced plasma sources (including the x-ray laser) are the more promising for applications.

For a decade, high-intensity subpicosecond lasers with chirped pulse amplification (CPA) [1] have opened a new field of study of laser matter interaction with solid targets [2,3]. Very short temporal ($< ps$) and spatial ($< 100 nm$) scale plasmas are produced with highly transient and non-equilibrium properties. These plasmas have attracted attention as potential sources for ultrafast pulsed x-rays in the sub-keV energy range [4–6], and the keV range [7–9]. Several experimental efforts for bringing picosecond time resolution in diffraction, spectroscopy, or microscopy of transient phenomena [10–12] have been reported. In this paper, after a short review of the emission properties of the various x-ray sources available, we will present experimental results of the ultrashort pulse K_α emission produced during the interaction of a p-polarized, a few $10^{16} W cm^{-2}$ intensity, 45° angle of incidence, laser pulse of $120 fs$ duration on solid targets. By modifying the electron density gradient scale length, we explore the highly complex transition between steplike gradient absorption and resonance absorption. This is done by varying the temporal separation between the main laser pulse and a prepulse of the same duration but with 1% of the intensity of the main interacting pulse. The electron density gradient scale length produced by the earlier pulse is measured as a function of time by dual-polarization (quadrature) frequency domain interferometry [13]. These measurements are correlated with Si K_α line emission measurements from the bulk of the target and electron emission measurements from the front of the target. We find an optimum delay between the prepulse and the main pulse for which the laser absorption and the K_α yield are greatly enhanced with respect to prepulse-free conditions. Doing so, the K_α emission yield is improved by a factor of 40 with respect to our previous measurements [8]. Hydrodynamic simulations with the codes FILM [14], MULTI [15] and particle-in-cell simulations with the code EUTERPE [16] are in satisfactory agreement with the experimental results.

SHORT PULSE DURATION X-RAY SOURCES

There are practical difficulties in comparing the relative benefits and drawbacks of different x-ray sources which reside, first, in their wide range of operational repetition rates and, second, in the fact that some sources are collimated while others emit in the whole 4π steradian solid angle. Repetition rates go from a few shots per hour with the high energy pulsed lasers, a few kHz for harmonic generation, to a few MHz for synchrotron radiation. We have chosen to compare the sources according to two their peak spectral brilliance B_{peak}, expressed as a number of x-ray photons per second per square millimeter per square milliradian in a spectral bandwidth of 0.1% of the photon energy. The peak spectral brilliance is proportional to the number of photons emitted by the source divided by the pulse duration. This favors bright and ultrashort pulsed sources. However, for the experimentalist, the most important source feature is the real number of x-ray photons reaching the sample per pulse. This quantity is highly constrained, of course, by the particular arrangement of the experiment (energy range and spectral bandwidth) and will dictate the choice of the detection technique (signal to noise ratio and repetition rate) to be used.

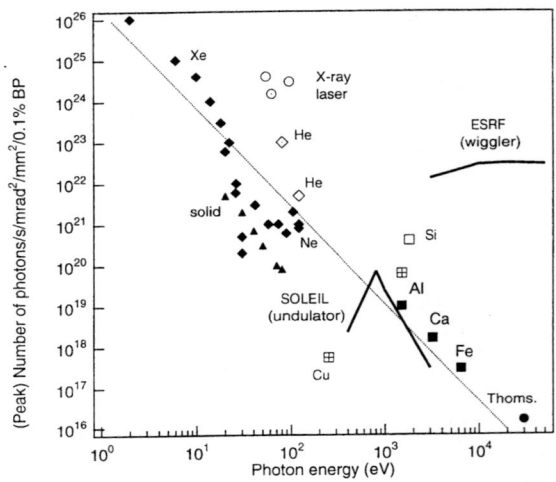

FIGURE 1. Peak spectral brilliance in photons/(s. mrd^2 mm^2 0.1%BW) as a function of photon energy. Diamonds: harmonic generation in gases; triangles: harmonic generation on solids; open circles: x-ray laser; solid lines: synchrotron radiation insertion devices; squares: laser plasma characteristics radiation; crossed squares: laser plasma thermal radiation; circle: Thomson scattering. The dotted line is an aid to the eye.

Third generation synchrotron machines

Here, we will concentrate on the performances which are obtained with third generation storage rings such as the European Synchrotron Radiation Facility (ESRF) in Grenoble [17] for the high energy range ($\geq 1000 eV$). At the ESRF, the x-ray pulse duration is about $150 ps$ at $15 mA$ current per bunch. For the SOLEIL proposal [18], a pulse duration of $52 ps$ is quoted. On both machines, insertion devices (undulators and wigglers) enhance the x-ray intensity over the bending magnet intensity by several orders of magnitude. In figure 1 is plotted the *peak* spectral brilliance as a function of photon energy for both the ESRF and the SOLEIL machines. The ESRF insertion device is a wiggler thus producing a broad and smooth spectrum around $10 keV$ whereas the SOLEIL undulator is highly brilliant in a narrow band around $1 keV$. Synchrotron radiation x-ray sources are at their disadvantage because of their rather long pulse duration in the tens of picoseconds range. This can be reduced to the femtosecond time scale by using collinear interaction of the electron bunch along an insertion device with an ultrashort pulse optical laser [19] or by electron Thomson scattering [20] of femtosecond laser light.

High order harmonic generation

High order harmonic generation provides another source of coherent radiation in the X-UV region of the spectrum. This type of radiation is obtained by focusing an intense short pulse laser in a gas. For intensities in the perturbative regime (up to $10^{12} W/cm^2$), the efficiency of the Nth harmonic obeys a simple power law. For higher intensities, the spectrum contains a plateau of approximately *constant efficiency* harmonics followed by a sharp cut-off at an energy given roughly by $\hbar\omega_{max} = I_p + 3U_p$ where I_p is the ionization potential of the atom and U_p is the electron oscillation energy. More details on the physics of the harmonic generation processes and of the phase matching between the fundamental and high order radiation can be found in recent reviews [21,22]. Shorter laser pulses generate higher harmonics and stronger laser pulses are more efficient. Figure 1 gives the peak brilliance of high order harmonic generation in neon (light diamonds) and xenon (darker diamonds). Very efficient harmonic generation in helium (open diamonds) with a high intensity Nd:glass laser at $1.05\mu m$ and $0.53\mu m$ has also been reported [21]. There are difficulties in obtaining wavelengths shorter than $\approx 10nm$ since ionization prevents the atom from experiencing high laser intensity. However, recent experiments have extended the range of application of harmonic generation down to the low energy edge of the so-called water window ($4.4 < \lambda < 2.3nm$) [23]. Time duration of harmonic emission has been measured to be of a few tens of femtoseconds (for $150 fs$ laser duration) by laser-assisted photoelectric effect in gases [24]. The size of the x-ray source is generally much smaller than the original laser focal spot diameter due to the high nonlinearity of the generation process. The angular spread of the harmonic beam is also very much narrower than the driving laser beam.

Other sources of high harmonics are produced from dense plasma surfaces [25]. In solids, the very large light pressure is responsible for the generation of harmonics. Pushing on the steep vacuum-solid interface at the laser frequency modulates the reflected beam at the same frequency. This "oscillating mirror" gives rise to a series of sidebands on the reflected spectrum, separated by ω or 2ω where ω is the fundamental laser pulsation. Accordingly, all harmonic orders are emitted, contrary to the gas phase where only odd orders are symmetry permitted. In addition, contrary to gas harmonics, the emission is diffuse (into 2π steradians). Figure 1 summarizes the results obtained at the Rutherford Appleton Laboratory [25] (line and solid circles). Harmonics appear more interesting as sources for applications at high repetition rate with "table-top" terawatt lasers.

Laser-produced plasma sources

The study of laser-plasma x-ray emission has been stimulated by the field of laser fusion [26]. Laser-plasmas have a number of characteristics [7,9,27,28] that make them valuable as x-ray sources: i) a wide range of pulse durations from a few hundred femtoseconds to hundreds of nanoseconds, ii) very bright x-ray sources can be generated with source sizes as small as a few microns, and iii) laser-plasma x-ray sources can be accurately synchronized with other events that can be driven, triggered or stimulated by the same laser light.

When a high intensity laser is focused onto a solid, the electron temperature can reach values of $100 - 1000eV$ depending on laser intensity. At these high temperatures, thermal X-rays at energies above a kilovolt are produced. For ultrafast pulses, due to the strong gradient and high density, rapid quenching of X-ray emission is expected by thermal conduction into the underlying cold material and by hydrodynamic expansion. Besides collisional absorption, non-linear absorption mechanisms have been shown to contribute efficiently to the overall deposition of laser light [29]. These non-linear mechanisms produce hot electrons which give rise to bremsstrahlung and K_α radiation from the target bulk. This non-thermal emission is thought also to be very short because, in principle, hot electrons are produced only during the laser pulse.

Thermal x-ray emission from laser-produced plasmas in the subpicosecond regime occurs with pulse durations of the order of a few tens of picoseconds [30–32], as measured by a streak camera. Peak brilliance results for laser-produced plasma thermal x-ray sources are given in Fig. 1 for a streak-camera-measured x-ray pulse duration of $4ps$ at $1500eV$ (open crossed-squares).

Suprathermal electrons produced by non-collisional absorption mechanisms have proved to be a convenient way of generating x-rays in the photon energy range above one keV [8,33–35]. In the $1 - 10keV$ energy range, efficient production of K_α radiation in aluminum, calcium, and iron has been demonstrated [8]. The x-ray throughput was controlled by varying the energy contrast ratio between the main ultrashort pulse and its nanosecond pedestal. Results are shown in Fig. 1 (filled squares) for the three target materials. Peak brilliance is comparable to SOLEIL synchrotron radiation but the streak-measured pulse duration is below $2ps$, instead of $\approx 50ps$. The measured source diameter is $10\mu m$, the repetition rate is $10Hz$ and the total number of photons per shot is about 3×10^8.

X-ray lasers have also a great potential for applications in time-resolved x-ray studies [36,37]. They emit partially coherent, linearly polarized light and they are particularly efficient in the $10 - 30nm$ range. The number of photons per pulse can be as large as 10^{14} and the pulse duration is about $70 - 80ps$. They have a peak brightness (see Fig. 1) very much larger than high order harmonic generation in the $10 - 100eV$ energy range. However, their very low repetition rate of several shots per hour limit them to applications where coherence (interferometry of plasmas, surface studies) or the x-ray flux (nonlinear studies in the X-UV) are the important factors.

THE OPTIMIZED K_α X-RAY SOURCE

Over the last years it has been recognized in several fields of laser plasma interactions that the temporal shape and the intensity contrast ratio of the laser pulse is of paramount importance in tailoring the plasma properties. For example, the importance of a controlled prepulse (or multipulses) in the pumping and amplification of X-ray lasers has been reviewed recently [37]. By varying the temporal separation between the main laser pulse and a prepulse, one can generate preformed plasmas with different density gradients scale lengths. The description of the numerous linear and non linear laser absorption mechanisms in the ultra short laser pulse regime for different values of L/λ (where L is the gradient scale length and λ is the laser wavelength) has been detailed in our previous paper [38]. A comprehensive review of state-of-the-art theory has been given recently [3].

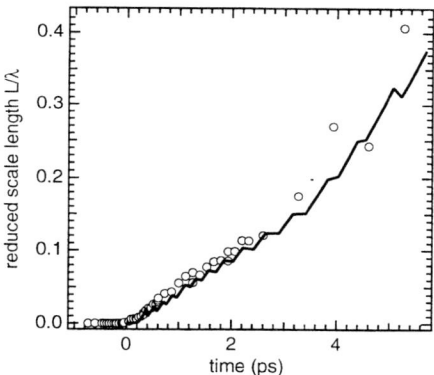

FIGURE 2. Measured (circles) reduced electron density gradient scale length as a function of time after the prepulse. The solid line is the result of FILM hydrocode simulation.

Optimization procedure and results

The experiments were carried out with the Laboratoire d'Optique Appliquée $Ti : Al_2O_3$ CPA laser in Palaiseau [39]. Experiments have been performed for a laser pulse incident on a silicon target at an angle of 45° with respect to the normal of the target. Laser intensities were kept constant at $4 \times 10^{16} W\,cm^{-2}$ for the main pulse and at $4 \times 10^{14} W\,cm^{-2}$ for the prepulse. Further details on the experimental procedures have been described in great detail elsewhere [35,38,40].

Our first step to optimize laser energy conversion into fast electrons was to characterize precisely the variation of the electron density gradient scale length as a function of the delay between the main pulse and the prepulse. To do this, we have used the technique of spectral (frequency-domain) interferometry to perform electron density gradient scale length measurements [41]. We have extended the technique to allow simultaneous measurements of the phase shift for the two (s and p) probe polarizations. Figure 2 shows the variation of the reduced scale length L/λ as a function of the delay. Assuming that the predominant absorption mechanism is resonance absorption for scale lengths of the order of a fraction of a laser wavelength (this hypothesis is supported by previous measurements [41] and by theory), we find [35] that the optimum value of $L/\lambda \approx 0.3$ is reached at a time $t \approx 5ps$.

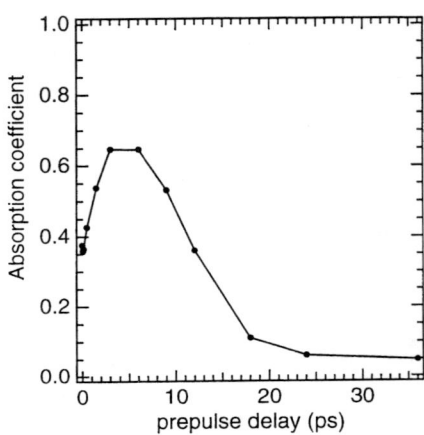

FIGURE 3. Laser absorption coefficient of the main pulse as a function of time after the prepulse. The range of delays is much larger than in Fig. 2.

Absorption measurements of the main pulse are shown in Fig. 3 as a function of the prepulse delay. Without a prepulse, i.e. for a very steep gradient, our measured absorption coefficient is in good agreement with Fresnel absorption at 45° incidence angle. In Fig. 3, the presence of a clear maximum of absorption can be seen for a delay of $\approx 5ps$. This corresponds to the optimum initial density gradient scale length shown in Fig. 2.

Up to now, we have determined the variation of the gradient scale length and of the laser absorption as a function of the time delay of the prepulse. For the optimization of the x-ray source, which relies on the production of energetic electrons, direct measurements of the electrons energies, both outside the target (towards the laser) and inside the target, would be very useful to give a complete view of the laser to electron energy conversion efficiency. For the electrons escaping the target, we have used an electron spectrograph [35] to determine the average "temperature" of the fast electrons.

Because we have only 6 electron channels, the determination of the electron temperature is quite imprecise. To improve the signal-to-noise ratio, we have integrated the electron signal over a range of angles excluding the specular direction, where

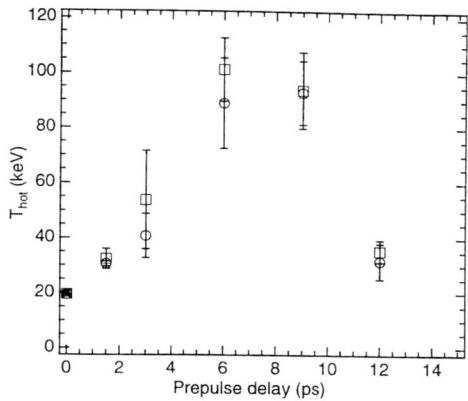

FIGURE 4. Hot electron temperature as a function of the prepulse delay.

electron jetting has been observed [38]. Results are shown in Fig. 4 where we have plotted the slope of the electron distribution function (the hot electron "temperature") as a function of the prepulse delay. Again, we clearly show that the highest temperature is obtained for the "optimum" delay of $\approx 6ps$. To determine the average energy of the electrons inside the target, we have made use of the multilayer technique [8] where a thin, variable thickness aluminum layer has been deposited over an iron substrate. By recording the relative intensities of the aluminum and iron K_α lines as a function of the aluminum thickness, the electron energy distribution can be unfolded from experiment with the help of Monte-Carlo simulations of the electron energy penetration and deposition inside the target [40]. Results are shown in Fig. 5 where we have plotted the ratio of Fe and Al K_α intensities for delays of 0, 3, and $6ps$ between the prepulse and the main pulse. The theoretical estimate of this ratio with an incident Maxwellian distribution of electrons at $12keV$ is shown for comparison. In contradiction with the temperature measurements of the electrons escaping the target, we find that the electrons depositing their energy inside the bulk of the target exhibit an average electron energy which is independent of the prepulse delay. We will discuss this finding in the theory section.

Finally, we have measured the K_α emission as a function of the delay between the prepulse and the main pulse. The results are shown in Fig. 6, where the open circles are the results of a single measurement and the solid circles are the averaged values. We have also given the values of the conversion efficiency from laser energy into x-ray energy, on the K_α line. The conversion efficiency from laser to electron energy can be estimated to be about 10%. We observe a strong increase of the K_α yield for the first $10ps$ and then a slow decrease for longer delays. We obtain an enhancement of about a factor 7 with respect to the case of an abrupt initial

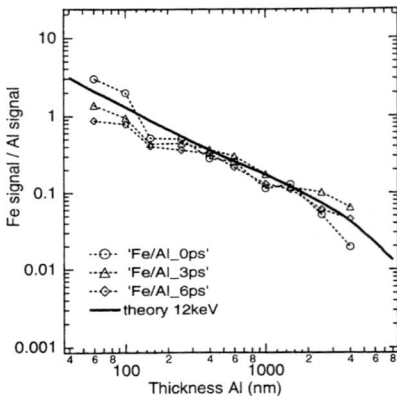

FIGURE 5. Intensity ratio of Fe and Al K_α signals as a function of the thickness of the Al layer. The heavy line is the theoretical estimate with an incident Maxwellian distribution with $12 keV$ temperature.

gradient for delays corresponding to maximum resonant absorption ($\approx 6ps$). Our new experimental brilliance is shown as an open square in Fig. 1. Around $2keV$, an improvement of a factor of ≈ 40 has been obtained over our previous results. The correlation between the detection at $\approx 6ps$ delay of fairly "hot" electrons in vacuum by the electron spectrometer and the detection of lower energy ($12keV$) electrons by K_α line emission points to the fact that the electron energy distribution function is highly anisotropic in directions going outwards and inwards of the target.

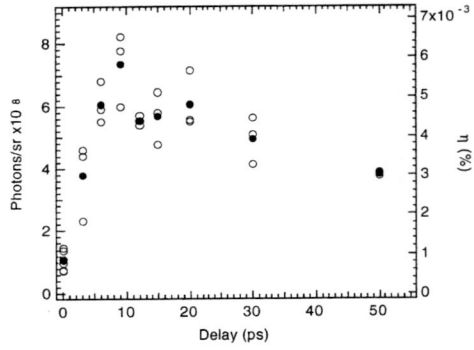

FIGURE 6. Si K_α emission (left) and conversion efficiency (right) as a function of the prepulse delay.

PIC simulations and discussion of experimental results

To get more insight into the relation between the scale length of the electron density gradient and such varied quantities as laser absorption, electron energy, electron distribution anisotropy, and optimum K_α emission (see Figs. 3-6), we have performed particle-in-cell simulations with the relativistic 1.5-D code EUTERPE. In this code, particle motion under the action of the electromagnetic fields (E_x, E_y, B_z) is described in the three-dimension phase space (x, p_x, p_y). To treat oblique laser incidence, we used the boosted-frame-of-reference method [42] in which the 2-D (x, y) periodic system is reduced to one spatial dimension by transforming to a frame in which $k'_y = 0$ (wave vector perpendicular to the electron density gradient). In our calculations, we used p-polarized light at an angle of incidence of 45° and we followed 1.4×10^5 particles over 100 optical cycles. The laser pulse duration (gaussian pulse) was $120 fs$ and the laser intensity $I\lambda^2 = 2.6 \times 10^{16} W/cm^2$.

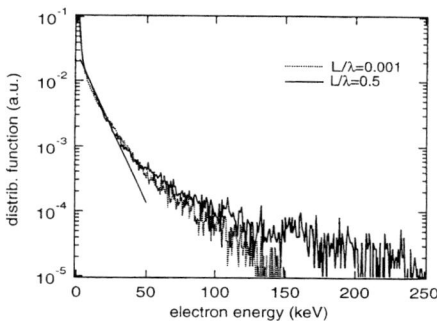

FIGURE 7. PIC electron energy distributions for two values of the reduced electron density gradient scale length. The straight line corresponds to an electron temperature of 12 keV.

We used mobile ions with $m_i/m_e = 3600$, the maximum electron density was $10 \times n_c$, the electron and ion temperatures $0.6 keV$ and $0.1 keV$, respectively. The reduced scale length L/λ of the initially linear electron density profile could be varied between 0.001 and 1. We note that for our experimental laser intensity, the electron oscillation energy in the vacuum laser field is $\approx 5 keV$. Laser absorption and electron energies in these modest intensities 1.5-D simulations were compared favorably with full blown 2-D simulations [43].

Time integrated energy distribution functions of electrons penetrating deeply inside the target bulk are shown in Fig. 7 for $L/\lambda = 0.001$ (steep density gradient) and $L/\lambda = 0.5$. Above $10 keV$, the distribution can be studied in two parts, according to electron energies. In the low energy range ($E \leq 50 keV$), the slope of the distribution function (the "hot" electron temperature) is almost independent of the reduced gradient scale length whereas above $50 keV$, the slope of the distribution varies with the reduced scale length. This variation is shown in Fig. 8 where we have plotted the hot electron temperature as a function of the scale length. Two features qualitatively in agreement with our experimental results can be pinpointed immediately. First, from the low energy part of the distribution function, we deduce an average "temperature" of the order of $\approx 12 keV$ (see the line drawn in Fig. 7) in good agreement with our K_α results shown in Fig. 5. Indeed, K_α measurements are sensitive only to those electrons with energies below $\approx 50 keV$ [8]. Second, the measured and calculated hot electron temperature show both a maximum as a function of the scale length (or prepulse delay, see Fig. 2). The optimum scale length of about $L/\lambda \approx 0.4$, corresponding to a $6ps$ delay, is consistent with our other observations.

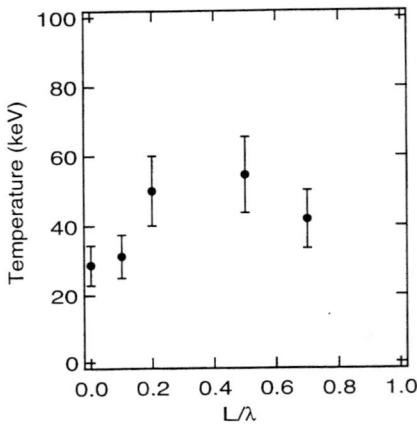

FIGURE 8. Hot electron temperature versus reduced scale length.

However, the absolute values of the electron energies disagree near the peak shown in Fig. 4. Hot electrons measured in front of the target have followed complex trajectories before reaching the electron spectrometer. These trajectories cannot be described by our collisionless code because once the electrons have penetrated the bulk of the target, they travel in straight lines for they do not feel any more the effect of the laser field. In reality, after penetrating the target, Monte Carlo electron energy deposition codes predict that about 10% of the electrons are backscattered towards the laser and, on average, these electrons lose less than half of their initial energy. Accordingly, the measured electron energies in Fig. 4 have to be considered as minimal values of the energies of the electrons which were originally accelerated towards the target. Thus, we get a factor of two discrepancy between the PIC code estimate and the measurements.

Amazingly, we note that our various estimates of the conversion of laser to electron energy are consistent. If 10% of the incoming electrons are backscattered, we would measure about 1% conversion efficiency on the electron spectrograph, in good agreement with our previous estimate [38]. We have also tried to explain how such large electron energies ($\approx 250 keV$) can be obtained in the simulations. To understand this, the labels of energetic electrons falling in an energy band above $150 keV$ were stored and the simulation was repeated to monitor the position and momentum of the labelled particles. Results show that some particles are trapped in the resonant field near critical density and accelerated towards the laser before wave-breaking damps out the resonance, undergo large vacuum excursions where they are submitted to the influence of the stationnary wave pattern of the laser field, and are accelerated back into the plasma by the ambipolar field produced by the less mobile ions.

CONCLUSIONS

For the higher photon energy range where structural informations could be recovered from time-resolved absorption spectroscopy or diffraction, despite several orders of magnitude differences in the number of photons per pulse with respect to synchrotron radiation, laser-produced plasmas are the x-ray sources of choice for applications. In this paper, we have characterized by interferometric, electron, and X-ray diagnostics the interaction of a subpicosecond, medium-intensity, p-polarized laser with a solid target. By applying a controlled prepulse, we have changed (and measured the change) of the electron density gradient scale length with which the main pulse interacts. The laser absorption coefficient and the "temperature" of the hot electrons backscattered towards the laser show a maximum as a function of prepulse delay which can be explained in terms of resonance absorption in a moderately long ($L/\lambda \approx 0.3$) electron density gradient. These observations are reasonably well interpreted with the help of 1.5-D particle-in-cell simulations. As a consequence of the increased high energy electron production, an optimum in the K_α emission yield has been shown; this optimized yield is about 40 times larger than the one obtained in our previous measurements. Our actual x-ray source performances are the following: a photon wavelength of 7.13Å, a number of photons per short per steradian of 8×10^8, and a pulse duration below $300 fs$ [12]. By increasing the laser energy to several hundreds of millijoules, photon yields in excess of 10^{10} photons per steradian could be reached in the near future [44].

ACKNOWLEDGMENTS

We would like to thank the generous help of J.-C. Adam for performing 2-D PIC simulations. Many contributions by E. Lefebvre, G. Bonnaud, and J. Delettrez have been helpful in improving the EUTERPE code. We thank L. Gremillet for helping us in Monte Carlo calculations of electron energy deposition. The present work was supported by the Centre National de la Recherche Scientifique and by the European Union contract ERBFMRX CT96-0080.

REFERENCES

1. Perry M.D. and G. Mourou, Science **264**, 917 (1994).
2. Gauthier J.C., in *Laser Interaction with Matter*, I.O.P. Conference Series Vol. 140, ed. S. Rose (IOP, Bristol, 1994), pp. 1-12.
3. P. Gibbon and E. Förster, Plasma Phys. Control. Fusion **38**, 769 (1996).
4. Pelletier J.F., M. Chaker, and J.C. Kieffer, Optics Lett. **21**, 1040 (1996).
5. Pelletier J.F., M. Chaker, and J.C. Kieffer, Appl. Phys. Lett. **69**, 2172 (1996).
6. Workman J., A. Maksimchuk, X. Liu, U. Ellenberger, J.S. Coe, X.-Y. Chien, and D. Umstadter, J. Opt. Soc. Am. B **13**, 125 (1996).
7. Kieffer J.-C., M. Chaker, J.P. Matte, H. Pépin, C.Y. Côté, Y. Beaudoin, C.Y. Chien, S. Coe, G. Mourou, and O. Peyrusse, Phys. Fluids B5, 2676 (1993).
8. Rousse A., P. Audebert, J. P. Geindre, F. Falliès, J.-C. Gauthier, A. Mysyrowicz, G. Grillon, and A. Antonetti, Phys. Rev. E **50**, 2200 (1994).
9. Kieffer J.-C., Z. Jiang, A. Ikhlef, and C.Y. Côté, J. Opt. Soc. Am. B **13**, 132 (1996).
10. Rousse A., P. Audebert, J.P. Geindre, F. Falliès, J.C. Gauthier, A Mysyrowicz, A. Dos Santos, G. Grillon, and A. Antonetti, J. Phys. B: Atom. Optical Mol. Phys. **27**, L697 (1994).
11. Ráksi F., K.R. Wilson Z. Jiang, A. Ikhlef, C.Y. Côté, and J.C. Kieffer, J. Chem. Phys. **104**, 6066 (1996).
12. Rischel C., A. Rousse, I. Uschmann, P.-A. Albouy, J.-P. Geindre, P. Audebert, J.-C. Gauthier, E. Frster, J.-L. Martin, and A. Antonetti, Nature **390**, 497 (1997).
13. Lepetit L., G. Chériaux, and M. Joffre, J. Opt. Soc. Am. B **12**, 2467 (1995).
14. Teubner U., P. Gibbon, E. Förster, F. Falliès, P. Audebert, J.P. Geindre, and J.C. Gauthier, Phys. Plasmas **3**, 2679 (1996).
15. Ramis R., R. F. Schmalz, and J. Meyer-ter-Vehn, Comput. Phys. Commun. **49**, 475 (1988).
16. Lefebvre E., *Mécanismes d'absorption et d'émission dans l'interaction d'une impulsion laser ultra-intense avec une cible surcritique*, PhD Thesis, Orsay University (1996).
17. Wulff M., D. Bourgeois, T. Ursby, L. Goirand, and G. Mourou, in *Time-resolved Diffraction*, Eds. J.R. Helliwell and P.M. Rentzepis (Oxford University Press, Oxford, 1997), pp. 195-228.
18. Nenner I., Annales de Physique, **22**, C1-11 (1997).
19. A.A. Zholents and M.S. Zolotorev, Phys. Rev. Lett. **76**, 916 (1996).
20. Leemans W.P., R.W. Schoenlein, P. Volfbeyn, A.H. Chin, T.E. Glover, P. Balling, M. Zolotorev, K.J. Kim, S. Chattopadhyay, and C.V. Shank, IEEE Jl. Quant. Electron. **33**, 1925 (1997).
21. L'Huillier A., T. Auguste, Ph. Balcou, B. Carré, P. Monot, P. Salières, P. Altucii, C. Gaarde, M.B. Larsson, E. Mevel, T. Starczewski, S. Svanberg, C.G. Wahlstrom, R. Zerne, K.S. Budil, T. Ditmire, and M. Perry, J. Nonlin. Opt. Phys. Mat. **4**, 647 (1995).
22. L'Huillier A., in *X-ray Lasers 1996*, Eds. S. Svanberg and C.G. Wahlström, (Institute of Physics Publishing, Oxford, 1996), pp. 444-451.
23. Chang Z., A. Rundquist, H. Wang, M.M. Murnane, and H.C. Kaypten,

Phys. Rev. Lett. **79**, 2967 (1997).
24. Chins J.M., P. Breger, P. Agostini, R.C. Constantinescu, H.G. Muller, A. Bouhal, G. Grillon, A. Antonetti, and A. Mysyrowicz, J. Opt. Soc. Am. B **13**, 197 (1996).
25. P. Norreys et al., Phys. Rev. Lett. **76**, 1832 (1996), and references therein.
26. Lindl J., Phys. Plasmas **2**, 3933 (1995).
27. Gauthier J.-C., in *Laser Interaction with Matter*, Ed. S. Rose, (Institute of Physics Publishing, Bristol, 1995), pp. 1-9.
28. Hauer A.A. and G.A. Kyrala, in *Time-resolved Diffraction*, Eds. J.R. Helliwell and P.M. Rentzepis (Oxford University Press, Oxford, 1997), pp. 71-105.
29. Gibbon P. and E. Förster, Plas. Phys. Contr. Fus. **38**, 769 (1996).
30. Workman J., A. Maksimchuk, X. Liu, U. Ellenberger, J.S. Coe, C.Y. Chien, and D. Umstadter, Phys. Rev. Lett. **75**, 2324 (1995).
31. Pelletier J.-F., *Etude et caractérisaton de la production d'impulsions de rayonnement X-UV ultra-courtes*, PhD Thesis, University of Montreal (1996).
32. Pelletier J.-F., M. Chaker, and J.C. Kieffer, Optics Lett. **21**, 1040 (1996).
33. Gauthier J.-C., J.P. Geindre, P. Audebert, S. Bastiani, C. Quoix, G. Grillon, A. Mysyrowicz, A. Antonetti, and R.C. Mancini, Phys. Plasmas **4**, 1811 (1997).
34. Bastiani S., A. Rousse, J.P. Geindre, P. Audebert, C. Quoix, G. Hamoniaux, A. Antonetti, and J.-C. Gauthier, Phys. Rev. E **56**, 7179 (1997).
35. Gauthier J.C., S. Bastiani, P. Audebert, J.P. Geindre, A. Rousse, C. Quoix, G. Grillon, A. Mysyrowicz, A. Antonetti, R. Mancini, et A. Shlyaptseva, SPIE Proceedings **3157**, 52 (1997).
36. Key M.H., T.W. Barbee, et al., in *X-ray Lasers 1996*, Eds. S. Svanberg and C.G. Wahlström, (Institute of Physics Publishing, Oxford, 1996), p. 9.
37. Jaeglé P., S. Sebban, A. Carillon, A. Jamelot, A. Klisnick, P. Zeitoun, B. Rus, F. Albert, and D. Ros, in *X-ray Lasers 1996*, Eds. S. Svanberg and C.G. Wahlström, (Institute of Physics Publishing, Oxford, 1996), p. 1.
38. Bastiani S., A. Rousse, J.P. Geindre, P. Audebert, C. Quoix, G. Hamoniaux, A. Antonetti, and J.-C. Gauthier, Phys. Rev. E **56**, 7179 (1997).
39. Le Blanc C., G. Grillon, J.P. Chambaret, A. Migus and A. Antonetti. Optics Lett. **18** 140 (1993).
40. Rousse A., *Génération d'électrons rapides et émission X dans l'interaction d'une impulsion laser sub-picoseconde et intense avec une cible solide*, Phd Thesis, University of Paris XI (1994), available upon request to the authors.
41. Blanc P., P. Audebert, F. Falliès, J. P. Geindre, Gauthier J.C., A. Dos Santos, A. Mysyrowicz and A. Antonetti, J. Opt. Soc. Am. B **13**, 118 (1996).
42. Bourdier, A., Phys. Fluids **26**, 1804 (1983); Gibbon P., A.R. Bell, Phys. Rev. Lett. **68**, 1535 (1992).
43. Guérin S., P. Mora, J.C. Adam, A. Héron, and G. Laval, Phys. Plasmas **3**, 2693 (1996).
44. Chambaret J.-P., C. Le Blanc, G. Chriaux, P. Curley, G. Darpentigny, P. Rousseau, G. Hamoniaux, A. Antonetti, and F. Salin, Optics Lett. **21**, 1921 (1996).

High Gain X-ray Lasers Pumped by Transient Collisional Excitation

J. Dunn, A.L. Osterheld, V.N. Shlyaptsev [†], J.R. Hunter, R. Shepherd, R.E. Stewart, and W.E. White

Lawrence Livermore National Laboratory, P.O. Box 808, Livermore, CA 94551
[†] *Permanent Address: P.N. Lebedev Physical Institute, 117924 Moscow, Russia.*

Abstract. We present recent results of x-ray laser amplification of spontaneous emission in Ne-like and Ni-like transient collisional excitation schemes. The plasma formation, ionization and collisional excitation can be optimized using two laser pulses of 1 ns and 1 ps duration at table-top energies of 5 J in each beam. High gain of 35 cm^{-1} has been measured on the 147 Å 4d→4p J=0→1 transition of Ni-like Pd and is a direct consequence of the nonstationary population inversion produced by the high intensity picosecond pulse. We report the dependence of the x-ray laser line intensity on the laser plasma conditions and compare the experimental measurements with hydrodynamic and atomic kinetics simulations for Ne-like and Ni-like lasing.

1. INTRODUCTION

The first laboratory demonstration of x-ray laser amplification of spontaneous emission was achieved in 1984 by pumping with high power laser drivers (1, 2). Since that time different x-ray laser schemes working on a range of ions, as described in recent proceedings (3, 4), have been investigated. In particular, progress has been rapid in small scale table-top x-ray lasers in the last three to four years. In the work by Rocca *et al.*, the fast capillary discharge collisional x-ray laser was produced on the 3p→3s J=0→1 transition of Ne-like Ar at 469 Å (5) and later the output was increased to the saturation limit (6). The development of compact, high power subpicosecond lasers based on chirped pulse amplification (CPA) (7) has also been very important for extending x-ray lasers to shorter wavelengths. This is because the pumping power required to produce the plasma formation, ionization and excitation for shorter wavelength collisional x-ray lasers has traditionally scaled approximately as λ^{-4}, where λ is the x-ray laser wavelength.

Short pulse, high power laser drivers at the Terawatt level or higher with high repetition rates can provide the necessary pump for a new class of x-ray laser (XRL). A novel scheme was reported in 1995 for a Pd-like Xe x-ray laser operating at 10 Hz with a gain length product of gL ~11 at 418 Å for 40 fs irradiation of a xenon gas cell (8). This scheme used a combination of field induced tunneling ionization followed by collisional excitation. A shorter wavelength scheme operating at 135 Å with a gain length product of gL ~6 has also been demonstrated for an H-like Li Ly-α inversion pumped by a 300 fs laser pulse (9) in a microcapillary. The transient collisional excitation using a picosecond pulse has been reported recently by Nickles et al. for the $3p \rightarrow 3s$ $J=0 \rightarrow 1$ transition of Ne-like Ti which lases at 326 Å with a gain of 19 cm^{-1} and a gain length product of gL ~9.5 (10). This scheme has also been reported to work with a $gL > 14$ for Ti and at 255 Å for Ne-like Fe (11, 12), and to achieve saturated output on Ne-like Ti and Ge at 196 Å (13).

The Ni-like ion has the similar benefits of the closed shell configuration as the Ne-like ion and therefore it is a robust XRL plasma medium. It also has the advantage of a larger x-ray photon energy to excitation energy ratio: the ratio is ~0.19 for Ni-like palladium. Therefore, while the Ne-like transient collisional scheme is interesting, the extension to the Ni-like ion sequence yields higher efficiency, higher output and shorter wavelength x-ray lasers. Previously, Ni-like collisional x-ray lasers for $3d^9 4d \rightarrow 3d^9 4p$ transitions in high-Z ions have required large energy laser drivers (14, 15). Using the NOVA laser, short wavelength lasing has been extended to 35 Å for the $4d \rightarrow 4p$ $J=0 \rightarrow 1$ transition of Ni-like Au (16). Lasing on Ni-like ions has also been measured on the Lanthanide series for example Neodymium and Lanthanum for x-ray lasers operating at wavelengths of 79 Å and 89 Å (17). Saturated output for the Ni-like scheme has been demonstrated very recently at 140 Å for Ag and at 73 Å for Sm by using 150 J of energy in a 70 ps pulse and pre-pulse combination (18).

We have recently observed lasing on the transient collisional excitation Ni-like Pd scheme pumped with less than 10 J of laser energy (19). This is a further reduction in the pumping energy by more than one order of magnitude. Gain in excess of 35 cm^{-1} and a gL product of ~12.5 have been measured on the $4d \rightarrow 4p$ $J=0 \rightarrow 1$ transition at 147 Å for this scheme. The next section describes the transient gain scheme and the experimental implementation. Section 3. summarizes experimental results on the transient Ni-like Pd scheme. Section 4. compares the Ni-like Pd results to simulations while Section 5. discusses the important issue of Ne-like Ti output as a function of short pulse delay. The final part of the paper, Section 6., discusses some future trends including a brief description of the purpose built table-top COMET laser driver for the next phase of experiments.

2. TRANSIENT GAIN SCHEME DESCRIPTION AND EXPERIMENTAL IMPLEMENTATION

The transient collisional excitation (TCE) or transient gain scheme as proposed by Afanasiev and Shlyaptsev (20, 21) describes the creation of a short-lived inversion on the timescale of femtoseconds to a few tens of picoseconds. This is characterized by a plasma lifetime dictated by the fast pumping source which in turn should be comparable with the relaxation timescales of the excited levels. Most collisional x-ray lasers reported in the last decade have been variations of the quasi-steady state (QSS) inversion scheme where the pumping source is much longer than the lifetime of the excited levels. The TCE scheme differs from the QSS inversion in a number of important areas but mainly because the risetime of the level excitation rates is shorter than the collisional excitation timescales. This produces a short-lived transient population inversion pumped directly from the ground state until collisions redistribute the populations among all levels achieving finally the quasi-steady state. During the time of population redistribution the inversion is not defined by the small difference in populations of upper and lower level as in the case of QSS but in fact solely by the the upper laser level population. Besides, since the plasma can be made sufficiently hot for maximal level population during this short transient time it is predicted that TCE will produce very high gains above 100 cm^{-1} and therefore very high efficiency x-ray lasers.

One proposed method for demonstrating the transient scheme x-ray laser is to use two sequential stages of laser irradiation (10, 21). A formation pulse of 1 ns heats a solid planar target at 10^{12} W cm^{-2} to produce a long scalelength plasma with a high fraction of ions in the Ni-like stage. It is essential for a delay before the second pulse to optimize the conditions within the plasma for maximum amplification. The short 1 ps pump pulse at 10^{15} W cm^{-2} produces rapid plasma heating with an increase in the electron temperature $T_e \geq \Delta E_e$, where ΔE_e is the upper laser level excitation energy. This generates the transient inversion and the scheme works most efficiently when the plasma formation conditions are at the correct ionization, with low initial electron temperature and low electron density gradients. The high gain conditions last for a few picoseconds then will quickly decay after ~10 ps as a result of collisional redistribution of the electron population among all excited levels, ionization and plasma cooling. Therefore, high gain is expected for short 0.1 to 0.3 cm targets with decreasing gain for longer lengths.

The Ni-like Pd experiments were performed at the Lawrence Livermore National Laboratory JANUS laser facilities. One arm of the JANUS laser provided an 800 ps (FWHM) pulse at 1064 nm wavelength with 5 - 6 J on target at a repetition rate of 1 shot/ 3 minutes. This produced the plasma forming beam. The short pulse needed to pump the inversion was provided by the 5 - 6 J hybrid chirped pulse amplification JANUS 500 fs system. This is based on a Ti:Sapphire oscillator and regenerative amplifier front end tuned to 1053 nm wavelength with Nd:phosphate glass power amplifiers. The pulse duration was lengthened to 1.1 ps (FWHM) by de-tuning the compressor gratings. The regenerative amplifiers of the two lasers were synchronized using the 80 Mhz radio-frequency output from the short pulse oscillator resulting in a relative timing jitter of 80 ps rms. The arrival of the short pulse was delayed by 1 to 2 ns relative to the peak of the long pulse to minimize refraction effects and to allow for sufficient plasma cooling. After amplification, the beams were enlarged to 8.4 cm diameter, aligned and co-propagated under vacuum to the target chamber. The combination of a long focus cylindrical lens and a paraboloid were used to produce a line focus of dimensions 70 μm × 12.5 mm. Slab palladium targets were used in the experiment. A flat-field grating spectrometer with a back-thinned CCD detector measured the axial spectral emission. Further experimental details are described in (11, 19).

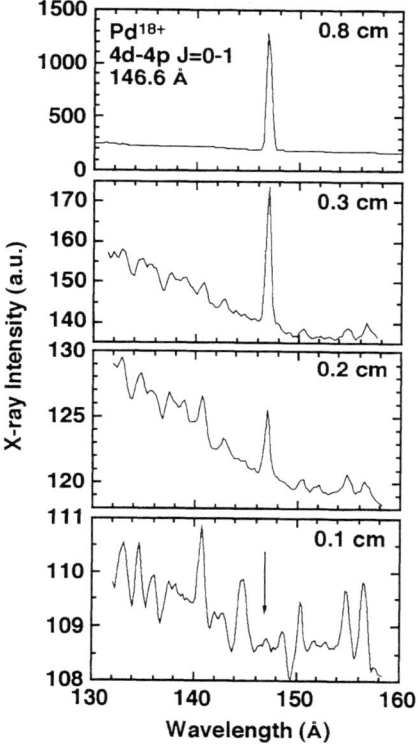

FIGURE 1. Axial spectra for various target lengths of Pd, from 0.1 to 0.8 cm, showing the exponential increase in the Ni-like Pd $4d \rightarrow 4p$ $J=0 \rightarrow 1$ x-ray laser line at 146.6 Å measured in second order. The transition, weak but visible, is arrowed in the 1 mm target length, bottom panel. The laser line experiences 4 orders of magnitude increase for a 0.8 cm target. Note the change in the intensity scale for each panel: there is a factor of 500 difference in the intensity scale between the bottom panel and top panel.

3. EXPERIMENTAL RESULTS

Figure 1 shows the axial spectrum for different target lengths. The strong exponential increase in the $4d \rightarrow 4p$ $J=0 \rightarrow 1$ transition is observed unambiguously in second order. Incident laser energy on target was approximately 4 J of long pulse and 5 J of short pulse. The lasing wavelength is measured to be 146.6 ± 0.9 Å by fitting a high order polynomial function to the spectrum using Ne-like transient gain 3→3 x-ray laser lines of Ti and Fe, observed in first order, as calibration lines. These Ne-like lasing lines have been previously reported in long pulse collisional excitation schemes using the pre-pulse technique (22). The Ni-like Pd wavelength is in good agreement with the calculated wavelength of 148 Å (23). (More accurate wavelengths have been recently measured and extended to other ions on the Ni-like isoelectronic sequence (24) using the table-top COMET laser driver.) As can be seen from Fig. 1 the output of the line increases rapidly with small increments of target length. The laser dominates the spectrum above 0.3 cm targets. The largest increases in x-ray laser output are observed for 0.05 and 0.1 cm steps in the shortest targets.

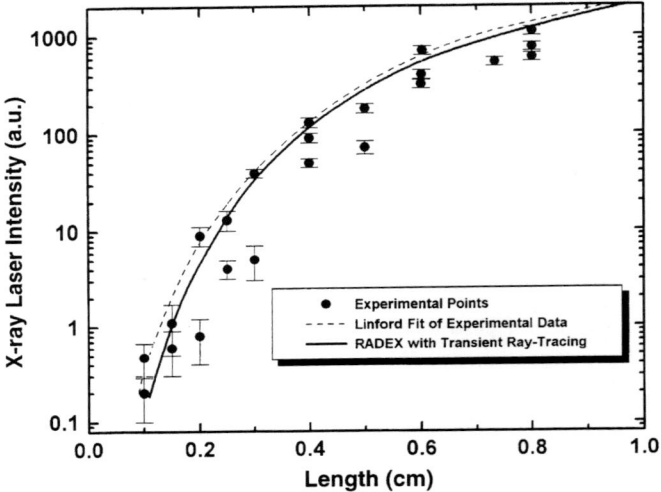

FIGURE 2. Intensity of 146.6 Å laser line, measured in second order, for 0.1 to 0.8 cm Pd target lengths. Full circles are experimental data. Dashed curve is Linford fit to experimental data. Solid curve is RADEX simulations with transient gain ray-tracing.

Figure 2 plots the measured laser line intensity as a function of target length. More than four orders of magnitude increase in the laser output is observed when the target is lengthened from 0.1 to 0.8 cm. All laser driver conditions, including the short pulse energy, long pulse energy, line focus and delay between the two pulses, are kept constant. Also shown are the local Linford fit to the experimental points (25) and the RADEX simulations with transient ray-tracing. The latter is discussed in more detail in the next section. The shape of the intensity versus length output of the x-ray laser indicates continually changing gain conditions with the highest gain of 35 cm^{-1} observed at the shortest target lengths of 0.1 to 0.2 cm. The gain drops at intermediate and longer target lengths. Although the shape is similar to saturation, this effect is explained by the transient gain timescale lasting for 5 to 15 ps. This is significantly shorter than the x-ray laser propagation time along the line focus (26) and so the laser experiences continually decaying gain conditions as it travels along the gain medium. The overall gL product is determined to be ~12.5 by integrating the measured gain along the full target length. The hydrodynamics, atomic kinetics modeling and ray-tracing are described in the next section.

4. COMPARISON OF NI-LIKE PALLADIUM RESULTS WITH SIMULATIONS

Very high gains in excess of 300 cm^{-1} have been predicted for transient gain in Ni-like Xe at 96 Å (21) and for Ni-like Mo 189 Å (27) at laser irradiances and pulse durations close to the work described in (19). High values have also been predicted for transient gain Ne-like schemes (28). The experimental gain of 35 cm^{-1} measured for Pd is high in comparison with previously reported x-ray gain results but still significantly lower than the predictions. The main reasons for the observed lower gain are the combined effects of refraction deflecting the x-ray laser out of the gain region of the plasma, short-lived high gain and collisional line broadening. We used the 1-dimensional numerical code RADEX (10, 20, 21) which treats the transient hydrodynamics, atomic kinetics and radiation transport self-consistently. An additional ray-tracing package, as a post-processor, is used to model the propagation of the x-ray laser along the gain medium and calculate the x-ray laser intensity. It is important to note that the hydrodynamics, atomic kinetics and ray-tracing have to be made using a transient approximation to simulate the experimental conditions. In particular, calculations show that if the x-ray laser line ray-tracing is made in a quasi-steady state approximation for gain described as transient this produces results inconsistent with the observed x-ray laser characteristics including the lasing intensity, pulse duration, deflection angle and effective gain. This can be explained mainly by the fast gain risetime of 1 - 3ps and short lifetime of 5 - 15 ps (for plasma densities n_e ~ 1 - 3 × 10^{20} cm^{-3}) compared to the propagation time L/c ~ 30 ps along the amplified medium. The major observed XRL characteristics are consistent with the experiment only when the effects of fast temporal evolution of the gain and photon transit time are included and properly described.

Figure 3. shows output from RADEX for simulations of the experimental conditions used to generate the Ni-like Pd x-ray laser mainly during or after the laser short pulse which produces the transient excitation. A general overview is that during the short pulse laser, the electron temperature exhibits dramatic changes near the critical density where locally most of the absorbed laser energy is deposited. The transient gain here reaches ultra-high values during the short pulse but also vanishes very rapidly. Two other regions, the over- and under-critical, also produce high gain: the former is on the front of the strong heat conduction wave. The latter, where much of the observed XRL gain is measured, is in the region of inverse bremsstrahlung absorption where the increase in the temperature is sufficiently high for efficient transient excitation $T_e \sim \Delta E_e$, where ΔE_e is the excitation energy.

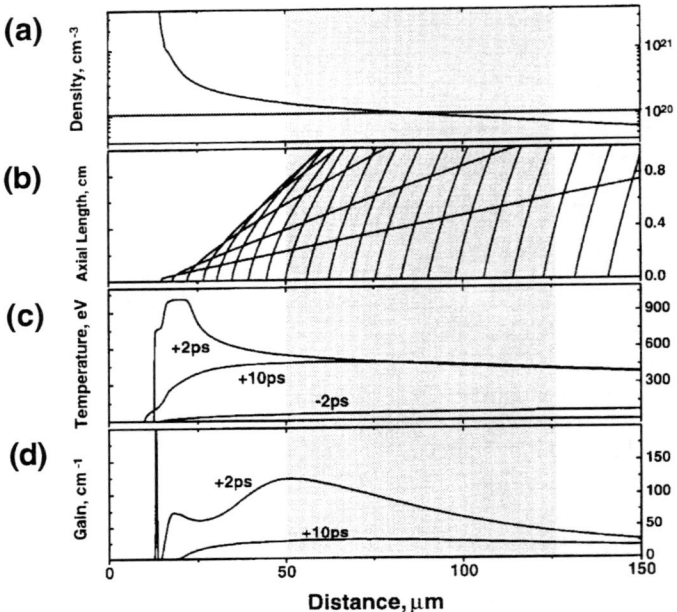

FIGURE 3. Simulations from the one-dimensional numerical code RADEX of the experimental lasing conditions for the Ni-like Pd $4d \rightarrow 4p$ $J=0 \rightarrow 1$ transition at 147 Å. The plasma formation pulse of 850 ps (FWHM) duration arrives 1.34 ns before the peak of the 1.1 ps short pulse. The conditions above are relative to the peak of the short pulse. The shaded area between 50 and 125 μm is region where strongest lasing is observed but is not confined to this region only. (a) Electron density profile at 2 ps after the peak of the short pulse laser. Horizontal line indicates 0.1 n_c position, 10^{20} cm^{-3}. (b) Ray-tracing of x-ray laser beam for target lengths up to 1 cm at 2 ps after peak of short pulse laser. (c) Electron temperature profile for -2 ps (-2 ps curve), +2 ps (+2 ps curve) and +10 ps relative to peak of short pulse. (d) Gain profile observed at +2 ps and +10 ps.

Looking at Fig. 3 in more detail, it is observed that refraction is caused by density gradients normal to the target surface, as shown in Fig. 3 (a) for +2 ps after the peak of the picosecond pulse. These deflect the laser out of the high density and high gain region as it propagates along the plasma column. Critical density is ~15 μm in front of the target. This is clearly shown for the XRL trajectory plot of Fig. 3 (b) at +2 ps, where refraction stops amplification in the high gain region at $n_e \sim 10^{21}$ cm^{-3} for plasma lengths $L < 0.1 - 0.3$ cm. This indicates that refraction is still very important for the shorter 147 Å wavelength here than in the transient Ne-like

Ti x-ray laser at 326 Å [10]. It is also applicable to the shorter target lengths of 0.8 cm than in the few centimeter long geometries QSS schemes. However, refraction is much less pronounced at lower densities $n_e \sim 1 - 3 \times 10^{20}$ cm^{-3} for 147 Å photons of Ni-like Pd compared to 326 Å Ne-like Ti x-ray laser.

Secondly, the conjunction of the fast transient nature of the atomic kinetics and the finite photon transit time L/c is significant. This transit time effect, when the XRL experiences gradually decreasing gain during propagation evident in Fig. 2, is basically not observed in most QSS lasers. This is illustrated in Fig. 3 (c) for the predicted temperature distribution in space at three different times. The long plasma formation pulse heats the plasma to a peak temperature of $T_e \sim$ 150 - 200 eV sufficient to ionize Pd to the Ni-like charge stage. At the end of the long pulse, the plasma expands and cools to less than ~90 eV as shown by the -2 ps curve, representing 2 ps prior to the short pulse arrival, without significant change in the Ni-like abundance. The absorption of the picosecond pulse energy occurs mostly at the critical density surface at $n_e = 10^{21}$ cm^{-3} rapidly heating the electron temperature there to ~2 keV in the first few picoseconds, curve +2 ps. There is partial heating in the under-dense corona to an optimal of 350 - 500 eV. As a result of strong heat flux, close to the free streaming limit, the high temperature region cools to 400 eV after 10 ps (curve +10 ps), and then continues to fall slowly. The transient $J=0 \rightarrow 1$ gain, Fig. 3 (d), reaches high values of 200 cm^{-1} during the first 1 - 3 ps and 50 - 80 cm^{-1} during a further 10 ps. The former occur near the critical density in the thin ablative layer, the isolated spike in Fig. 3 (d) at 15 μm, while the latter are in areas of relatively flat lower density profile, shaded region. As indicated by Fig. 3 (b) and (d), the short inversion lifetime near the high density critical region decreases the photon transit path in the high gain region to an axial length of ~500 μm.

The third important phenomenon is the dominance of collisional line broadening over Doppler broadening of the excited levels at electron densities above 0.3× critical. As a consequence of all these effects, the gain-length product at high density and gain region is substantially decreased. The outer plasma regions at lower density $n_e \sim 0.9 - 2 \times 10^{20}$ cm^{-3} are more optimal for amplification resulting in larger local gL. Hence, the unusual properties of the TCE scheme such as the high gain and saturation-like behaviour find consistent explanation by a combination of the above main plasma and transient kinetics effects. Thus, returning to Fig. 2 the saturation-like behavior for the intensity of Ni-like Pd x-ray laser with length obtained in the experiment, is well reproduced in the RADEX calculations. The dashed curve on this figure represents the Linford fit to the experimental points at highest intensity for a given length. These estimations and numerical investigations indicate that x-ray laser intensity saturation is in fact very close and can be achieved by several methods including just simply increasing the plasma length. Calculations also suggest that the high efficiency obtained in this work can be further improved with the use of a pre-pulse, low density targets or traveling wave irradiation [21].

5. NE-LIKE TI LASING FOR DELAYED SHORT PULSE

An explanation of the long pulse-to-short pulse delay was one of the important questions to be answered for effective laser operation in the transient regime. This was resolved in the current experiments. It was found experimentally for Ne-like Ti x-ray laser plasmas that lasing action had a maximum intensity if the two pulses were separated by between one and two nanoseconds (12). This issue was known but not investigated systematically in the experiments of other groups and required better understanding. Therefore, numerical modeling of this Ne-like laser scheme represented a way to explaining the optimum lasing, the strong lasing at ~1.6 ns delay and the role of the underlying key processes.

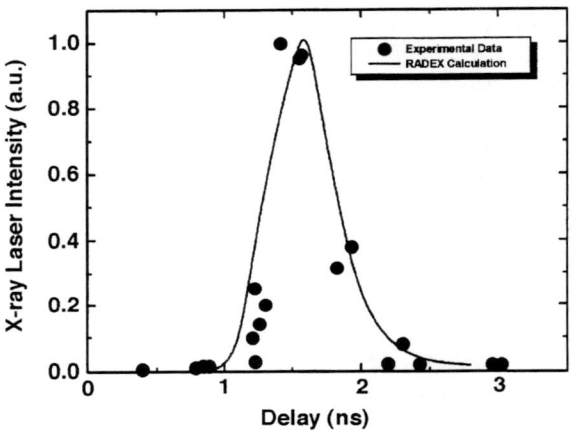

FIGURE 4. Experimental measurements of the 326 Å Ne-like Ti $3p \rightarrow 3s$ $J=0 \rightarrow 1$ x-ray laser intensity (full circles) as a function of delayed arrival of the short pulse relative to the peak long pulse. Strong lasing is observed at a delay of 1.6 ns. Full curve shows comparison from the one-dimensional numerical code RADEX for the experimental lasing conditions.

There exists a similarity here with previous experiments for pre-pulse formed x-ray laser plasmas in the quasi-steady state regime. It was found in many experiments mostly conducted by Nilsen and co-workers, see for example (22), that the QSS lasing could be maximized with different kinds of low density pre-pulses, multiple pulses of different duration, with optimal energies and delays. The main conclusion was that enhanced laser action was a result of substantial suppression of refraction. With the transient excitation scheme there are several new aspects in the fundamental operation and hence the optimized parameters for it are different from the previous QSS approach. A number of comparisons have been made for the experimental data and the code RADEX simulations at the same conditions for different atomic elements.

Fig. 4 shows the intensity of Ne-like Ti $3p \rightarrow 3s$ $J=0 \rightarrow 1$ laser at 326 Å as a function of delay introduced for the short pulse relative to the peak of the 800 ps plasma forming pulse. It can be seen, that lasing does not occur if the picosecond pulse arrives earlier than 1.0 - 1.2 ns after the long pulse. There is a window where good lasing is observed, centered at ~1.6 ns, followed by a fast decrease for delays of more than 2.2 ns. This non-lasing early delay behavior is more than just strong refraction effects at the beginning of expansion. This is confirmed in the simulations by artificially reducing the effects of refraction by an order of magnitude: the time of optimal delay and width of lasing window remain relatively unchanged. The additional reason why lasing does not appear prior to a specific moment lies in the physics of the transient inversion. To achieve substantial transient gain, the initial plasma temperature before the picosecond temperature jump must be low enough to empty all excited level populations. After the 800 ps pulse has finished, the laser plasma is allowed to cool down by expansion and radiation to reach less than 80 eV for Ti and less than 90 eV for Pd. Unless these conditions are achieved the transient gain is small. The fall in XRL intensity for large delays is due to substantial plasma expansion and the resultant drop in the density. This causes in turn a decrease in the short pulse laser absorption and hence reduced plasma temperatures during the short pulse. At these late delays the density and corresponding gradients are reduced by more than 50% which is beneficial for refraction effects but is not sufficient to compensate for the lower gain coefficient.

6. DISCUSSION AND FUTURE DIRECTIONS

The transient gain x-ray laser scheme opens up rich and diverse opportunities for atomic physics and plasma physics research. In particular, there are plans to study energy level measurements, short pulse laser plasma interactions and sub-picosecond non-stationary atomic kinetics in plasmas. There are many possible applications which can be pursued with table-top x-ray lasers but have previously required larger laser facilities to pump the inversion. These include x-ray microscopy (29), interferometry of plasmas (30), radiography of plasmas (31), and interferometry of materials (32). X-ray laser applications are enhanced by the high repetition shot rate and inherent short pulse duration available with table-top transient gain x-ray laser systems. However, high output of the x-ray laser line is essential. Gains as high as 50 - 100 cm^{-1} may be achieved with careful optimization of the target and plasma conditions.

The original CPA short pulse laser has been re-designed with some enhancements not previously available in the first experiments to meet these objectives. The new COMET (COmpact Multi-pulsE Terawatt) laser system is based around the 15 TW short pulse laser, maximum of 7.5 J in 500 fs, but has a long pulse arm sharing the same oscillator. The laser is a table-top system and occupies two standard optical tables with area less than 10 m^2. The long pulse of 500 - 800 ps duration has 12 - 15 J energy available on target and replaces the plasma forming beam previously provided by the JANUS laser. A pre-pulse network has also been installed to produce a pre-formed plasma in front of the two main pulses to improve the coupling of laser energy into the gain medium. We have recently looked at lower Z Ni-like materials studied previously, for example Nb and Mo (33), and observed very strong lasing on Ni-like Y through Mo from 240 Å to 189 Å (34). The COMET laser is a versatile system which will allow the further detailed study of the transient gain scheme.

ACKNOWLEDGMENTS

We thank B. Sellick for technical assistance. Thanks to Mark Eckart for continuing support and encouragement for this research. One of us (V.N.S.) acknowledges support from H. Baldis of ILSA and Yu. Afanasiev of LPI. This work was performed under the auspices of the U.S. Department of Energy by the Lawrence Livermore National Laboratory under Contract No. W-7405-Eng-48.

REFERENCES

1. D.L. Matthews, P.L. Hagelstein, M.D. Rosen, M.J. Eckart, N.M. Ceglio, A.U. Hazi, H. Medecki, B.J. MacGowan, J.E. Trebes, B.L. Whitten, E.M. Campbell, C.W. Hatcher, A.M. Hawryluk, R.L. Kauffman, L.D. Pleasance, G. Ramback, J.H. Scofield, G. Stone, and T.A. Weaver, *Phys. Rev. Lett.* **54**, 106 (1985).
2. S. Suckewer, C.H. Skinner, H. Milchberg, C. Keane, and D. Voorhees, *Phys. Rev. Lett.* **55**, 1753 (1985).
3. Proceedings of the 5th International Conference on X-ray Lasers 1996, Lund, Sweden. *IOP Conf. Series No.* **151**, ed. S. Svanberg and C.-G. Wahlström (1996).
4. Proceedings of the conference on Soft X-ray Lasers and Applications II, ed. J.J. Rocca and L.B. Da Silva, *SPIE Proceedings Vol.* **3156** (1997).
5. J.J. Rocca, V.N. Shylaptsev, F.G. Tomasel, O.D. Cortázar, D. Hartshorn, and J.L.A. Chilla, *Phys. Rev. Lett.* **73**, 2192 (1994).
6. J.J. Rocca, D.P. Clark, J.L.A. Chilla, and V.N. Shylaptsev, *Phys. Rev. Lett.* **77**, 1476 (1996).
7. D. Strickland and G. Mourou, *Opt. Commun.* **56**, 219 (1985).
8. B.E. Lemoff, G.Y. Yin, C.L. Gordon III, C.P.J. Barty, and S.E. Harris, *Phys. Rev. Lett.* **74**, 1574 (1995).
9. D.V. Korobkin, C.H. Nam, S. Suckewer, and A. Goltsov, *Phys. Rev. Lett.* **77**, 5206 (1996).
10. P.V. Nickles, V.N. Shylaptsev, M. Kalachnikov, M. Schnürer, I. Will, and W. Sandner, *Phys. Rev. Lett.* **78**, 2748 (1997).
11. J. Dunn, A.L. Osterheld, R. Shepherd, W.E. White, V.N. Shylaptsev, A.B. Bullock, and R.E. Stewart, *SPIE Proceedings Vol.* **3156**, 114 (1997).
12. J. Dunn, A.L. Osterheld, R. Shepherd, W.E. White, V.N. Shylaptsev, and R.E. Stewart, Internal Report, Lawrence Livermore National Laboratory, CA UCRL-ID-127872 (1997).
13. P.V. Nickles, M. Schnürer, M.P. Kalachnikov, W. Sandner, V.N. Shylaptsev, C. Danson, D. Neely, E. Wolfrum, J. Zhang, A. Behjat, A. Demir, G. Tallents, P.J. Warwick, and C.L.S. Lewis, *SPIE Proceedings Vol.* **3156**, 80 (1997).

14. B.J. MacGowan, S. Maxon, P.L. Hagelstein, C.J. Keane, R.A. London, D.L. Matthews, M.D. Rosen, J.H. Scofield, and D.A. Whelan, *Phys. Rev. Lett.* **59**, 2157 (1987).
15. S. Maxon, P. Hagelstein, B. MacGowan, R. London, M. Rosen, J. Scofield, S. Dalhed, and M. Chen, *Phys. Rev. A* **37**, 2227 (1988).
16. B.J. MacGowan, L.B. Da Silva, D.J. Fields, C.J. Keane, J.A. Koch, R.A. London, D.L. Matthews, S. Maxon, S. Mrowka, A.L. Osterheld, J.H. Scofield, G. Shimkaveg, J.E. Trebes, and R.S. Walling, *Phys. Fluids B* **4**, 2326 (1992).
17. H. Daido, Y. Kato, K. Murai, S. Ninomiya, R. Kodama, G. Yuan, Y. Oshikane, M. Takagi, H. Takabe, and F. Koike, *Phys. Rev. Lett.* **75**, 1074 (1995).
18. J. Zhang, A.G. MacPhee, J. Nilsen, J. Lin, T.W. Barbee Jr., C. Danson, M.H. Key, C.L.S. Lewis, D. Neely, R.M.N. O'Rourke, G.J. Pert, G.J. Tallents, J.S. Wark, and E. Wolfrum, *Phys. Rev. Lett.* **78**, 3856 (1997); J. Zhang, A.G. MacPhee, J. Lin, E. Wolfrum, R. Smith, C. Danson, M.H. Key, C.L.S. Lewis, D. Neely, J. Nilsen, G.J. Pert, G.J. Tallents, and J.S. Wark, *Science* **276**, 1097 (1997).
19. J. Dunn, A.L. Osterheld, R. Shepherd, W.E. White, V.N. Shylaptsev, and R.E. Stewart, *Phys. Rev. Lett.* **80**, 2825 (1998).
20. Yu.V. Afanasiev and V.N. Shylaptsev, *Sov. J. Quant. Electron.* **19**, 1606 (1989).
21. V.N. Shylaptsev, P.V. Nickles, T. Schlegel, M.P. Kalashnikov, and A.L. Osterheld, *SPIE Proceedings Vol.* **2012**, 111 (1993).
22. J. Nilsen, B.J. MacGowan, L.B. Da Silva, and J.C. Moreno, *Phys. Rev. A* **48**(6), 4682 (1993); E.E. Fill, Y. Li, G. Pretzler, D. Schlögl, J. Steingruber, and J. Nilsen, Physica Scripta **52**, 158 (1995).
23. J.H. Scofield and B.J. MacGowan, *Phys. Scr.* **46**, 361 (1992).
24. Y. Li *et al*, submitted for publication in *Phys. Rev. A* (1998).
25. G.J. Linford, E.R. Peressini, W.R. Sooy, and M.L. Spaeth, *Appl. Opt.* **13**(2), 379 (1974).
26. J. Dunn *et al*, submitted to *Opt. Lett.* (1998).
27. J. Nilsen, *J. Opt. Soc. Am. B* **14**, 1511 (1997).
28. J. Nilsen, *Phys. Rev. A.* **55**, 3271 (1997); S.B. Healy *et al*, *Opt. Commun.* **132**, 442 (1996); K.G. Whitney *et al*, *Phys. Rev E* **50**, 468 (1994).
29. L.B. Da Silva, J.E. Trebes, R. Balhorn, S. Mrowka, E. Anderson, D.T. Attwood, T.W. Barbee Jr., J. Brase, M. Corzett, J. Gray, J.A. Koch, C. Lee, D. Kern, R.A. London, B.J. MacGowan, D.L. Matthews, and G. Stone, *Science* **258**, 269 (1992).
30. L.B. Da Silva, T.W. Barbee Jr., R. Cauble, P. Celliers, D. Ciarlo, S. Libby, R.A. London, D. Matthews, S. Mrowka, J.C. Moreno, D. Ress, J.E. Trebes, A.S. Wan, and F. Weber, *Phys. Rev. Lett.* **74**, 3991 (1995).
31. D.H. Kalantar, M.H. Key, L.B. Da Silva, S.G. Glendinning, J.P. Knauer, B.A. Remington, F. Weber, and S.V. Weber, *Phys. Rev. Lett.* **76**, 3574 (1996).
32. G. Jamelot, F. Albert, A. Belsky, M. Boussoukaya, A. Carillon, A. Egbert, S. Hubert, P. Jaeglé, D. Joyeux, I. Kamenskikh, A. Klisnick, S. Meyer, D. Phalippou, D. Ros, B. Rus, S. Sebban, B. Wellegehausen, T. Woyvod, Ph. Zeitoun, and A. Zeitoun-Fakiris, *SPIE Proceedings Vol.* **3156**, 124 (1997).
33. S. Basu, J.G. Goodberlet, M.H. Muendel, S. Kaushik, and P.L. Hagelstein, in X-ray Lasers 1992, ed. E. E. Fill, *IOP Conf. Series No.* **125**, 71 (1992); S. Basu, P.L. Hagelstein, J.G. Goodberlet, M.H. Muendel, and S. Kaushik, *App. Phys. B* **57**, 303 (1993).
34. J. Dunn *et al*, submitted to *Opt. Lett.* (1998).

ELECTRON EXCITATION
OF ATOMIC IONS

Recent Developments in the Theory of Electron–Ion Collisions

Klaus Bartschat

Drake University[1]
Des Moines, Iowa 50311, USA

Abstract. Until a few years ago, calculations for electron scattering from atoms and ions were typically performed using non-perturbative close-coupling-type approximations for low impact energies and perturbative Born-type methods in the high-energy regime. Especially for singly and doubly ionized targets, however, neither one of those methods is particularly reliable for the "intermediate energy regime", i.e., impact energies near and up to about five times the ionization threshold. This gap has recently been closed by the convergent close-coupling (CCC), the R-Matrix with pseudo-states (RMPS), the intermediate-energy R-matrix (IERM), and time-dependent lattice methods. The principal ideas behind these approaches are introduced and some example cases, illustrating the need for using such sophisticated approaches, are discussed.

INTRODUCTION

The accurate knowledge of electron–ion collision cross sections is of great importance for modeling astrophysical and laboratory plasmas. Due to the difficulties associated with experimental determinations of these cross sections, as well as the enormous amount of data needed for the modeling programs, the vast majority of currently available input data for these programs consists of theoretical predictions. In most cases, these have have only been checked against a few experimental benchmark results or, more often, against predictions from other theoretical models. Prominent examples of extensive, world-wide efforts in the field include the "Opacity Project" [1] and the "Iron Project" [2].

In light of this situation, it is obviously very important to assess the reliability of the various theoretical approaches. This assessment should not only provide theoretical benchmark results, obtained with the most sophisticated numerical models, but it should also indicate under what circumstances simpler methods can provide results of sufficient accuracy, thus requiring much less computational effort.

This paper will concentrate on the benchmark aspect of the theoretical models and focus on relatively simple (quasi-)one-electron targets, such as He^+ as well as

---

[1] work supported by the National Science Foundation under grant PHY-9605124.

Li-like and Na-like ions, while some references to current work on more complex targets will be provided.

NUMERICAL METHODS

We now summarize the principal ideas behind three recently developed close-coupling methods that account for the effect of the target continuum states in the expansion and keep the full coupling between all the physical discrete as well as the bound and continuum pseudo-channels. Furthermore, an approach to solve the time-dependent Schrödinger equation more directly, i.e., without the use of a basis function expansion, will be discussed.

Time-independent close-coupling methods

The convergent close-coupling method (CCC)

The details of the CCC theory have been given by Bray and Stelbovics [3]. The method can be thought of as a standard close-coupling approach where, in addition to the discrete target states, the target continuum is treated with the aid of positive-energy, square-integrable pseudo-states. All states are obtained by diagonalising the target Hamiltonian in an orthogonal Laguerre basis. The usage of such a basis ensures that "completeness" is, in principle, approached by increasing the basis size until the results of interest become sufficiently stable.

A key feature of the formalism, as developed by Bray and Stelbovics, lies in the fact that the coupled equations are formulated in momentum space, where they take the form of coupled Lippmann-Schwinger equations for the T-matrix. These are solved separately, upon partial-wave expansion, for each collision energy of interest. As such the method is not ideally suited for the study of detailed scattering behavior as a function of incident energy, as needed, for example, in the vicinity of Rydberg resonance series. However, modern computational resources involving large clusters of workstations allow for the parallel execution of the CCC program at many energies, thereby somewhat reducing this problem. Furthermore, the method has to date only been applied to (quasi-)one-electron and (quasi-)two-electron targets in a non-relativistic framework, although the formalism, as presented by Fursa and Bray [4], is more general.

The R-matrix with pseudo-states method (RMPS)

The low-energy R-matrix method (for details, see Burke and Robb [5]) is another method to solve the close-coupling equations, this time in coordinate space. The important difference to the standard formulation is the division of configuration space into two regions, $r \leq a$ and $r > a$, where the R-matrix radius a is chosen in such a way that exchange effects between the projectile and the target electrons can be neglected in the external region. Here the coupled equations (without exchange) are solved for each collision energy and matched, at the boundary $r = a$, to the solution in the inner region. However, instead of solving a set of coupled integro-differential equations in the internal region for each collision energy, the $(N+1)$-electron wavefunction at energy E is expanded in terms of an energy-independent basis set, ψ_k, as

$$\Psi_E = \sum_k A_{Ek} \psi_k . \tag{1}$$

The basis states ψ_k are constructed as

$$\psi_k = \mathcal{A} \sum_{ij} \Phi_i(1,\ldots,N)\, u_j(N+1)\, a_{ijk} + \sum_i \chi_i(1,\ldots,N+1)\, b_{ik} , \tag{2}$$

where the Φ_i are N-electron target states, the u_j are members of a complete set of numerical continuum orbitals used to describe the motion of the scattered electron inside the box, and the χ_i are $(N+1)$-electron configurations. The latter are formed from the one-electron orbital basis used to describe the N-electron target. They are included to allow for electron correlation effects when the scattered electron is close to the nucleus, and also to ensure completeness of the total trial wavefunction if the continuum orbitals are constructed orthogonal to the bound orbitals.

Each of the N-electron target states is expanded in terms of a sum of orthonormal configurations

$$\Phi_i = \sum_{j=1}^{m} c_{ij} \phi_j . \tag{3}$$

The configurations ϕ_j are built up from one-electron orbitals, coupled together to give a function which is completely antisymmetric with respect to the interchange of the space and spin coordinates of any two electrons.

The one-electron orbitals used in constructing the ϕ_j configurations may be of two types: i) physical (Hartree-Fock) orbitals and ii) suitably chosen non-physical pseudo-orbitals. In standard R-matrix calculations, performed over many years by the Belfast group and their collaborators worldwide [6], the number of pseudo-orbitals is very small (one or at most two per angular momentum), and generally their sole purpose is to improve the description of the discrete target spectrum.

Following the work on the "Opacity Project" [1], very large calculations are currently being performed within the "Iron Project" [2], mostly using a version of the computer code **RMATRX I** of Berrington et al. [7]. One of many examples is the work of Zhang and Pradhan [8] on electron impact excitation of Fe^{2+}. In addition, the newly developed program **RMATRX II** of Burke et al. [9] has been used for other complex targets, such as Ni^+ and Ni^{2+} [10,11]. While **RMATRX II** is more efficient than **RMATRX I** in many aspects, it is currently restricted to non-relativistic calculations, whereas **RMATRX I** can allow for some relativistic effects through the inclusion of the one-electron terms in the Breit-Pauli Hamiltonian. In addition, some fully relativistic R-matrix calculations have been performed using the DARC program [12,13] based on the Dirac-Breit Hamiltonian.

The R-matrix with pseudo-states (RMPS) method was described in detail by Bartschat et al. [14]. The principal idea behind this method is the same as in the CCC approach, i.e., one includes a large set of pseudo-orbitals which can then be used both (i) to improve the description of the physical target states and (ii) to approximate the effects of the infinite number of discrete and continuum states of the target atom or ion that cannot be included explicitly in the calculation. As in the CCC method, these effects are only represented accurately in a certain region of configuration space (usually the R-matrix box), due to the finite range of the pseudo-orbitals. However, this is expected to be sufficient if one is interested in transitions between discrete states whose range is also restricted to within this box. In fact, the method also works for ionization if the box is made sufficiently large and some averaging over effective box sizes, determined by different ranges of the pseudo-orbitals, is included [15,16].

It should be pointed out that special care has to be taken in the RMPS method due to the overcomplete basis represented by the true continuum orbitals and the pseudo-orbitals. Bartschat et al. [14] used an explicit orthogonalization procedure to address this problem, while Badnell and Gorczyca [17] suggested an alternative scheme based on matrix transformations between the various basis sets. Also, one needs to ensure that the Buttle correction [18] to the R-matrix, which accounts for the part of the continuum basis omitted in the expansion, does not yield an over-correction due to double-counting part of the spectrum.

The RMPS method method has recently been implemented in updated versions of RMATRX I and RMATRX II and can thus be used, in principle, for complex targets as well as single and even double photoionization [19]. An example for treating a complex open-shell target is the recent work on electron scattering from neutral boron [20].

The intermediate energy R-matrix method (IERM)

The intermediate energy R-matrix method (IERM) differs from the RMPS approach in that the target continuum orbitals and the motion of the projectile inside the box are described in terms of the *same* one-electron numerical continuum functions whose radial terms satisfy the differential equation

$$\left(\frac{d^2}{dr^2} - \frac{\ell(\ell+1)}{r^2} - 2V(r) + k_{n\ell}^2\right) u_{n\ell}(r) = 0 \qquad (4)$$

for each angular momentum ℓ, subject to the R-matrix boundary conditions

$$u_{n\ell}(0) = 0; \qquad (5)$$

$$\frac{a}{u_{n\ell}(a)} \cdot \left.\frac{du_{n\ell}}{dr}\right|_{r=a} = b. \qquad (6)$$

The potential $V(r)$ is chosen in such a way that, provided a suitable R-matrix boundary radius is selected, the first few $u_{n\ell}(r)$ will correspond to the bound target orbitals. (For example, $V(r) = -Z/r$ would be suitable for e–H scattering.) The reason for this choice of basis stems from the work of Burke et al. [21] who discussed the problem of including two continuum electrons within the R-matrix box and allowing for the escape of both electrons from this region. The two-electron IERM approach, for which results are shown in the next section, grew out of this work, since the two-electron case was the simplest example to study.

One advantage of the IERM approach over the RMPS method lies in the fact that the atomic pseudo-orbitals are automatically orthogonal to the physical bound orbitals and there is no need to use any numerical orthogonalization procedure. Therefore, it is possible to generate a large number of pseudo-orbitals without encountering linear-dependence problems. On the other hand, the range of the pseudo-orbitals generated in the IERM approach is determined by the value of the R-matrix boundary being used. For example, if one chooses a boundary radius to envelope all the excited target states of interest, convergence of the cross section with respect to the number of pseudo-states will be most rapid for transitions involving the outermost orbitals, but usually slow for transitions involving the ground state, particularly the elastic scattering cross section.

Comparison of CCC and R-matrix approaches

There are several important differences between the CCC and the two R-matrix based approaches that should be pointed out. For the CCC method, as a straightforward way to solve the close-coupling equations, the computational effort required is roughly the same for each collision energy; some variation is possible depending on the number of states and the number of partial waves required for convergence of the observables of interest. On the other hand, the computational effort in an R-matrix approach consists of two parts: (i) a one-time effort to solve the internal region problem by diagonalizing the $(N+1)$-electron Hamiltonian and thereby determining, for example, the coefficients a_{ijk} and b_{ik} in Eq. (2) for the RMPS case, and (ii) the solution of the (no-exchange) coupled equations outside the box for each collision energy. Since the latter can be done very efficiently using newly developed packages such as **FARM** [22], the R-matrix method is most suitable if results for a large number of collision energies are required, as for example in the study of resonances. On the other hand, the size of the Hamiltonian matrix to be diagonalized grows quickly with the size of the box and the maximum collision energy that needs to be treated. Consequently, the standard R-matrix method works best for transitions between low-lying discrete states and collision energies below the ionization threshold. However, intermediate energies up to several times the ionization energy can now be reached by both the RMPS and the IERM methods, and transitions involving higher-lying Rydberg states can be calculated by subdividing the R-matrix internal region into a number of smaller sub-regions, as done in the two-dimensional R-matrix propagator approach [21,23].

Recently, a benchmark calculation using the CCC, RMPS, and IERM methods was performed for electron scattering from atomic hydrogen. Excellent agreement between the predictions from all three approaches as well as experiment was found [24].

Time-dependent lattice approaches

With the recent development of massively parallel supercomputers, the use of more direct numerical approaches has become very popular. One example is the time-dependent close-coupling method described by Pindzola *et al.* [25]. For a target with one electron outside a closed shell, they derive the time-dependent close-coupling equations for the radial parts of the wavefunction describing the projectile and the active target electron. For each partial-wave symmetry, these equations are propagated in time over a two-dimensional lattice.

The initial wavefunction is usually written as the properly symmetrized product of a bound orbital, represented on the lattice, and a Gaussian wavepacket for the projectile. The width of the package is chosen in such a way that any overlap between the projectile and the target wavefunction before the collision can be neglected. The two-electron wavefunction is then propagated for a sufficiently long time to ensure that the final state after the collision has been reached. Note, however, that the propagation time has to be small enough to ensure that boundary effects, due to collisions with the wall of the lattice, do not disturb the result. Cross sections for elastic scattering, excitation and ionization are obtained by projecting the final-state wavefunction onto the various scattering channels.

The principal advantage of the method lies in the fact that the radial part of the wavefunction, which contains all the dynamics of the collision, is not described by a finite basis set, as in the time-independent close-coupling approaches discussed above. On the other hand, the computational resources needed for such calculations are relatively extensive. Hence, the method has so far only been applied to (quasi-)one-electron targets. In addition, just the lowest few partial waves are typically treated in this way, while contributions from higher partial waves are handled by much simpler methods, such as the first-order distorted-wave Born approximation (DWBA).

SELECTED EXAMPLES

We now present a few examples of transitions and energy ranges for which very sophisticated methods are required if the desired accuracy of the results is higher than about 50%. In addition, we show some cases where much simpler methods are apparently sufficient.

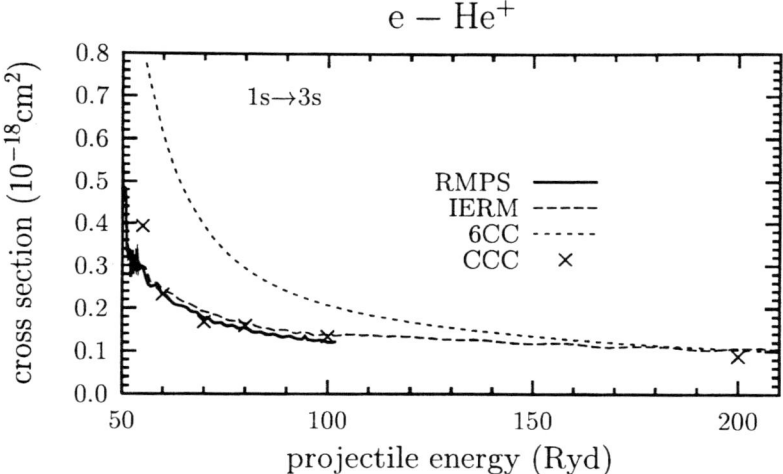

FIGURE 1. Results for electron impact excitation of the 1s→3s transition in He$^+$. The various theoretical models are described in the text.

Electron impact excitation of He$^+$

Figure 1 shows theoretical results from a standard 6-state (1s,2s,2p,3s,3p,3d) calculation for electron impact excitation of the 1s→3s transition in He$^+$ in comparison with those obtained with the CCC method [26] as well as in preliminary IERM [27] and RMPS calculations. As one would expect, the 6-state model (6CC) clearly overestimates the cross section in the intermediate energy range between 54.4 eV (the ionization threshold) and about 150 eV incident energy. For higher energies, however, the coupling to continuum states becomes much less important and the simpler model can provide results of similar accuracy. The agreement between our RMPS results and the few available CCC data points is very encouraging, while the preliminary IERM results are still a few percent higher at intermediate energies. However, extended IERM calculations are currently being performed [28] and are expected to resolve the remaining discrepancies with the CCC and RMPS predictions.

Electron impact excitation of Be$^+$

The e-Be$^+$ 2s→2p excitation cross section results are shown in Figure 2 for an extended energy range (top) and the resonance region below the $n = 5$ thresholds (bottom). For this transition, the close-coupling predictions of Mitroy and Norcross (MN88) [29], extending to about 21 eV, are only marginally higher than those of the RMPS and CCC calculations [15], and of the 9-state R-matrix calculation (RM9) presented by Berrington and Clark [30]. Although the distorted-wave

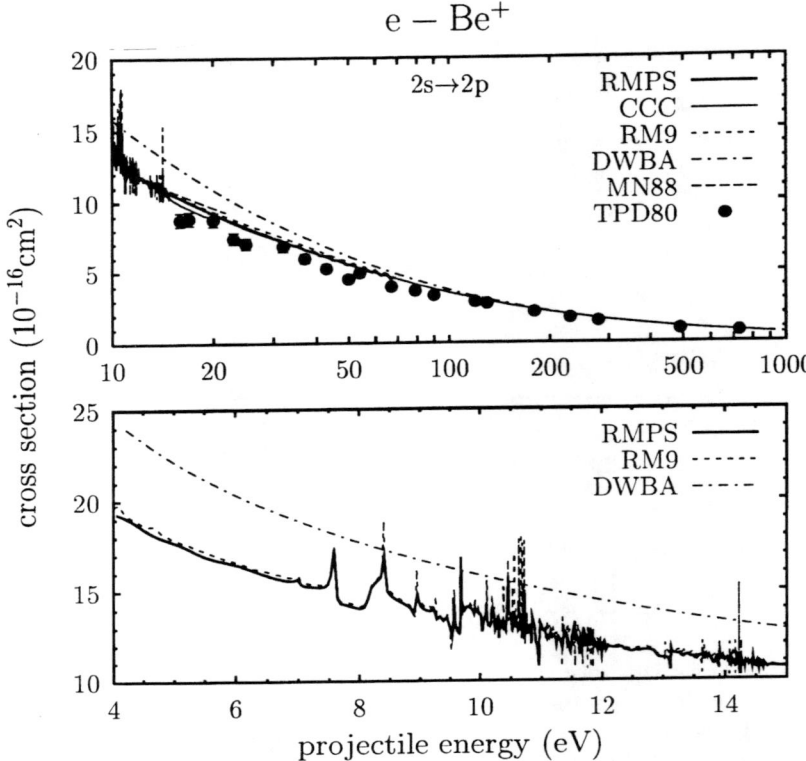

FIGURE 2. Electron-impact excitation cross section for the 2s→2p resonance transition in Be$^+$. The experimental data of Taylor *et al.* (TPD80) are compared with predictions from various theoretical models described in the text.

Born predictions of Clark and Abdallah [31] are substantially too high at low and intermediate energies, this discrepancy is much lower than what it would likely be in the case of neutral targets. The agreement with the experimental data of Taylor *et al.* [32] is generally satisfactory, although the measured results lie systematically, by a few percent, below the RMPS/CCC curves which we believe to represent the most accurate theories.

The situation, however, changes dramatically when one looks at an optically forbidden transition, such as 2s→3d shown in Fig. 3. To begin with, we note substantial differences between the close-coupling results of Mitroy and Norcross [29] and the 9-state R-matrix results given by Berrington and Clark [30]). Since the latter agree fairly well with the first-order DWBA calculations of Clark and Abdallah [31] for incident energies above 20 eV, one might be tempted to conclude that, within a few percent, the transition is accurately described by the RM9 results below 20 eV and the DWBA results for higher energies. However, this conclusion is apparently

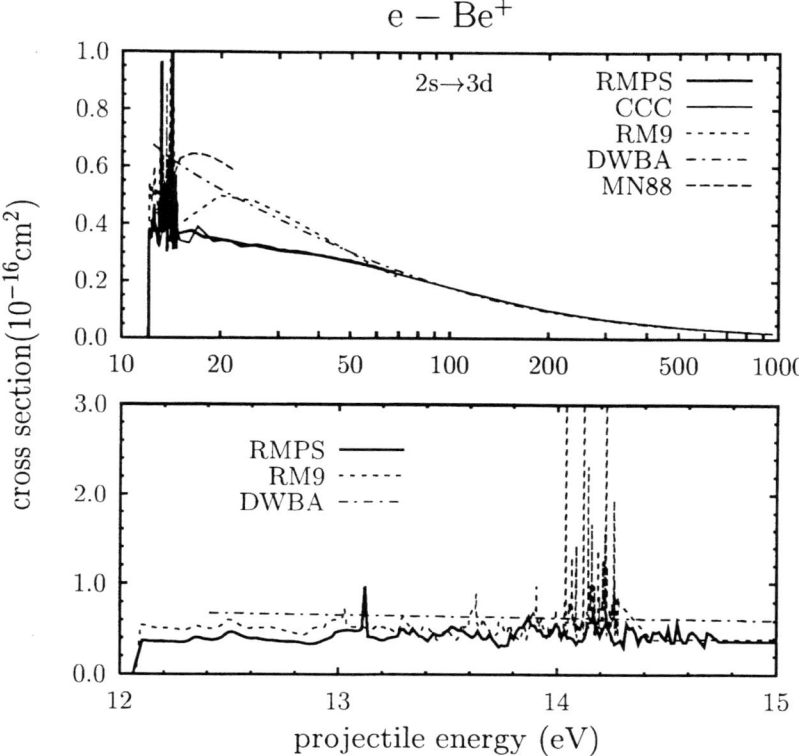

FIGURE 3. Electron-impact excitation cross section for the 2s→3d forbidden transition in Be$^+$. The various theoretical models are described in the text.

faulty, as demonstrated by comparing the RM9 and DWBA predictions with the RMPS and CCC results. Since a common error in the latter entirely independent calculations, which nevertheless agree extremely well with each other, seems rather unlikely, the cross section values provided by Berrington and Clark [30] were recently re-assessed [34], with the recommended data being reduced substantially in the energy range between 15 and 70 eV.

Electron impact excitation of B^{2+}

It is generally assumed that with increasing charge of the target, the level of sophistication in calculations of electron–ion collision cross sections can be reduced. This philosophy has been widely used, for example, in the original calculations for the Opacity Project [33]. Indeed, the trend is confirmed in Fig. 4 where the differences between the RM9/DWBA predictions on one side and the CCC/RMPS

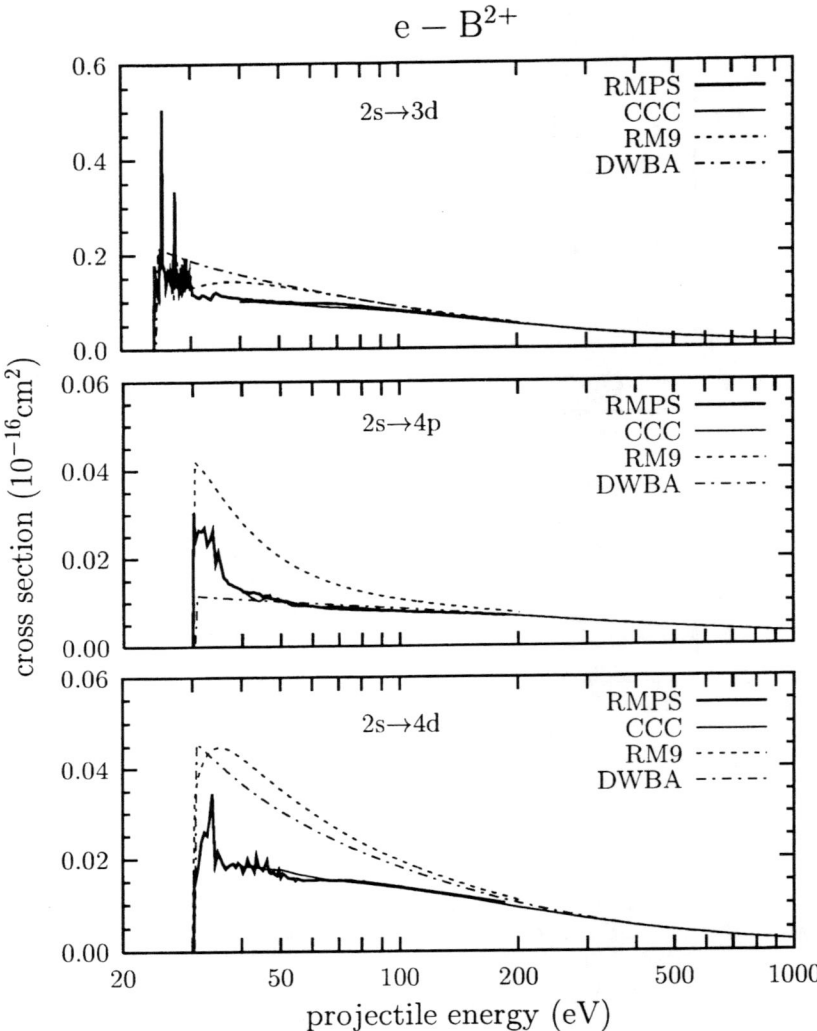

FIGURE 4. Electron-impact excitation cross section for the 2s→3d, 2s→4p, and 2s→4d transitions in B^{2+}. The various theoretical models are described in the text.

results on the other side are less pronounced for the 2s→3d transition in B^{2+} [16] than for the corresponding transition in Be^+ discussed above. Note, however, that the disagreement increases dramatically with the principal quantum number of the states involved. This is also demonstrated in Fig. 4 by the comparison of results from the different models for excitation of the 2s→4p and 2s→4f transitions. For the optically allowed 2s→4p transition, the RM9 model seems to severely overestimate this cross section, while the DWBA results, except for the very low energies, agree very well with the RMPS/CCC predictions. On the other hand, for the 2s→4d transition, RM9 and DWBA agree well with each other, but differ by up to a factor of two from the RMPS/CCC predictions.

Finally, we note that the 2s→2p excitation cross section has recently been measured over a small energy range of approximately 2 eV above the excitation threshold. The agreement between experiment and the RMPS predictions [16] was found to be excellent [35].

Electron-impact ionization

As mentioned above, the CCC, RMPS, IERM, and time-dependent methods can also be used in a straightforward way to obtain ionization cross sections, essentially by re-interpreting the results for excitation of positive-energy pseudo-states as transitions into the continuum. An example is shown in Fig. 5 for electron-impact ionization of sodium-like Al^{2+}. There is very good agreement between the predictions from a time-independent RMPS model and a time-dependent hybrid (i.e., time-dependent close-coupling for $L \leq 6$ supplemented by DWBA for higher partial waves) approach, and satisfactory, though not perfect, agreement with the CCC results [36]. Based on the fact that all the theoretical predictions are substantially higher than the experimental data of Crandall *et al.* [37], it was suspected that an experimental normalization problem may have existed. This conclusion

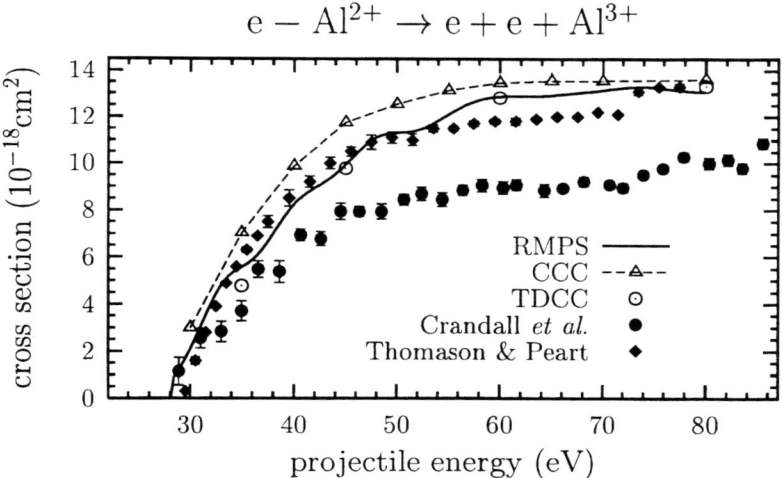

FIGURE 5. Electron-impact ionization cross section for Al^{2+}. The experimental data of Crandall *et al.* and of Thomason and Peart are compared with predictions from various theoretical models described in the text.

is, indeed, supported, by the most recent experimental results of Thomason and Peart [38].

A similar problem exists for electron-impact ionization of lithium-like Be^+, except that in this case the probably best theoretical values are substantially smaller than the experimental data of Falk and Dunn [39]. On the other hand, there is extremely good agreement between several theoretical predictions for this problem, namely time-dependent close-coupling, CCC, and two independent RMPS calculations. For a detailed discussion, see Bartschat and Bray [15] and Pindzola et al. [25].

SUMMARY

We have presented benchmark examples, demonstrating the need for highly sophisticated theoretical methods not only in the treatment of low-energy and intermediate-energy electron scattering from neutral atoms but also for singly and doubly ionized targets. Agreement between predictions from simpler low-energy and high-energy methods in the intermediate-energy regime does *not* necessarily mean that those results are highly accurate. In general, good agreement is obtained between results from the most sophisticated theoretical approaches that account for channel coupling within both the discrete and the continuous target spectrum, independent of the collision energy. The agreement with the few available experimental data for resonance transitions is satisfactory, but large discrepancies remain for ionization of several quasi-one-electron targets. In light of the good agreement between the various theoretical datasets, new experiments to independently check some of the earlier measurements seem highly worthwhile.

ACKNOWLEDGMENTS

The RMPS method was developed in collaboration with P.G. Burke, V.M. Burke, E.T. Hudson and M.P. Scott at the Queen's University of Belfast. I would like to thank N.R. Badnell, I. Bray, B.M. McLaughlin, and M.P. Scott for providing some of their results in electronic form. Financial support from the United States National Science Foundation under grant PHY-9605124 is gratefully acknowledged.

REFERENCES

1. Seaton M.J., Yu Y., Mihalas D., and Pradhan A.K., *Mon. Not. R. Astr. Soc.* **266**, 805 (1994).
2. Hummer D.G., Pradhan A.K., Saraph H.E., and Tully J.A., *Astron. Astrophys.* **279**, 298 (1993).
3. Bray I. and Stelbovics A.T., *Adv. Atom. Mol. Phys.* **35**, 209 (1995).
4. Fursa D.V. and Bray I., *Phys. Rev. A* **52**, 1279 (1995).
5. Burke P.G. and Robb W.D., *Adv. At. Mol. Phys.* **11**, 143 (1975).

6. Burke P.G. and Berrington K.A, *Atomic and Molecular Processes – An R-Matrix Approach*, Bristol: Institute of Physics (1993).
7. Berrington K.A., Eissner W., and Norrington P.H., *Comp. Phys. Commun.* **92**, 290 (1995).
8. Zhang H.K. and Pradhan A.K., *J. Phys. B* **28**, 3403 (1995).
9. Burke P.G., Burke V.M., and Dunseath K.M., *J. Phys. B* **27**, 5341 (1994).
10. Watts M.S.T., Berrington K.A., Burke P.G., and Burke V.M., *J. Phys. B* **29**, L505 (1996).
11. Watts M.S.T. and Burke V.M., *J. Phys. B* **31**, 145 (1998).
12. Norrington P.H. and Grant I.P., *J. Phys. B* **20**, 4869 (1987).
13. Wijesundera W.P., Parpia F.A., Grant I.P., and Norrington P.H., *J. Phys. B* **24**, 1803 (1991).
14. Bartschat K., Hudson E.T., Scott M.P., Burke P.G., and Burke V.M., *J. Phys. B* **29**, 115 (1996).
15. Bartschat K. and Bray I., *J. Phys. B* **30**, L571 (1997).
16. Marchalant P.J., Bartschat K., and Bray I., *J. Phys. B* **30**, L435 (1997).
17. Badnell N.R. and Gorczyca T.W., *J. Phys. B* **30**, 2011 (1997).
18. Buttle P.J.A., *Phys. Rev.* **160**, 719 (1967).
19. Marchalant P.J. and Bartschat K., *Phys. Rev. A* **56**, R1697 (1997).
20. Marchalant P.J. and Bartschat K., *J. Phys. B* **30**, 4373 (1997).
21. Burke P.G., Noble C.J., and Scott M.P., *Proc. Roy. Soc.* **A410**, 289 (1987).
22. Burke V.M. and Noble C.J., *Comp. Phys. Commun.* **85**, 471 (1995).
23. Le Dourneuf M., Launay J.M., and Burke P.G., *J. Phys. B* **23**, L559 (1990).
24. Bartschat K., Bray I., Burke P.G., and Scott M.P., *J. Phys. B* **29**, 5493 (1996).
25. Pindzola M.S., Robicheaux F., Badnell N.R., and Gorczyca T.W., *Phys. Rev. A* **56**, 1994 (1997).
26. Bray I., McCarthy I.E., Wigley J., and Stelbovics A.T., *J. Phys. B* **24**, L831 (1993).
27. McLaughlin B.M., Scott M.P., Burke P.G., and Dahler J.S., *Book of Abstracts ICPEAC XX* (TH054), eds. Aumayr F., Betz G., and Winter HP., Vienna (1997).
28. McLaughlin B.M., Scott M.P., and Burke P.G., private communication (1998).
29. Mitroy J. and Norcross D.W., *Phys. Rev. A* **37**, 3755 (1988).
30. Berrington K.A. and Clark R.E.H., *Nuclear Fusion (supplement)* **3**, 87 (1992).
31. Clark R.E.H. and Abdallah J., *Physica Scripta* **T62**, 7 (1996).
32. Taylor P.O., Phaneuf R.A., and Dunn G.H., *Phys. Rev. A* **22**, 435 (1980).
33. Berrington K.A., in: *Photon and Electron Collisions with Atoms and Molecules*, eds. Burke P.G. and Joachain C.J., New York: Plenum (1997).
34. Bartschat K., Berrington K.A., Bray I., Stephens J.A., and Janev R.K., *Report INDC(NDS)-369*, International Nuclear Data Committee, Vienna (1997).
35. Bannister M.E., private communication (1998).
36. Badnell N.R., Pindzola M.S., Bray I., and Griffin D.C., *J. Phys. B* **31**, 911 (1998).
37. Crandall D.H., Phaneuf R.A., Falk R.A., Belić D.S., and Dunn G.H., *Phys. Rev. A* **15**, 143 (1982).
38. Thomason J.W.G. and Peart B., *J. Phys. B* **31**, L201 (1998).
39. Falk R.A. and Dunn G.H., *Phys. Rev. A* **27**, 754 (1983).

Low-Energy Electron Collisions With Multiply-Charged Positive Ions

A. Chutjian, J. B. Greenwood and S. J. Smith

*Jet Propulsion Laboratory, California Institute of Technology
Pasadena, CA 91109 USA*

Abstract. Cross sections for a variety of electron-ion collision phenomena are the backbone for understanding energy balance in high electron temperature plasmas. Such plasmas include such seemingly disparate objects such as the Io torus around Jupiter, solar and stellar atmospheres, the interstellar medium, and fusion devices. Several experimental approaches used with multiply-charged ions (MCIs) will be reviewed. These include measurement of excitation cross sections using the electron energy-loss method, measurement of ionic lifetimes using a Kingdon trap, and measurement of dissociative recombination cross sections using ion storage rings. New JPL results will be presented of e-S^{2+} inelastic scattering, relevant to the problem of ion density and radiated energy in the Io torus; and metastable-state lifetimes in C^+, N^+, and Ar^{2+} relevant to stellar absorption by the interstellar medium.

INTRODUCTION

Electron-ion interactions are present in a variety of astronomical objects such as the sun, stars, quasars, planetary nebulae, the interstellar medium, comets, planetary magnetospheres and ionospheres. The electron energies and ionic charge states will depend on the object: a violent solar flare can produce emissions in Fe^{24+} (IP of 8828 eV), while a dense interstellar cloud can have a high density of co-existent singly-charged atomic and molecular ions, and thermal electrons as cold as 10-100 K.

Excitation cross sections are needed to convert, for example, the rich optical emissions observed by the Hubble Space Telescope (especially from the GHRS and STIS instruments), the Extreme Ultraviolet Explorer, and from the many ground-based telescopes, into ion densities. In the equations of statistical equilibrium the collision strength and the radiative emission rate play key roles in determining an excited-state ion population (1). In the simple case of coronal equilibrium one has the expression useful in determining the electron temperature T_e for the excited-state population N_i,

$$N_i = N_e N_g C(g \to i)/A(i \to g) \tag{1}$$

where g refers to the ion ground state, and $C(g{\rightarrow}I)$, $A(I{\rightarrow}g)$ are the collisional excitation rate (cm^3/sec) and the spontaneous radiative decay rate (sec^{-1}), respectively. The collision rate is an average of the collision strength or cross section over the (usually) Maxwellian electron distribution function of the astronomical plasma described by T_e. Low-energy electron excitation cross sections are difficult to calculate, and any calculation has to be benchmarked against reliable measurements. As evident from Eq. (1) Einstein A-values are also required in large numbers. The majority have to be calculated, but a smaller fraction need to be measured experimentally in order to calibrate theory.

Finally, the broader issue of ionization equilibrium in hot solar, stellar, and fusion plasmas involves phenomena such as direct and dielectronic recombination; photoionization, electron ionization (both direct and indirect); and charge exchange (1). A vast array of theoretical and experimental cross sections, with and without superimposed electric fields (critical in the case of dielectronic recombination) are required.

THE ELECTRON ENERGY-LOSS METHOD

The electron energy-loss method, when applied to singly- and multiply-charged ion targets, can provide a more versatile path for exciting spin- and symmetry-forbidden transitions than can, for example, excitation-optical emission methods. The process can be represented as,

$$e(E,0°) + A(nl)^{m+} \rightarrow e(E-\Delta E,\theta) + A(n'l')^{m+} \qquad (2)$$

where ΔE is the electron energy loss corresponding to the energy difference between the nl and $n'l'$ states, and θ is the electron polar scattering angle. As seen in Eq. (2), one is able to measure *angular distributions* of elastically- and inelastically-scattered electrons. These differential cross sections (DCSs) provide a more stringent test of theory, since the angular results probe both the short-range (large scattering angles) and long-range (small scattering angles) portions of the e-ion scattering potential. Calculated integral cross sections, while the preferred input to plasma-modeling calculations, tend to integrate over approximations to the scattering potential, wavefunction, number of states, *etc*.

Recent methods applied to energy-loss scattering can be divided into those using crossed electron-ion beams (3,4) and merged beams (5,6). Crossed beams have been used to measure elastic, superelastic, and inelastic DCSs, and merged beams to detect inelastically-scattered electrons collected over the angular range $\{0,\pi\}$. Representative results for both classes of experiments will be reviewed in the following two sections.

Inelastic Scattering

The first application of energy-loss techniques to ions was in a crossed-beams geometry (7). A highly-collimated, non-monochromatized incident electron beam was used, with a 180° hemispherical analyzer for electron energy analysis. The resonance $4^2S \rightarrow 4^2P$ in Zn^+ was detected, and relative DCS with comparison to a five-state close-coupling theory, reported (8). Further DCS were obtained on low-lying allowed and forbidden transitions in Mg^+, Zn^+, and Cd^+ (9).

As mentioned above, the primary need in both astronomical and fusion plasmas is the *integral* excitation cross section (10). Two similar methods of approach have been developed, at JPL (5,11-13) and at ORNL (6,14-16), using energy-loss and merged-beams methods. In these approaches, one is able to handle issues of full angular collection, good signal-to-backgrounds, efficient electron detection, and be able to accommodate MCIs as the target.

Details of the ORNL apparatus are given by Bannister, *et al.* in this volume. A brief description will be given here of the JPL apparatus, with an indication of differences between the JPL and ORNL methods. Shown in Fig. 1 is a schematic of the JPL MCI facility (13). The facility is designed to generate high charge states using the *Caprice* 14 GHz electron-cyclotron resonance ion source. Separate beam lines are dedicated to measuring excitation cross sections in a merged-beams section, ion lifetimes in a Kingdon-trap section, and charge-exchange cross sections with a neutral gas cell. Recent additions to the merged-beams sections are a multipole "electronic aperture" **EA** to separate elastically-scattered electrons, with their larger gyroradii, from the inelastically-scattered electrons (17); a new vane system with finer sampling of the electron and ion beam profiles; and a new beam-pulsing unit using fast MOSFET switches (18), with PC control of all pulsing, beams-profile monitoring, and data acquisition. Some differences between the JPL and ORNL merged-beams system are: (a) metallic beam profile monitors to monitor profiles at four locations in a 20.0±0.3 cm merged pathlength (microchannelplate-CCD camera combination for ORNL which monitors continuously along a 6.35 cm pathlength), (b) location of a PSD along the magnetic-field direction (orthogonal in the ORNL setup) allowing discrimination against energetic electrons using retarding grids, an electron mirror MI to reflect into the forward PSD electrons which are backscattered ($\vartheta > 90°$) in the LAB frame, (d) an electronically-variable multipole aperture (five physical apertures in the ORNL system)

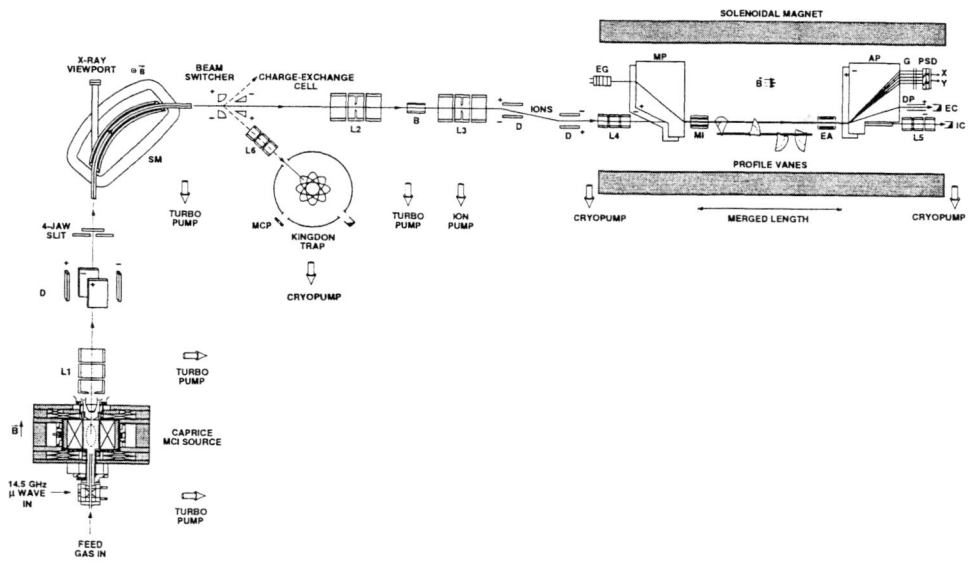

FIGURE 1. Present configuration of the JPL multiply-charged ion facility. (**L1-L5**), three-element focusing lenses, (**B**) differential pumping baffle, (**D**) deflector plates, (**MP**) merging trochoidal plates, (**AP**) analyzing trochoidal plates, (**MI**) electron mirror, (**EA**) electronic aperture, (**DP**) trochoidal deflection plates to deflect parent electron beam out of the scattering plane, (**PSD**) position-sensitive detector.

to filter unwanted elastically-scattered electrons having the same axial velocity as inelastically-scattered electrons.

As is well known, ion beams produced in discharges can contain a significant fraction f of metastable states. The population of these states will depend on lifetime, ion transit time from source to target, and the source operating conditions (microwave power, gas pressure, *etc.*). The presence of metastables means that only a fraction *(1-f)* of the total beam current is available for excitation, while the entire beam is counted in measurement of the ion current. Hence any measured cross section will be smaller than its true value by a factor *1/(1-f)*. In the JPL approach an ion-beam attenuation method is used, wherein a section of the beam line is filled with a charge-exchange (*ce*) gas (He, N_2, or Ar). Assuming that different excited states will have a different *ce* cross section, one will observe breaks in the slope of the transmitted current *vs* pressure. Extrapolation of the high-pressure slope (corresponding to *ce* by the ground state) to zero pressure will give the metastable fraction. Fractions as high as 35% have thus far been observed in the JPL *Caprice* source. (Interestingly, one would like to *minimize f* for the excitation measurements, and *maximize f* for metastable *f*-value measurements! See below.)

Shown in Figs. 2 and 3 are several recent JPL results on excitation of singly-charged C^+ (5) and S^+ (12). These ions are strong emitters in solar and stellar atmospheres, and in the Io torus (S^+). It is interesting to note that the torus is supplied by SO_2 from volcanic action on Io (19). The SO_2 is dissociated and ionized by the energetic particle environment at Jupiter, and ions are trapped in Jupiter's magnetosphere. Successive ionizations of the trapped singly-charged ions lead to higher charge states of S and O. Shown In Fig. 4 are the first JPL results in MCIs, in this case excitation of the $3s^2 3p^2$ $^3P \rightarrow 3s 3p^3$ $^5S^\circ$ transition in e-S^{2+} (20). Comparison with a 17-state close-coupling calculation (21) shows good agreement between experiment and theory, especially for the broad resonances at 11.5 eV. It is essential to filter out as completely as possible the elastically-scattered electrons. A schematic of the so-called electronic aperture, used in the results of Fig. 5 to eject the unwanted elastically-scattered electrons, is shown in Fig. 5. It consists of sixteen 2.00-mm dia rods, each 25.0 mm long. Potentials of equal and opposite sign are impressed on the rods, forming a null potential in the center of the structure through which the merged electron and ion beams

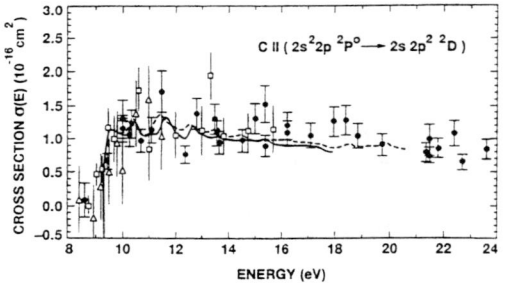

FIGURE 2. Experimental (filled circles) and an 8-state *R*-matrix cross sections (solid line, convoluted with a 250 meV FWHM resolution) in C^+ (5). Other data: Luo and Pradhan (22) theory (dashed line), Lafyatis and Kohl experiment (open square)(23), and Greenwood *et al.* experiment (open triangle)(24).

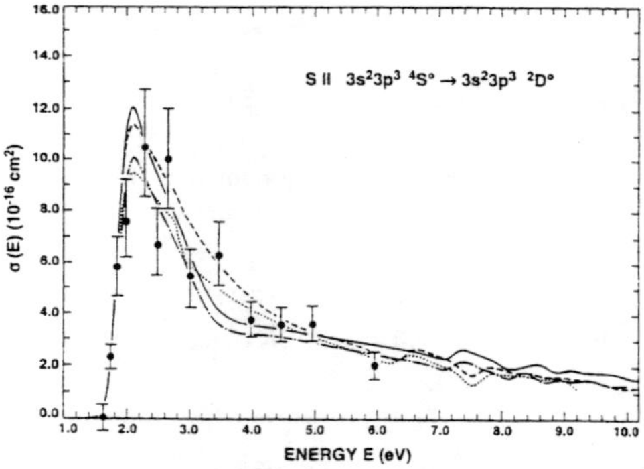

FIGURE 3. Experimental (filled circles) and 19-state close-coupling cross sections (solid line, convoluted with a 250 meV FWHM resolution) in S^+ (12). Other theories: 19-state *R*-matrix [dash-dot line (25)], 12-state *R*-matrix [dashed line (26)], 6-state *R*-matrix [dotted line (27)].

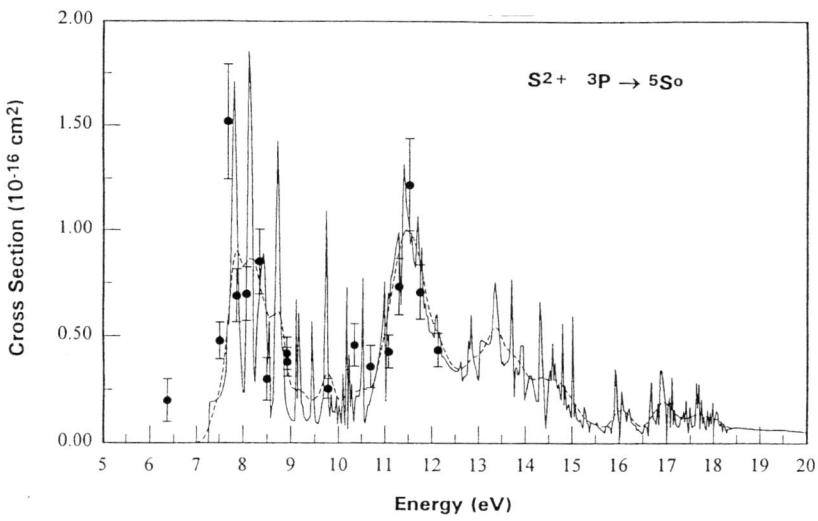

FIGURE 4. Experimental (filled circles) (20) and theoretical cross sections [solid line (21)] for the $3s^2 3p^2\ ^3P \to 3s3p^3\ ^5S°$ transition in S^{2+}. Dashed line is theory line convoluted with a 250 meV FWHM resolution.

pass. An elastically-scattered electron with its larger gyroradius will make an excursion close to the poles and be ejected into the shield. An inelastically-scattered electron with its smaller perpendicular energy will stay closer to the null center and traverse the aperture. The diameter of the "iris" of the aperture is set by the potentials of the rods. Further details will be presented elsewhere (17).

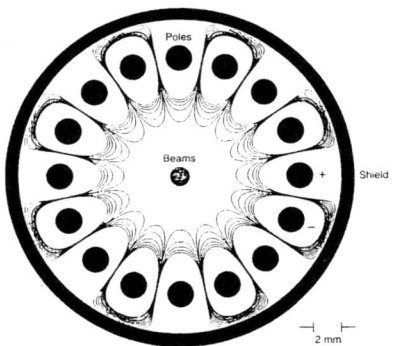

FIGURE 5. Schematic of an "electronic aperture" used to filter elastically-scattered electrons with their larger gyroradii (17). This end-on view shows sixteen rods (dark circles) symmetrically placed about the merged beams axis (shaded central region). Light lines are calculated equipotentials. Opposite potentials are placed on adjacent rods, and unwanted electrons are ejected into the positively-biased shield.

Elastic Scattering

It has been noted in the previous section that elastic scattering can be an unwelcome addition to electron-ion excitation signal. For typical collision energies the elastic scattering signal can be order of magnitude greater than the inelastic signal, especially for high charge states. Although various techniques are applied to remove this corrupting signal an estimate of the elastic contribution may be necessary. For this reliable elastic differential and integral cross sections are required.

In plasmas elastic electron-ion scattering determines properties such as electron transport coefficients and ion heating rates. In modeling these parameters the classical Rutherford Coulomb DCS is often used. This DCS is strongly forward peaked. However Greenwood and Williams (28) have demonstrated deviations from the Coulombic formula which can have significant effect on the momentum-transfer cross section for certain types of plasmas. Therefore, the accurate angular distribution has to be taken into account to obtain an accurate plasma model.

Elastic scattering from a Coulombic potential can be solved analytically (29), and the result is equivalent to the classical Rutherford scattering model,

$$\frac{d\sigma}{d\Omega} = \frac{q^2}{16 E^2 \sin^4(\theta/2)}, \tag{3}$$

where $d\sigma/d\Omega$ is the elastic DCS, E the electron energy, q the ionic charge, and θ the scattering angle. For partially-stripped ions the scattering potential is only Coulombic for large distance, so the potential can be written as $V = V_c + V_s$, where V_c is the Coulombic (long-range) potential and V_s is the short-range potential. Similarly, the scattering amplitude may be represented as a Coulombic and a short-range part (29) by $f = f_c + f_s$, to give a DCS as,

$$\frac{d\sigma}{d\Omega} = |f_c|^2 + |f_s|^2 + 2\mathrm{Re}[f_c^* f_s].$$

This formula indicates that there is a Coulombic contribution to the DCS, a short-range part, and an interference term, respectively. The interference term can give dramatic deviations from the usual Coulomb formula in Eq. (3), especially at large scattering angles where the long range contribution is weak. Examples of this interference can be seen in recent elastic DCS measurements, shown here in Fig. 6 (4).

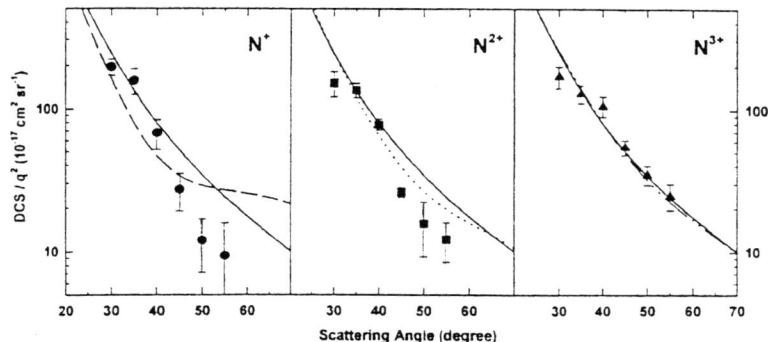

FIGURE 6. DCS, reduced by q^2, for elastic electron scattering from N^{q+} (q = 1-3) at 10 eV (4). Solid line in each case is the Rutherford formula [Eq.(3)], and broken lines are from Ref. 31.

The scattering amplitudes can be determined by calculating the Coulomb and short-range phase shifts. The former are given by gamma functions, and the latter can be calculated numerically given an assumed form for V_s. Manson (30) and Szydlik et al. (31) have done so for singly-charged ions, and some MCIs. A database of short-range phase shifts for all ions has recently been completed (32).

Elastic scattering was first investigated by an indirect method of fast ion-atom scattering through analysis of scattered binary-encounter electrons. Electrons scattered at $\vartheta = 0°$ in the LAB frame correspond to $\theta = 180°$ scattering in the projectile frame. Larger LAB angles correspond to smaller CM angles. Energy broadening of the emitted electron spectra occurs because the electrons are not stationary in the atom, but have a characteristic momentum distribution, the Compton profile. To assess the experimental data it is necessary to fold this profile into theoretical results. However, this limits the usefulness of the technique to large electron-ion scattering energies (E > 100 eV). A summary of experimental and theoretical investigations of binary-encounter electrons has been given by Liao et al. (33). Large enhancements in the DCS at $\theta = 180°$ have been observed. These increase with lower charge states of the partially-stripped ions, in contradiction to the Rutherford formula. Indications of interference structure in the DCS appear in some angular measurements, best shown in Ref. 34.

Due to the inherently difficult nature of electron-ion scattering measurements (low ion densities, requiring ultrahigh vacuum techniques and beams modulation) there are few direct scattering measurements, but the existence of interference structure in the DCS provides a sensitive test of theory. Basically, the interference is due to close-in (low partial waves) electron-ion interactions, where the electron can better "feel" the details of the (non-Coulomb) ion scattering potential. Agreement between experimental measurements (3,30,34-38) and partial-wave calculations (3,30-32,35) has been reasonably good. However, Srigengan *et al.* (39) find poorer agreement in energy-dependent measurements at fixed scattering angles. Other differences occur in the more sophisticated many-body perturbation theory (32), which is in accord with measurements of Huber *et al.*, but not with Srigengan *et al.* (38). Possible effects of polarization at large scattering angles has been suggested (40,41), but *R*-matrix calculations (42) with and without polarized pseudostates indicate that this is not a significant effect for the case of Ar^+.

These diverse results indicate that further e-ion elastic DCS measurements are required, especially at large scattering angles. Deviations from the Rutherford formula are dramatic, even for highly-charged ions. For plasma modeling the use of the short-range phase shifts of Manson and Turner to calculate DCSs is recommended.

MEASUREMENTS OF LIFETIMES AND *f*-VALUES

Because of the high sensitivity and optical resolution of present and planned space missions weak absorption and emissions, previously undetected, now assume new importance. One example is detection in the long path length through the interstellar medium (ISM) of weak spin- and symmetry-forbidden transitions from metastable levels in singly- and multiply-charged positive ions (43-45). Strong electric-dipole transitions lead to absorption lines in the ISM with equivalent widths (EW) that lie on the saturated part of the curve of growth. Hence reliable column densities require transitions (metastable) with small *f*-values, where EW's now lie on the linear part of the curve. Because rates for collisional excitation and de-excitation of ions in diffuse plasmas are comparable to radiative decay rates from metastable levels, the intensity ratios between forbidden and allowed transitions are density- and temperature sensitive (46) and critically depend on the lifetimes of the metastable or intersystem line (47,48). However, calculated and measured lifetimes in some cases can differ by a factor of three.

Use of the Kingdon Trap for Metastable Lifetime Measurements

For measuring the long radiative lifetimes of metastable states, long storage times are required: hence the use of an ion trap under ultrahigh vacuum operation for minimizing collisional de-excitation effects. One can store the ions, and access prominent intersystem and forbidden decays (45). Decay channels include intercombination, electric quadrupole E2, magnetic dipole M1 and double photon decay 2E transitions. The use of a Kingdon trap was first reported by Prior (49), and a number of lifetime measurements using it have been reported by Smith, Parkinson, and co-workers (46,50-55). Lifetime measurements have been made for C^{2+} 190.9 nm (46), and O^{2+} 166.1, 166.6 nm (54). Kingdon traps have been used to measure lifetimes of Ar^{2+}, $Kr^{2+,3+}$, $Xe^{2+,3+}$, B^+, N^+, and Cu^+ (49,56). Lifetime measurements carried out in Paul, Penning and Kingdon ion traps have been reviewed through 1993 (56).

The lifetime measurements in metastable MCIs reported herein use the JPL *Caprice* electron-cyclotron resonance ion source (Fig. 1). The metastable population in the beam is measured by filling a gas cell region with a charge exchanging gas such as Ar, N_2 or He and monitoring the fraction of transmitted ion beam *vs* target gas pressure. Populations can range up to 35% of the total beam current, depending on ion charge state and operating parameters of the source such as gas pressure and microwave power.

The mean lifetime of an upper level k is related to the transition rates or decay rates A_{jk} to N lower levels i the by,

$$\tau_k^{-1} = \sum_{i=1}^{N} A_{ik} . \tag{4}$$

The oscillator strength f_{ik} and the A-value are connected *via* the standard expression,

$$A_{ki}(\text{sec}^{-1}) = \frac{6.670 \times 10^{13}}{\lambda^2} \frac{g_i}{g_k} f_{ik} . \tag{5}$$

Here g_i and g_k are the statistical weights of levels i and k, and λ is in nanometer. It is apparent from Eqs. (4) and (5) that when there is only one channel branch for the decay of the metastable state, the lifetime yields the A-value or f-value of the transition directly. When more than one decay channel is open, the branching ratios of the transitions must be known (experimentally or theoretically) in order to obtain individual A- or f- values. The JPL ion trap, constructed in collaboration with Texas A & M University (57) is on a dedicated beam line. In the Kingdon ion trap ions are launched into orbits of defined angular momentum about a charged electrode (wire) which is pulsed once during the trapping cycle. The trap is formed by two coaxial cylinders, terminated at each end by cap electrodes. The ion trap is 15 cm long and 10 cm diameter. Four 2-cm dia apertures gird the middle of the cylinder. One aperture is for the ion beam entrance, one for the beam exit into its Faraday cup, one is for the UV fluorescence-detection optics, and one for dumping the trap contents onto an ion-counting CEM detector. The UV photon emissions are detected by an interference filter and phototube using a UV grade quartz optical system. For transitions shorter than 180 nm a UV-enhanced, CsI-coated CEM is used. A cryopump provides residual gas pressures in the vacuum chamber on the order 8×10^{-10} torr.

A multichannel analyzer records the decay N(t) of the upper state as

$$N(t) = N_o \exp[-t/\tau_m] + B. \quad (6)$$

where τ_m is the measured lifetime and B is a background term. The measured decay rate τ_m^{-1} is the sum of two rates,

$$\tau_m^{-1} = \tau^{-1} + \tau_Q^{-1}. \quad (7)$$

The quantity τ_Q^{-1} is the rate of loss of metastable ions due to collisional quenching within the trap and τ^{-1} is the natural decay rate of the metastable ion. The collisional quenching decay rate τ_Q^{-1} can be expressed as the product kn_p of the collision rate coefficient k and n_p, the density of the background gas. Experimentally, the total decay rate τ_m^{-1} is measured as a function of background pressure P at about 10 different pressures. A plot is made of τ_m^{-1} vs P. The intercept at zero pressure yields τ^{-1} and the slope yields k. At each P the data-taking procedure requires about 2500 scans (fill-dump cycles) and each scan last about 250-500 ms. The accuracy of the lifetimes measured are in the range 4-10% (56), which is the accuracy required by astronomers.

JPL Radiative Lifetime Measurements

Ar^{2+}: The trapping technique was first tested with the Ar^{2+} $^1S_o \rightarrow \, ^1D_2$ transition at 519.5 nm. This transition was measured previously (57), with results comparing favorably with the available calculations (58,59). JPL preliminary results are shown in Fig. 7. The lifetime of $\tau = 127$ ms compares with the measured results of 121.3 ms. This measured lifetime must be corrected for ion storage lifetimes, as shown in Eq.(7). Extrapolation to zero pressure resulted in a finite ion storage time constant of 511 ms for Ar^{2+}. The extrapolated lifetime is $\tau = 169$ ms, which is good agreement with the calculated work.

C$^+$: Boron-like ions of carbon, nitrogen and oxygen are astrophysical-abundant and spectra are observed in a number of astronomical sources. The $2s2p^2\ ^4P \rightarrow 2s^22p\ ^2P^e$ intersystem transitions at 232.5 nm were measured in C^+. The upper $2s2p^2\ ^4P$ state has three components, each decaying to the ground $^2P^e$ term. The emissions consist of five closely-spaced lines (60). Measurements (61) and calculations (52,63) also exist. The emission signal is seen to consist of three exponential components, corresponding to the radiative decay of the $J = 1/2$, $3/2$ and $5/2$ fine-structure levels. For the JPL effort, a narrow-band interference filter centered at a wavelength of 232 nm was used. Shown in Fig. 8 are preliminary C^+ results. The base pressure was 6×10^{-9} torr with the ECR beam on. Nitrogen was also deliberately added to the vacuum chamber to produce a set of plots of decay rate vs N_2 pressure, with pressure ranging from $(6\text{-}115) \times 10^{-9}$ torr. The true rates were obtained by extrapolating to zero N_2 pressure.

N^+: JPL measurements were recently started on the radiative lifetime of the $2s2p^3\ ^5S_2$ metastable state of N^+. This state decays by the E1 transition $2s2p^3\ ^5S_2 \rightarrow 2s^2 2p^2\ ^3P_{0,1,2}$ at 214 nm. These emissions are readily detected in auroral spectra from satellite observations. Three theoretical studies (64-66) and three measurements (67-69) of the N^+ $^5S_2$ lifetime have been reported, hence N^+ is a good benchmark to test the JPL apparatus. One measurement (69) is 5.4 ± 0.3 ms, while the JPL preliminary lifetime is $\tau = 5.0 \pm 1.5$ ms. The final experimental error is expected to be at the 5% level.

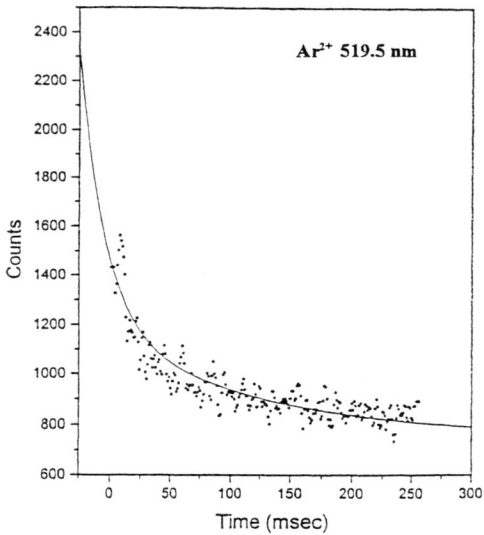

FIGURE 7. Decay of the Ar^{2+} 519.5 nm emission as function of time. Experimental data points are shown as circles and the solid line along with a two-term exponential fit shown as a solid line. The longer term measures 127 ms.

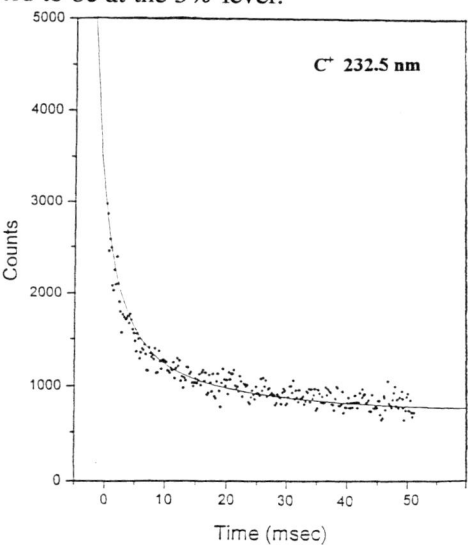

FIGURE 8. Decay of the C^+ 232.5 nm emission as a function of time at a base pressure of 6×10^{-9} torr. Experimental points are shown along with a three-term exponential fit.

ELECTRON-ION RECOMBINATION

With the use of storage rings in the various European laboratories the study of electron-atom and electron-molecular ion recombination has become a highly-developed field. A recent example of the dissociative recombination (DR) in the ions H_2O^+, H_3O^+, and CH_3^+ may be found in Vejby-Christensen et al.(70) using the ASTRID storage ring. With large energies in the LAB frame, CM energies in the merged cooler region as low as 10^{-4} eV can be attained. This correspond to a lower electron temperature of about 1 K, hence data are relevant for ionization balance and molecular formation in interstellar clouds where temperatures are of the order 10-100 K. Several reviews, including that of the operation of the CRYRING, may be found in Refs. 71 and 72. Another recent example of dissociative recombination is that of e-O_2^+ recombination which is relevant to a planet within our own solar system — the earth. Shown in Fig. 9 are the projected distances for neutral atoms formed in the DR process $e + {}^{18}O{}^{16}O^+ \to 2O$, where the states of O can be ^3P, ^1D, or ^1S (73). (Long ion storage times and the use of mixed isotopes ensured that the metastable $a^4\Pi_u$ and vibrational levels had decayed.) From the product yields of Fig. 9 one finds that the total quantum yield for $O(^1S)$ production is 0.05 ± 0.02, about an order of magnitude larger than the previously-assumed values of 0.0012-0.0016. This higher yield helps explain the bright emissions observed in the $O(^1S) \to O(^3P)$ 557.7 nm green line as arising from DR, without having to invoke a new source of $O(^1S)$ in the atmosphere.

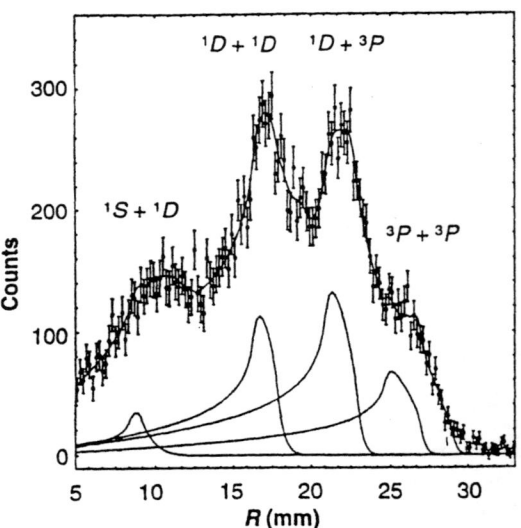

FIGURE 9. Product-state distance distribution arising from the process $e + {}^{18}O{}^{16}O^+ \to 2O$, with the atomic states of O as indicated (73).

ACKNOWLEDGMENTS

JBG thanks the National Research Council for a fellowship though the NASA-NRC program. This work was carried out at the Jet Propulsion Laboratory, California Institute of Technology, and was supported by the National Aeronautics and Space Administration.

REFERENCES

1. Mason, H. E., and Monsignori Fossi, B. C., *Astron. Rev.* 6, 123-179 (1994).
2. Smith, S. J., Chutjian, A., Mitroy, J., Tayal, S. S., Henry, R. J. W., Man, K-F., Mawhorter, R. J., and Williams, I. D., *Phys. Rev. A* 48, 292-309 (1993).
3. Belenger, C., Defrance, P., Friedlein, R., Guet, C., Jalabert, D., Maurel, M., Ristori, C., Rocco, J. C., and Huber, B. A., *J. Phys. B* 29, 4443-56 (1996).
4. Williams, I. D., Srigengan, B., Greenwood, J. B., Newell, W. R., Platzer, A., and O'Hagan, L., *Physica Scripta* T73, 119-20 (1997).
5. Smith, S. J., Zuo, M., Chutjian, A., Tayal, S. S., and Williams, I. D., *Astrophys. J.* 463, 808-17 (1996).
6. Bannister, M. E., Chung, Y.-S., Djurić, N., Wallbank, B., Woitke, O., Zhou, S., Dunn, G. H., and Smith, A. C. H., *Phys. Rev. A* 57, 278-81, (1998).
7. Chutjian, A., and Newell, W. R., *Phys. Rev. A* 26, 2271-3 (1982).
8. Chutjian, A., Msezane, A. Z., and Henry, R. J. W., *Phys. Rev. Letters* 50, 1357-60 (1983).
9. Williams, I. D., Chutjian, A., and Mawhorter, R. J., *J. Phys. B* 10, 2189-98 (1986), and refs. therein.
10. Models to date do not require knowledge of the trajectory and energy of an electron incident on a plasma. One application of DCSs would be in the earth's upper atmosphere: e-neutral DCS can give a better description of the magnetically-confined electrons raining down at the earth's poles to generate the aurora.
11. Smith, S.J., Man K-F., Mawhorter, R. J., Williams, I. D., and Chutjian, A., *Phys. Rev. Letters* 67, 30-3 (1991).
12. Liao, C., Smith, S. J., Hitz, D., Chutjian, A., and Tayal, S. S., *Astrophys. J.* 484, 979-84 (1997).
13. Liao, C., Smith, S. J., Chutjian, A., and Hitz, D., *Phys. Scripta* T73 382-3 (1997).
14. Wåhlin, E. K., Thompson, J. S., Dunn, G. H., Phaneuf, R. A., Gregory, D. C., and Smith, A. C. H., *Phys. Rev. Letters* 66, 157-60 (1991).
15. Bell, E. W., Guo, X. Q., Forand, J. L., Rinn, K., Swenson, D. R., Thompson, J. S., Bannister, M. E., Gregory, D. C., Phaneuf, R. A., Smith, A. C. H., Müller, A., Timmer, C. A., Wåhlin, E. K., DePaola, B. D., and Belić, D. S., *Phys. Rev. A* 49, 4585-96 (1994).
16. Wallbank, B., Djurić, N., Woitke, O., Dunn, G. H., Smith, A. C. H., Bannister, M. E., *Phys. Rev. A* 56, 3714-8 (1997).
17. Greenwood, J. B., Smith, S. J., and Chutjian, A., electronic aperture, unpublished results.
18. Bernius, M. T., and Chutjian, A., *Rev. Sci. Instr.* 60, 779-82 (1989); Bernius, M. T., and Chutjian, A., *ibid.* 61, 925-7 (1990).
(19) Ballester, G. E., McGrath, M. A., Strobel, D. F., Zhu, X., Feldman, P. D., and Moos, H. W., *Icarus* 111, 2-17 (1994).
20. Greenwood, J. B., Smith, S. J., and Chutjian, A., e-S^{2+} excitation, unpublished results.
21. Tayal, S. S., *Astrophys. J.* 481, 550-6 (1997).
22. Luo, D., and Pradhan, A. K., *Phys. Rev. A* 41, 165-73 (1990).
23. Lafyatis, G. P., and Kohl, J. L., *Phys. Rev. A* 36, 59-65 (1987).
24. Greenwood, J. B., O'Neill, R. W., Hughes, I. G., and Williams, I. D., unpublished.
25. Ramsbottom, C. A., Bell, K. L., and Stafford, R. P., *At. Data Nucl. Data Tables* 63, 57-91 (1996).
26. Cai, W., and Pradhan, A. K., *Astrophys. J. Suppl. Ser.* 88, 329-356 (1993).
27. Tayal, S. S., Henry, R. J. W., and Nakazaki, S., *Astrophys. J.* 313, 487-93 (1987).
28. Greenwood, J. B., and Williams, I. D., *Physica Scripta* T73, 108-9 (1997).
29. Burke, P. J., and Joachain, C. J., *Theory of Electron-Atom Collisions*, New York: Plenum,1994.
30. Manson, S. T., *Phys. Rev.* 182, 97-103 (1969).
31. Szydlik, P. P., Kutcher, G. J., and Green, A. E. S., *Phys. Rev. A* 10, 1623-32 (1974).
32. Turner, C. S., and Manson, S. T., to be published.

33. Liao, C., Richard, P., Grabbe, S. R., Bhalla, C. P., Zouros, T. J. M., and Hagmann, S., *Phys. Rev. A* **50**, 1328-34 (1994).
34. Liao, C., Hagmann, S., Bhalla, C. P., and Grabbe, S. R., *Physica Scripta* **T73**, 225-6 (1997).
35. Huber, B. A., Ristori, C., and Kuchler, D., *AIP Conf. Proc. Ser.* **271**, 218 (1993).
36. Huber, B., Ristori, C., Guet, C., Küchler, D., and Johnson, W., *Phys. Rev. Lett.* **73**, 2301-3 (1994).
37. Srigengan, B., Williams, I. D., and Newell, W. R., *Phys. Rev. A* **54**, R2540-2 (1996).
38. Srigengan, B., Williams, I. D., and Newell, W. R., *J. Phys. B* **29**, L897-900 (1996).
39. Srigengan, B., Williams, I. D., and Newell, W. R., *J. Phys. B* **29**, L605-L610 (1996).
40. Johnson, W. R., and Guet, C., *Phys. Rev. A* **49**, 1041-8 (1994).
41. Greenwood, J. B., Williams, I. D., and McGuinness, P., *Phys. Rev. Lett.* **75**, 1062-5 (1995).
42. Griffin, D. C., and Pindzola, M. S., *Phys. Rev. A* **53**, 1915-8 (1996).
43. Shull, J. M., *Physica Scripta T47*, 165-70 (1993).
44. Savage, B. D., *Physica Scripta T47*, 171-5 (1993).
45. Linsky, J. L., in *Atomic and Molecular Data for Space Astronomy* (eds. P. L. Smith and W. L. Wiese) New York: Springer Verlag, 1992.
46. Kwong, V. H. S., Fang, Z., Gibbons, T. T., Parkinson, W. H., and Smith, P. L., *Astrophys. J.* **411**, 431-7 (1993).
47. Keenan, F. P., Conlon, E. S., Harris, K. M., Aggarwal, K. M., and Widing, K. G., *Astrophys. J.* **389**, 440-2 (1992).
48. Henry, R. J. W., *Physics Reports* **68** 1-91 (1981).
49. Prior, M. H., *Phys. Rev. A* **30**, 3051-6 (1984).
50. Smith, P. L., *Atomic and Molecular Data Required for Ultraviolet and Optical Astrophysics*, Cambridge: Harvard College Observatory, 1989.
51. Calamai, A. G., Han, X. F., and Parkinson, W. H., *Phys. Rev. A* **45**, 2716-22 (1992).
52. Calamai, A. G., and Johnson, C. E., *Phys. Rev. A* **42**, 5425-32 (1990).
53. Johnson, C. E., *J. Appl. Phys.* **55**, 3207-14 (1984).
54. Johnson, B. C., Smith, P. L., and Knight, R. D., *Astrophys. J.* **281**, 477-81 (1984).
55. Kwong, H. S., Johnson, B. C., Smith, P. L., and Parkinson, W. H., *Phys. Rev. A* **27**, 3040-3 (1983).
56. Church, D., *Physics Reports* **228**, 253-358 (1993).
57. Yang, L., and Church, D. A., *Phys. Rev. Lett.* **70**, 3860-3 (1993).
58. Doschek, G. A., in *Autoionization* (ed. A. Temkin) New York: Plenum, 1985.
59. Feldman, U., Doschek, G. A., and Rosenberg, F. D., *Astrophys. J.* **215**, 652-65 (1977).
60. Edlen, B., *Physica Scripta* **23**, 1079-86 (1981).
61. Fang, Z., Kwong, H. S., and Wang, J., *Phys. Rev. A* **48**, 1114-22 (1993).
62. Lennon, D. L., Dufton, P. L., Hibbert, A., and Kingston, A. E., *Astrophys. J.* **294**, 200-6 (1985).
63. Nussbaumer, H., and Storey, P. J., *Astron. Astrophys.* **96**, 91-5 (1981).
64. Hibbert, A., and Bates, D. R., *Planet. Space Sci.* **29**, 263-8 (1981).
65. Dalgarno, A., et al. *Geophys. Res. Lett.* **8**, 603-5 (1981).
66. Cowan, R. D., Hobbs, L. M., and York, D. G., *Astrophys. J.* **257**, 373-5 (1982).
67. Knight, R. D., *Phys. Rev. Lett.* **48**, 792-5 (1982).
68. Johnson *et al.*, private communication to Calamai and Johnson.
69. Calamai, A. G., and Johnson, C. E., *Phys. Rev. A* **44**, 218-22 (1991).
70. Vevby-Christensen, L., Andersen, L. H., Heber, O., Kella, D., Pedersen, H. B., Schmidt, H. T., and Zajfman, D. *Astrophys. J.* **483**, 531-40 (1997).
71. Graham, W. G., Fritsch, W., Hahn, Y., and Tanis, J. A. eds., *Recombination of Atomic Ions*, New York: Plenum Press, 1992.
72. Larsson, M., *Int. J. Mass. Spectrom. Ion Processes* **149/150**, 403-14 (1995).
73. Kella, D., Vevby-Christensen, L., Johnson, P. J., Pedersen, H. B., and Andersen, L. H. *Science* **276**, 1530-3 (1997).

Recent Experiments on Near-Threshold Electron-Impact Excitation of Multiply Charged Ions

M. E. Bannister[*], N. Djurić[†], O. Woitke[†], G. H. Dunn[†], Y.-S. Chung[‡], A. C. H. Smith[§], and B. Wallbank[∥]

[*] *Physics Division, Oak Ridge National Laboratory,*
Oak Ridge, Tennessee 37831-6372

[†] *JILA, University of Colorado and National Institute of Standards and Technology,*
Boulder, Colorado 80309-0440

[‡] *Department of Physics, Chungnam National University,*
Gung-Dong 220, 3005-764 Daejon, South Korea

[§] *Department of Physics and Astronomy, University College London,*
London WC1E 6BT, United Kingdom

[∥] *Department of Physics, St. Francis Xavier University,*
Antigonish, Nova Scotia, Canada, B2G 2W5

Abstract. Some recent measurements of excitation of multiply charged ions by electrons studied in beam-beam experiments are highlighted. The emphasis is on absolute total cross sections measured with the merged electron-ion beams energy-loss (MEIBEL) technique, although some results obtained with the crossed-beams fluorescence method are also presented. The MEIBEL technique allows the investigation of optically-allowed and forbidden transitions with sufficient energy resolution, typically about 0.2 eV, to resolve resonance structures in the cross sections. Results from the JILA/ORNL MEIBEL experiment on dipole-allowed transitions in several ions demonstrate the success of various theoretical methods in predicting cross sections in the absence of resonances. Comparisons of R-matrix calculations and measured cross sections for spin-forbidden transitions in Mg-like Si^{2+} and Ar^{6+}, however, show that further refinements to the theory are needed in order to more accurately predict cross sections involving significant contributions from dielectronic resonances and interactions between neighboring resonances.

INTRODUCTION

Vitally important in almost all plasma environments are collisions between electrons and ions, particularly those that involve the transfer of energy. Cross sections

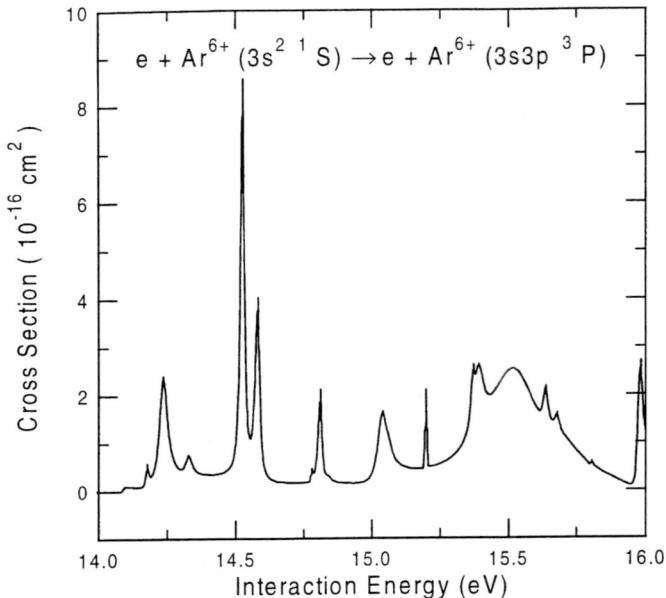

FIGURE 1. Cross sections for electron-impact excitation of the $3s^2\,^1S \rightarrow 3s3p\,^3P$ transition in Ar^{6+}. The curve represents the close-coupling R-matrix calculations of Ref. 1. The small step at the threshold (14.1 eV in the calculation of Ref. 1) is the contribution of direct excitation.

and rate coefficients for these processes are essential for modeling and diagnosing plasmas in areas of research such as controlled fusion, plasma processing, lighting discharges, and astrophysics. Theoretical efforts have produced much of the existing data, especially for electron-impact excitation of ions. Excitation can be a direct process or a result of resonant dielectronic capture followed by autoionization leaving the target ion in an excited state. Resonant enhancements to the cross section are significant in optically-forbidden transitions and can dominate the direct contribution by an order of magnitude or more near threshold, as shown in Fig. 1 for the $3s^2\,^1S \rightarrow 3s3p\,^3P$ transition in Ar^{6+} calculated [1] with the close-coupling R-matrix (CCR) method.

Direct configuration interaction (CI) and indirect interactions with a common continuum have been shown [2] to produce interference between nearby resonances and have a strong effect on the resonance contributions to the cross sections as calculated in the close-coupling formulation. It has also been found [2] that the resonance structure is sensitive to the exact energies of the individual resonances. Only in the past few years have experimentalists been able to provide theorists with benchmark cross sections for transitions dominated by resonances.

EXPERIMENTAL TECHNIQUES

Two primary beam-beam methods have been used to investigate electron-impact excitation of ions. The older method, the crossed-beams fluorescence method, is based on detecting photons emitted following excitation to a radiating state of the ion. This technique has been employed by a number of researchers [3-7] using similar experimental arrangements, but the details presented here are specific to Savin et al. [7]. A multiply charged ion beam is extracted from a Penning ion source, mass-to-charge ratio analyzed, and passed through an electrostatic charge purifier. The ion beam then intersects a magnetically confined electron beam at an angle of 55° (other experiments typically use 90°). After leaving the collision region, the ions are electrostatically charge analyzed, with the primary ions collected in a Faraday cup. A mirror subtending slightly over π steradians placed below the collision volume reflects emitted photons back through the collision volume and into a photomultiplier tube (PMT) oriented perpendicular to the collision plane. The bandpass of the photon detector is set by the lower-wavelength cutoff of a quartz filter and the higher-wavelength cutoff of the PMT. The total photon detection efficiency is about 1%, and the largest experimental uncertainties derive from the radiometric calibration of the photon detection system. This method must account for the angular distribution of the emitted photons over the solid angle of the detector, the polarization of the photons, and the radiative lifetime of the upper level. The two beams are modulated in a four-way chopping scheme in order to extract the signal from the background and their spatial overlap is measured with a movable beam probe.

The second method relies on detecting electrons that have lost most of their energy during inelastic collisions with ions. This merged electron-ion beams energy-loss (MEIBEL) technique has been used by researchers at the Jet Propulsion Laboratory [8] for singly charged ions and in the JILA/ORNL collaboration [9], but the details given here will be specific to the latter group. The JILA/ORNL MEIBEL technique [10], shown in Fig. 2, employs trochoidal analyzers with crossed magnetic and electric fields to merge and demerge an electron beam with an ion beam extracted from an electron-cyclotron resonance ion source. The demerger serves as an energy analyzer, separating inelastically scattered electrons from unscattered or elastically scattered electrons. Inelastically scattered electrons are deflected onto a calibrated position sensitive detector, while the unscattered primary electrons and those elastically scattered at small angles are collected in a Faraday cup since they are deflected less by the trochoidal fields. A series of apertures at the entrance of the demerger blocks electrons elastically scattered through large enough angles to reach the detector. By measuring the beam overlaps at several points along the merge path using a two-dimensional video beam probe [11], the cross sections are put on an absolute scale. The measured cross sections at higher interaction energies may be corrected for backscattering losses by using a three- dimensional trajectory modeling program [12].

The MEIBEL technique has three distinct advantages over the crossed-beams

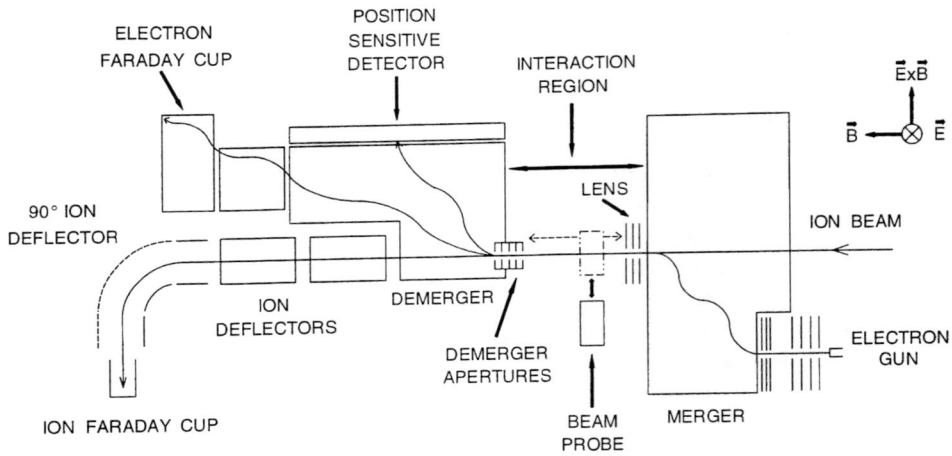

FIGURE 2. Schematic drawing of the JILA/ORNL merged electron-ion beams energy loss (MEIBEL) apparatus.

fluorescence method. Since the MEIBEL method involves the detection of low energy electrons, complete collection of the signal with detection efficiencies of 50-70% betters the capabilities of the fluorescence method by more than an order of magnitude. Secondly, by employing the merged-beams geometry, the energy resolution of the MEIBEL technique is typically 0.2 eV, which is 6-10 times better than that of the fluorescence method [7]. Finally, the fluorescence technique requires excitation to a radiating state, whereas the MEIBEL technique may be applied equally to excitation to radiating and non-radiating states.

RESULTS

The experimental results will be presented in two separate parts, one for dipole-allowed transitions measured with both the crossed-beams fluorescence and MEIBEL methods and the other for spin-forbidden transitions measured with the MEIBEL technique. The experimental data will be compared to theoretical predictions of the close-coupling R-matrix (CCR) and distorted-wave (DW) approaches.

Dipole-Allowed Transitions

The 2s → 2p transition in Li-like C^{3+} has been investigated in two different crossed-beams fluorescence experiments [7,13] as shown in Fig. 3. In addition, the transition was studied [14] with the MEIBEL technique, with the results also

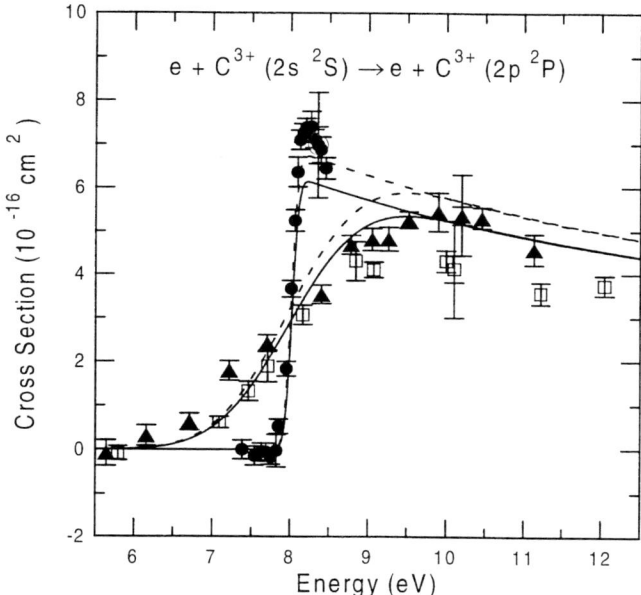

FIGURE 3. Cross sections for electron-impact excitation of the 2s → 2p transition in C^{3+}. The circles represent the MEIBEL results of Ref. 14 and the triangles and squares the crossed-beams results of Refs. 13 and 7, respectively, with relative error bars at the 90% confidence level. The outer error bars on one point in each set represent the total expanded uncertainty. The solid curves are the 9-state CCR calculations of Ref. 15 convoluted with 0.17 eV and 1.7 eV FWHM Gaussians representing the energy distributions of Refs. 14 and 7, respectively. The dashed curves are the DW results of Ref. 16 convoluted with the same two Gaussians.

shown in Fig. 3. The theoretical predictions of both Burke [15], a 9-state CCR calculation, and Clark et al. [16], a DW calculation, are shown convoluted with 0.17 eV and 1.7 eV FWHM Gaussians representing the energy distributions of Bannister et al. [14] and Savin et al. [7], respectively. There is good agreement between both calculations and the experimental data of Gregory and co-workers [13] and of Bannister et al. [14], but the measurements of Savin et al. [7] lie about 25% below the other data sets, barely within the limits of their total experimental uncertainty.

The MEIBEL technique was also used to measure cross sections [17] for excitation of the $3s^2\,^1S \to 3s3p\,^1P$ transition in Ar^{6+} as shown in Fig. 4. The 8-state CCR calculations of Griffin et al. [1] and the DW calculations of Clark et al. [16], convoluted with a 0.24 eV FWHM Gaussian representing the experimental energy distribution, are also shown in Fig. 4. There is excellent agreement between all three sets of data for this transition. The CCR theory includes some resonances

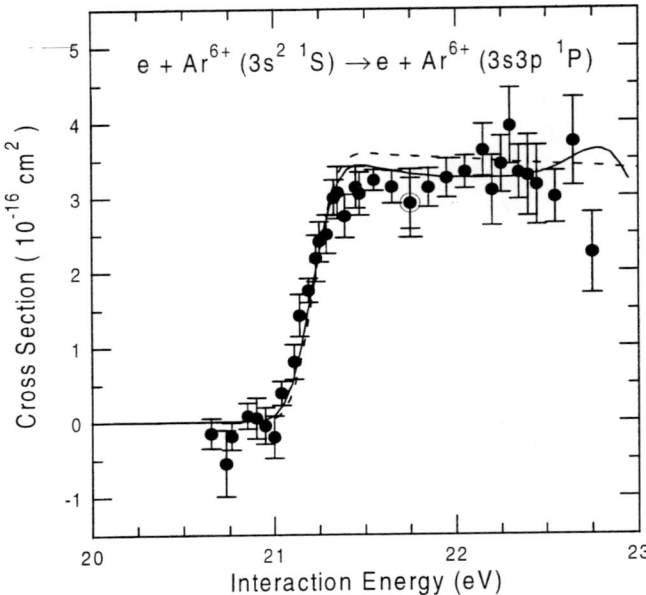

FIGURE 4. Cross sections for electron-impact excitation of the $3s^2\,^1S \to 3s3p\,^1P$ transition in Ar^{6+}. The circles are the MEIBEL results of Ref. 17 with relative error bars at the 90% confidence level. The outer bars on the point at 21.75 eV represent the total expanded uncertainty. The solid curve is the 8-state CCR calculation of Ref. 1 convoluted with a 0.24 eV FWHM Gaussian representing the experimental energy distribution. The dashed curve is the DW calculation of Ref. 16 convoluted with the same Gaussian.

near 22.7 eV, but their contributions are so small that the enhancement is only a small feature on the convoluted curve that is smaller than the relative uncertainties of the measurements. The measurement of this transition also serves to establish the absolute energy scale of the experiment, which is crucial for the measurement of the spin-forbidden transition in Ar^{6+} discussed below.

For the two dipole-allowed transitions presented above, dielectronic resonances make very minor contributions to the excitation cross section in the near-threshold region. This is not the case for the $3s^2\,^1S \to 3s3p\,^1P$ transition in Si^{2+}, as predicted by the 12-state CCR calculation of Griffin *et al.* [1]. Dielectronic resonances in this transition make significant contributions as shown in Fig. 5. There is reasonable agreement between the MEIBEL results [18] and the convoluted CCR curve about the average value of the excitation cross section, but the experimental data show a sharper drop from the peak cross section, perhaps due to a resonance just above threshold not predicted by theory. There is also disagreement about the magnitude of the resonance structure predicted near 11.7 eV. The DW results of Clark *et al.*

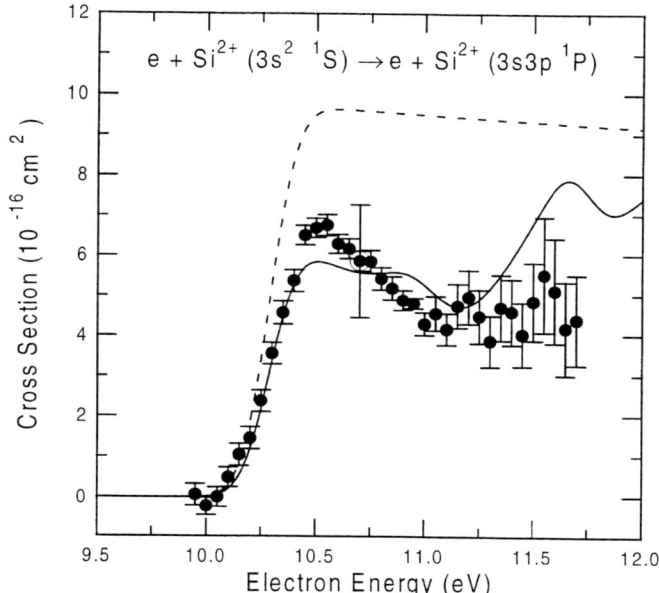

FIGURE 5. Cross sections for electron-impact excitation of the $3s^2\,^1S \to 3s3p\,^1P$ transition in Si^{2+}. The circles are the MEIBEL results of Ref. 18 with relative error bars at the 90% confidence level. The outer bars on the point at 10.7 eV represent the total expanded uncertainty. The solid curve is the 12-state CCR calculation of Ref. 1 convoluted with a 0.24 eV FWHM Gaussian representing the experimental energy distribution. The dashed curve is the DW calculation of Ref. 16 convoluted with the same Gaussian.

[16] overestimate the non-resonant cross section by a factor of nearly two and do not include any resonance contributions. The data emphasize the difficulty that *ab initio* calculations have in predicting cross sections in the presence of significant resonance contributions, even for this dipole-allowed transition. Interference between nearby resonances may account for some of the differences between the measured cross sections and those predicted by the CCR method. Ongoing measurements [19] using the crossed-beams fluorescence technique could shed more light on this discrepancy.

Spin-Forbidden Transitions

Since the non-resonant contributions are usually small for spin-forbidden transitions, the cross sections are often dominated by dielectronic resonances as discussed in the introduction. This is clearly demonstrated by the experimental excitation cross sections [18] for the $3s^2\,^1S \to 3s3p\,^3P$ transition in Si^{2+} that are shown in Fig.

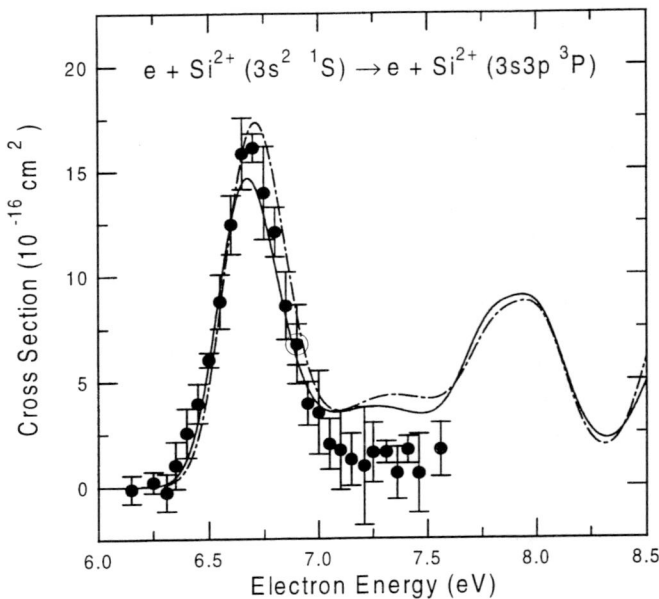

FIGURE 6. Cross sections for electron-impact excitation of the $3s^2\,{}^1S \to 3s3p\,{}^3P$ transition in Si^{2+}. The circles are the MEIBEL results of Ref. 18 with relative error bars at a 90% confidence level. The outer bars on the point at 6.9 eV represent the total expanded uncertainty. The curves are convolutions of a Gaussian (0.24 eV FWHM) with CCR calculations from Ref. 1 (solid) and Ref. 20 (dash-dot).

6 along with separate 12-state CCR calculations of Baluja et al. [20] and Griffin et al. [1], which are each convoluted with a Gaussian of 0.24 eV FWHM representing the experimental energy distribution. There is very good agreement between the three sets of data on the large resonance peak near 6.7 eV that dominates the non-resonant contributions to the cross section by almost an order of magnitude. Ion energy limitations of the experiment prevented measurements beyond 7.6 eV so that the second resonance peak predicted by theory could not be investigated.

Experimental excitation cross sections [17] for the $3s^2\,{}^1S \to 3s3p\,{}^3P$ transition in Ar^{6+} are also dominated by dielectronic resonances, as predicted by the 8-state CCR calculations of Griffin et al. [1] shown in Fig. 1. The measurements are shown in Fig. 7 along with the CCR calculations [1] convoluted with a Gaussian of 0.24 eV FWHM. The calculation agrees very well with the experiment for the resonance feature near 15.5 eV. The agreement for the peak near 14.4 eV is not good, indicating that the theory has difficulty calculating the precise energies of the contributing resonances and their interference. Comparison of the CCR predictions with those of the independent-processes isolated-resonance distorted-wave (IPIRDW) approx-

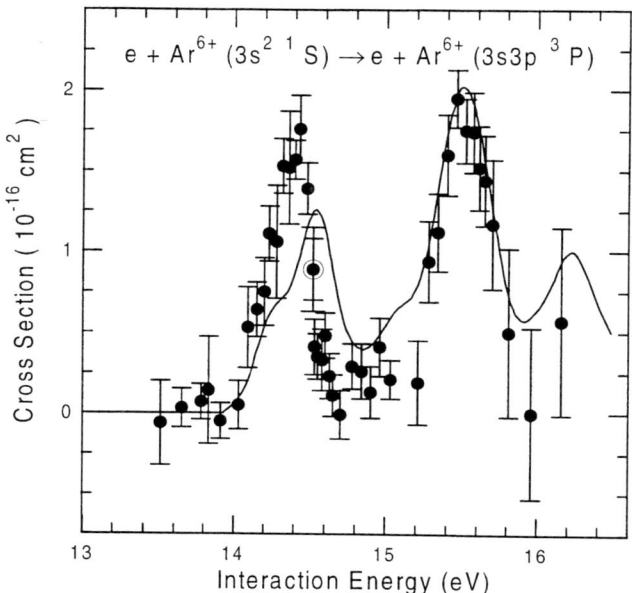

FIGURE 7. Cross sections for electron-impact excitation of the $3s^2\ ^1S \to 3s3p\ ^3P$ transition in Ar^{6+}. The circles are the MEIBEL results of Ref. 17 with relative error bars at a 90% confidence level. The outer bars on the point at 14.52 eV represent the total expanded uncertainty. The solid curve is a convolution of a Gaussian (0.24 eV FWHM) with CCR theory from Ref. 1.

imation [21], which neglects interactions between resonances, shows that the lower-energy peak is strongly influenced by interference effects in the CCR calculation.

CONCLUSIONS

The MEIBEL technique is a powerful tool [22] for investigating near-threshold electron-impact excitation of ions, particularly for forbidden transitions which are commonly dominated by dielectronic resonances. Both the close-coupling R-matrix (CCR) and distorted-wave (DW) methods are fairly successful in predicting cross sections for dipole-allowed transitions in the absence of significant resonance contributions. The agreement between experiment and theory is not as good when resonances dominate, and varies greatly even for different resonances in the same transition. The present experimental cross sections serve as crucial benchmarks for the close-coupling R-matrix theory and indicate that some refinements are required for the calculations to accurately reproduce the resonance positions and cross section contributions.

ACKNOWLEDGMENTS

The authors thank N. R. Badnell, T. W. Gorczyca, D. C. Griffin, M. S. Pindzola, and J. S. Shaw for providing unpublished differential and total cross section calculations. The JILA/ORNL MEIBEL research was supported by the Office of Fusion Energy Sciences of the U. S. Department of Energy, under Contract No. DE-AC05-96OR22464 with Lockheed Martin Energy Research Corp. and Contract No. DE-A105-86ER53237 with the National Institute of Standards and Technology.

REFERENCES

1. D. C. Griffin, M. S. Pindzola, and N. R. Badnell, *Phys. Rev.* A **47**, 2871-2880 (1993).
2. D. C. Griffin, M. S. Pindzola, F. Robicheaux, T. W. Gorczyca, and N. R. Badnell, *Phys. Rev. Lett.* **72**, 3491-3494 (1994).
3. D. F. Dance, M. F. A. Harrison, and A. C. H. Smith, *Proc. Roy. Soc.* A **290**, 74-93 (1966).
4. K. T. Dolder and B. Peart, *J. Phys.* B **6**, 2415-2426 (1973).
5. P. O. Taylor and G. H. Dunn, *Phys. Rev.* A **8**, 2304-2321 (1973).
6. I. P. Zapesochnyi, A. I. Dashchenko, V. I. Frontov, A. I. Imre, A. N. Gomonai, V. I. Len'del, V. T. Navrotskii, and E. P. Sabad, *JETP Lett.* **39**, 51-55 (1984).
7. D. W. Savin, L. D. Gardner, D. B. Reisenfeld, A. R. Young, and J. L. Kohl, *Phys. Rev.* A **51**, 2162-2168 (1995).
8. S. J. Smith, K.-F. Man, R. J. Mawhorter, I. D. Williams, and A. Chutjian, *Phys. Rev. Lett.* **67**, 30-33 (1991).
9. E. K. Wåhlin, J. S. Thompson, G. H. Dunn, R. A. Phaneuf, D. C. Gregory, and A. C. H. Smith, *Phys. Rev. Lett.* **66**, 157-160 (1991).
10. E. W. Bell, X. Q. Guo, J. L. Forand, K. Rinn, D. R. Swenson, J. S. Thompson, G. H. Dunn, M. E. Bannister, D. C. Gregory, R. A. Phaneuf, A. C. H. Smith, A. Müller, C. A. Timmer, E. K. Wåhlin, B. D. DePaola, and D. S. Belić, *Phys. Rev.* A **49**, 4585-4596 (1994).
11. J. L. Forand, C. A. Timmer, E. K. Wåhlin, B. D. DePaola, G. H. Dunn, D. Swenson, and K. Rinn, *Rev. Sci. Instrum.* **61**, 3372-3377 (1990).
12. SIMION 3D Version 6.0; David A. Dahl, Idaho National Engineering Laboratory.
13. D. Gregory, G. H. Dunn, R. A. Phaneuf, and D. H. Crandall, *Phys. Rev.* A **20**, 410-420 (1979); P. O. Taylor, D. Gregory, G. H. Dunn, R. A. Phaneuf, and D. H. Crandall, *Phys. Rev. Lett.* **39**, 1256-1259 (1977).
14. M. E. Bannister, Y.-S. Chung, N. Djurić, B. Wallbank, O. Woitke, S. Zhou, G. H. Dunn, and A. C. H. Smith, *Phys. Rev.* A **57**, 278-281 (1998).
15. V. M. Burke, *J. Phys.* B **25**, 4917-4928 (1992).
16. R. E. H. Clark, N. H. Magee, J. B. Mann, and A. L. Merts, *Astrophys. J.* **254**, 412-418 (1982).
17. Y.-S. Chung, N. Djurić, B. Wallbank, G. H. Dunn, M. E. Bannister, and A. C. H. Smith, *Phys. Rev.* A **55**, 2044-2049 (1997).

18. B. Wallbank, N. Djurić, O. Woitke, S. Zhou, G. H. Dunn, A. C. H. Smith, and M. E. Bannister, *Phys. Rev.* A **56**, 3714-3718 (1997).
19. D. B. Reisenfeld, P. H. Janzen, L. D. Gardner, D. W. Savin, and J. L. Kohl (private communication).
20. K. L. Baluja, P. G. Burke, and A. E. Kingston, *J. Phys* B **13**, L543-L545 (1980).
21. N. R. Badnell, D. C. Griffin, T. W. Gorczyca, and M. S. Pindzola, *Phys. Rev.* A **50**, 1231-1239 (1994).
22. Further experimental details and tabulations of the data are available at the World-Wide Web site http://www-cfadc.phy.ornl.gov/meibel/.

ATOMIC PROCESSES IN ASTROPHYSICAL PLASMAS

Quasars, the early universe, and plasma simulations

Gary J. Ferland & Dmitri A. Verner

Physics Department, University of Kentucky, Lexington, KY 40506 USA

Abstract. Quasars are among the most luminous objects in the universe, and the most distant for which we can obtain good spectroscopy. These spectra carry with them information about the formation of the first large structures in the universe and the first generations of star formation in newly born galaxies. The spectrum of a quasar has strong broad emission lines, formed by a non-equilibrium low-density photoionized gas. This spectrum can best be interpreted by reference to large-scale simulations of the full environment. This in turn is strongly coupled to many basic plasma physics questions. Here we discuss the current status of quasar research, numerical plasma simulations, and the atomic database.

1. Introduction

Quasars offer several puzzles – they are among the most luminous objects in the universe, so the nature of their energy source is an important question. They exist mostly at large redshifts, meaning that they were associated with some of the first large structures that formed in the universe, but also that they are no longer active. Finally, their spectra depend on their luminosity. Once these correlations are understood quasars will be used as standard candles to probe the redshift z = 5 universe, and measure fundamental cosmological parameters. Peterson (1997) gives a comprehensive review of the family of active nuclei, of which quasars are the brightest members.

The spectra of distant objects are Doppler-shifted to longer wavelengths due to the expansion of the universe. This is parameterized by the redshift z, the ratio of the wavelength displacement to the lab wavelength. Increasing redshift corresponds to increasing distance and look-back time – the time it took light to reach us. Redshift can also be converted into time since the creation of the universe. Figure 1 (taken from Hamann & Ferland 1993) shows the age of the universe as a function of redshift for two values of q_o (proportional to the mass density of the universe) and H_o, the Hubble constant. Spectra can be obtained of quasars at redshifts of nearly 5, and this gives us the chance to directly observe conditions when the universe was between ½ and 4 billion years old.

Deducing reliable abundances and luminosities is a central theme of quasar research. Their emission lines are produced by warm (~10^4 K) photoionized gas with moderate to low density ($n \leq 10^{12}$ cm^{-3}). Such gas is far from thermodynamic

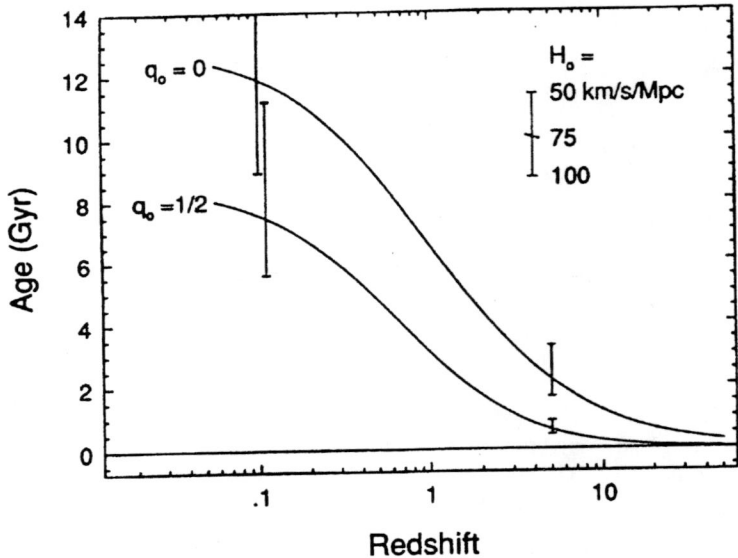

Figure 1 The age of the Universe as a function of the redshift z, for various cosmologies. From Hamann & Ferland (1993).

equilibrium, and its physical conditions cannot be known from analytical theory. Rather, the observed spectrum is the result of a host of microphysical processes that must be numerically simulated in detail. In the following sections we outline some of the bigger questions about quasars, show how this relates to large scale numerical plasma simulations, and outline the uncertainties in the underlying atomic database.

2. Quasar Emission Line Regions

Figure 2 shows a typical spectrum, taken from data in Francis et al. (1991). Many strong and broad emission lines are present, superimposed on a non-thermal continuum that extends from the ratio to at least the hard X-Rays. As a reference point, for a redshift $z=5$ quasar the Lyα transition of hydrogen, normally seen in the vacuum ultraviolet, is redshifted beyond the optical into the near infrared. The breadth of the emission lines is due to high-speed cloud motion since pressure broadening is unimportant at these densities, and most investigators interpret this as virialized clouds near a super-massive black hole. Matter falling into this potential well is believed to be the fundamental energy source.

2.1. Rapid star formation in young galaxies

The Big Bang produced predominantly hydrogen and helium, and successive generations of star formation are responsible for the heavier elements. Many of the emission lines shown in Figure 2 come from elements heavier than helium, and these elements are plentiful in even the highest redshift objects.

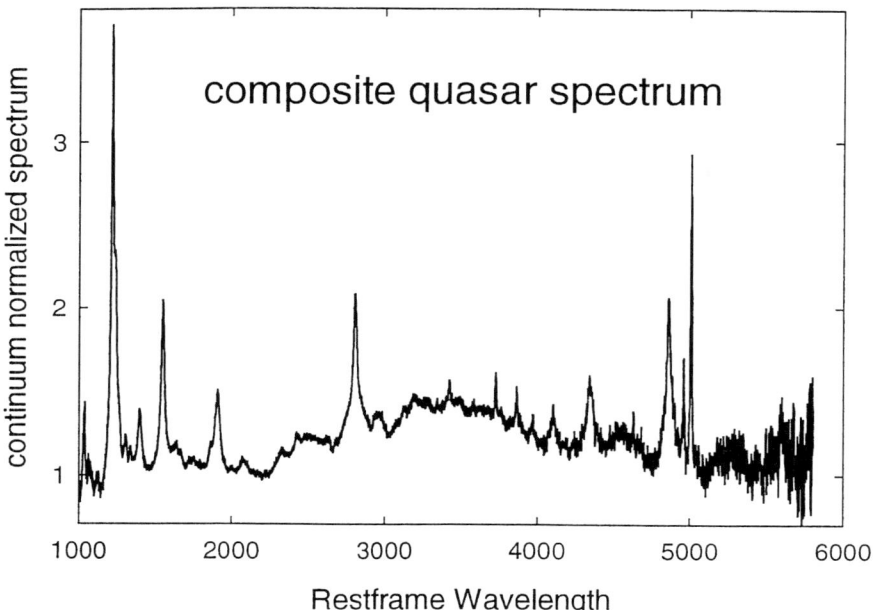

Figure 2 A composite spectrum of high redshift quasars. After Francis et al. (1991).

This spectrum clearly tells us a great deal about the first generations of star formation in very young galaxies. The elements heavier then helium are created by successive generations of stars forming from interstellar matter, undergoing nuclear processing, the expelling enriched material back into the interstellar medium. One might think that the amount of nuclear enrichment should decrease when going to higher redshift objects, since we observe them when the universe was far younger.

Figure 3 shows two abundance indicators developed by Hamann and Ferland (1993). The upper panels show NV $\lambda1240$ relative to C IV $\lambda1549$. Carbon is formed by the triple alpha reaction, after hydrogen burning forms helium. This process can happen in gas with primordial (H and He only) abundances, so carbon is referred to as a primary element. The lower panels show NV λ 1240 relative to HeII $\lambda1640$. Helium is created in the Big Bang and its abundance does not change by large factors as a galaxy ages. Nitrogen is fundamentally different. It has an odd number of protons, and so is not directly produced in high abundance from primordial material. Rather it is produced as a secondary element by CNO burning, which can only take place after C and O are produced by first generations of stars. Nitrogen is referred to as a secondary element, and complete simulations of evolving star clusters show that the abundance of nitrogen goes up as the square of the carbon abundance. Strong nitrogen lines are a sign of a system that has undergone extensive nuclear enrichment by many generations of star formation.

Surprisingly, Figure 3 shows that nitrogen lines tend to be stronger in the highest redshift objects. This is the opposite of initial expectations, which are that the abundances of the elements should go down with increasing look-back time. The two solid lines in the left panels shows the expected abundance ratios for the slow evolution that occurs here in the solar neighborhood (the lower lines) and the rapid evolution

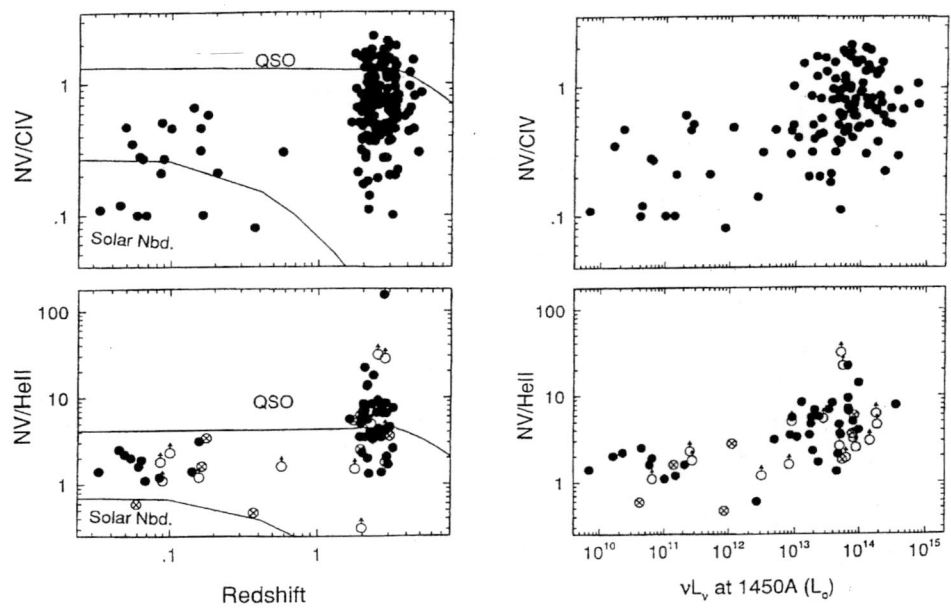

Figure 3 Two abundance indicators are shown as a function of redshift and luminosity. Taken from Hamann and Ferland (1993)

expected in the core of a very massive galaxy like M87 (the upper curve, marked QSO). At face value this figure makes no sense!

The right panels of the figure show the same data set but plotted against luminosity instead of redshift. Because of the way quasars are discovered, at the highest redshift we only find the very brightest ones. Posed this way, there is a luminosity – metallicty relation, in the sense that the highest luminosity objects have the largest abundances. This can be understood in the context of galactic evolution: more massive galaxies tend to be brighter, and can hold onto their interstellar gas better. This means that recycling can occur more often, leading to higher metallicities. This also means that the very highest redshift quasars were associated with extensive blasts of star formation during the birth of galaxies. Careful studies of the emission lines in the very highest redshift quasars can tell us something about our origins.

2.2. Quasars as cosmological probes

The redshift is an observed quantity. If we could independently measure a quasar's luminosity, then we could use them as standard candles and infer distances from the apparent brightness. We could then directly chart Figure 1 and measure q_o and H_o.

The spectrum of a quasar does indeed have a luminosity dependence, the so-called Baldwin Effect (Figure 4). There have been many extensive studies of it, such as those in Baldwin, Wampler, and Gaskell (1989), and Osmer et al. (1994). Basically the observation is that the lines tend to be weaker relative to the continuum for higher luminosity objects. The equivalent width of an emission line is the measure of the line to continuum ratio most often used, and is the y-axis in Figure 4.

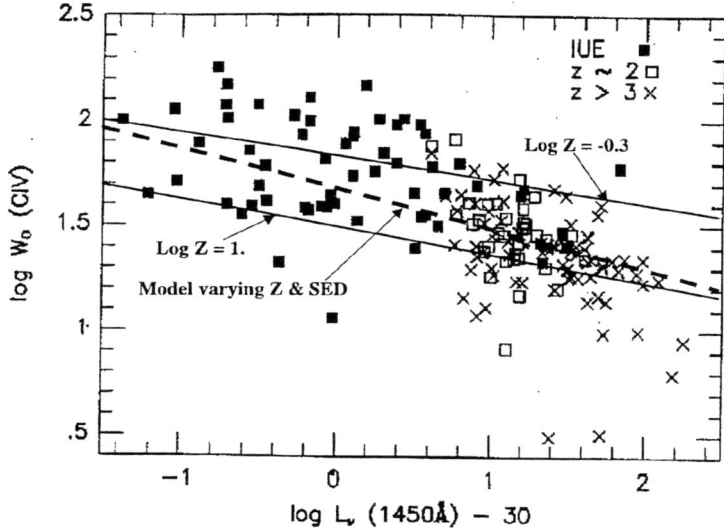

Figure 4 The correlation between the luminosity of a quasar (the x-axis) and the intensity of one emission line relative to the continuum (the y-axis). The figure is from Korista, Baldwin, and Ferland (1998), and is adopted from Osmer, et al (1994).

Luminosity – equivalent width correlations are complicated by other correlations. The x-ray to optical ratio α_{ox} and metallicity may both correlate with luminosity (Avni and Tananbaum 1986; Worral et al. 1987, Hamann and Ferland 1993; Ferland et al. 1996). All of this is further compromised by the plasma physics issues discussed below.

Our approach has been to generate large data cubes of plasma code predictions, and then integrate over this data set using cloud distribution functions. Details were given in a recent review (Ferland 1998) and will not be repeated here. The basic result is that the plasma physics introduces very powerful emission line selection effects that help determine the observed spectrum of a quasar fairly independently of cloud properties.

Figure 4 shows some recent calculations that attempt to reproduce the Baldwin effect (Korista, Baldwin, & Ferland 1998). The two solid lines show how the equivalent width is predicted to change due to changes in the shape of the ionizing continuum (the Spectral Energy Distribution or SED). Changes in the SED are observed as changes in the ratio of the x-ray to UV continuum. The two lines are for two different gas metallicities relative to the sun's metallicity – one is half the solar metallicity and the second is ten times larger. The dashed line shows the combined effects of changes in the SED and the tendency for high luminosity objects to have higher metallicities (Figure 3). The implication is that the observed correlation is due to the combined effects of changing continuum shape and gas metallicity.

The result is that we have a good idea of the underlying physics that produces the observed luminosity indicators. The next step is to use this knowledge to reduce the scatter in the correlations. If successful, we will have a direct probe of the redshift $z \approx 5$ universe.

3. Simulations of plasmas and their spectra

All of the discussion of the physics of quasars outlined above depended on the analysis and interpretation of their emission lines. Ferland has been developing a large-scale numerical code, Cloudy, to simulate non-equilibrium plasmas and predict their spectra. The most recent version of Cloudy is described by Ferland et al. (1998) and it is available on the web at http://www.pa.uky.edu/~gary/cloudy. There are nearly a dozen similar plasma codes designed to simulate such environments - Ferland et al. (1994) summarize a set of such codes and compare predictions. The number of codes is an indication of their importance to the interpretation of astronomical spectroscopy.

The basic problem is that the emitting plasma is not in equilibrium. In strict thermodynamic equilibrium the photons and particles are in equilibrium and Boltzmann statistics describe the particles, and Planck's law the light. This is seldom the case in astrophysics. Local thermodynamic equilibrium is the case where the particle distributions are in equilibrium but the light is not. Even this case is seldom found in plasmas. Instead, the ionization, level populations, and electron temperature are set by microphysical processes, and can be determined by self-consistently solving the equations of statistical and thermal equilibrium. The only assumptions are that the system has had time to reach steady state, and (usually) that the frequent collisions between free electrons cause their velocity distribution to become Maxwellian. The electron temperature is the only temperature used to characterize a photoionized plasma. Osterbrock (1989) is the definitive discussion of the physics of this situation.

Predictions of the intensities of thousands of lines and the column densities of all constituents result from the specification of only the incident continuum, gas density, and its composition. By their nature, such calculations involve enormous quantities of atomic/molecular data describing a host of microphysical processes, and the codes involved are at the forefront of modern computational astrophysics. Although the task is difficult the rewards are great, since numerical simulations make it possible to interpret astrophysical spectra on a physical basis.

4. The atomic data base

All of this work rests on a foundation of basic atomic and molecular data. Cross sections or rate coefficients are needed for all physical processes that can affect the excitation or ionization of an atom. These data needs are vast. Different processes are discussed below, and a more complete review is given by Ferland et al. (1998).

4.1. Ionization processes

This part of the database is fairly complete.

4.1.1. Heavy element photoionization cross sections

For many years the extensive tables of Reilman and Manson (1979) were the only complete data set. These calculations are quite accurate for highly ionized species, but were increasingly approximate for atoms or first ions. We now use the calculations of Verner & Yakovlev (1995), with extensions to more recent Opacity Project (OP) data (Verner et al. 1996). In particular, Verner et al. (1996) use the most recent results from the Iron Project (Hummer et al. 1993; Nahar & Pradhan 1994; Bautista & Pradhan 1995) for the ions they calculated. For some species, including O^+ and most neutral atoms and single ions of the third and fourth row, the results can be quite different from the Reilman and Manson work, and this has had a moderate affect on some predictions. The fits given in Verner's work average over sharp resonances and do not try to fit them. This is appropriate because the position of most resonances is uncertain by well more than their width. The photoionization rate, heating rate, and photoelectron yield is evaluated independently for each electronic subshell. There are a total of seven possible subshells for the heaviest elements. Fluorescence yields for each subshell come from electronic forms of the tables of Kaastra & Mewe (1993).

Suprathermal Auger electrons can often produce further collisional ionizations before they are thermalized. We use efficiencies from Xu & McCray (1991).

4.1.2. Collisional ionization

Rate coefficients for collisional ionization from the ground states of the first 28 elements are compiled and fitted by Voronov (1997). The previous large compilation is of Arnaud & Rothenflug (1985) with the updates for iron of Arnaud & Raymond (1992).

4.2. Recombination processes

The description of the recombination physics is the biggest uncertainty in the atomic database today. The total electron-ion recombination process involves three distinct physical channels - radiative, low-temperature dielectronic, and high-temperature dielectronic recombination. These are discussed in turn below.

Two limiting situations, collisional and photoionization equilibria, are found in nature. In collisional equilibrium the electron temperature is about equal to the ionization energy of the species, while in photoionization equilibrium kT is much less than the ionization energy. Radiative recombination can be significant at any temperature, low-temperature dielectronic recombination usually dominates in photoionization equilibrium, and high-temperature dielectronic recombination dominates at collisional temperatures.

4.2.1. Radiative recombination rates

Radiative recombination rates are obtained by integrating over the photoionization cross section using the Milne relation, a form of detailed balance. We have calculated radiative recombination rates for recombination to all H-like, He-like, Li-like and Na-like ions of elements from H through Zn, and fitted them with analytical formulae (Verner & Ferland 1996). For these cases the parent ion is a closed shell and autoionizing levels are not expected to occur close to the ionization threshold. Radiative recombination should dominate for most temperatures for these ions. For other isoelectronic sequences of C, N, O, and Ne we use radiative recombination rates from Pequignot et al. (1991), refitted by the asymptotically correct analytical formula. For the rest of ions, we still apply power-law fits from Aldrovandi & Pequignot (1973), with the adjustments from Arnaud & Raymond (1992) for iron ions.

Radiative recombination rate coefficients can be obtained to high accuracy for nearly all species. Unfortunately, for most species the recombination process is dominated by the other, far more uncertain, mechanisms described next.

4.2.2. Low-temperature dielectronic recombination

This is a process that occurs through low-lying autoionizing stages. This is usually an important, often dominant, recombination process for those species whose parent ions do not have closed shells. Nussbaumer & Storey (1983, 1984, 1986, 1987, hereafter NS) calculated low-temperature dielectronic recombination rates for some ions of C, N, O, Ne, Mg, Al, and Si using available experimental data on energies of autoionizing levels.

Uncertainties in energies of the autoionizing resonances are crucial. The autoionizing levels are assumed to be in LTE with the continuum due to rapid dielectronic recombination / autoionization. The recombination rate is then the population of the level multiplied by its rate of decay to bound levels. This population depends on the Boltzmann factor and so depends exponentially on the energies of these levels. It will be difficult to improve upon NS since they did all species with accurate experimental energies.

Within the framework used by NS there is a further uncertainty regarding which autoionizing levels will be in LTE with the continuum. Levels with many spins exist, but only those where the spin changes by one are connected by an allowed transition in the continuum. Only these levels were included in the early papers in the NS series. In later papers two rates were given for the two cases, one with only directly connected levels included, and another with all possible levels. The rates differ by a factor of 2 in some cases. Recent experimental work shows that channels not populated in LS coupling do in fact contribute (Savin et al. 1997).

The lack of reliable dielectronic rate coefficients for most third row and higher elements is the dominant uncertainty in the ionization balance in photoionization equilibrium. Most ions simply do not have complete spectroscopy and measured autoionization level energies. Theoretical structure calculations can be done, but uncertainties in the positions of the resonances remain. The question concerning populations of levels not coupled by an LS-allowed transition to the continuum is central. This is likely to remain an uncertainty for some time to come, and points to the need for complete basic experimental data.

4.2.3. *High-temperature dielectronic recombination*

We use published fits for high-temperature dielectronic recombination (HTDR) rates based on calculations made in the low-density limit (Aldrovandi & Pequignot 1972; 1974; Shull & Van Steenberg 1982; Arnaud & Rothenflug 1985; Arnaud & Raymond 1992). This process mainly affects gas in collisional, not photoionization, equilibrium, because it is efficient when the electron energy approaches the ionization potential of the species. These rates are largely based on the Burgess (1965) formula. Where detailed calculations have been made (Badnell 1991; Savin et al. 1997) rates have only been within 0.5 dex of the Burgess estimate. The Burgess formula seems to overestimate the more accurate rates. This remains a major uncertainty.

4.2.4. *Charge exchange*

All neutralization and ionization reactions between hydrogen and the first four ions of all species are included using the fits listed by Kingdon & Ferland (1996) and Ferland et al. (1997). Charge transfer between heavy elements and helium are from a variety of sources. Rates for this processes are uncertain, and an area of future needs.

5. Summary and acknowledgements

Quasars are the most distant objects we can directly observe. Their emission lines reveal the chemical composition of the emitting gas, and the luminosity of the quasar. Once we can fully interpret the spectrum we will be able to map out the very first generations of star formation and chart the early expansion of the universe. This is difficult since the emitting plasma is far from equilibrium. A number of plasma codes have been developed to simulate conditions and predict the resulting spectrum. All of these codes, and our ability to fully interpret spectroscopic data, rest on a foundation of atomic and molecular physics. The biggest uncertainty today is in recombination processes. Rates for dielectronic recombination and charge transfer are notoriously hard to compute, and in many cases even rough calculations do not exist.

We are grateful to the NSF and NASA for supporting this research.

6. References

Aldrovandi, S. M. V., & Pequignot, D. 1973, A&A, 25, 137
Arnaud, M., & Raymond, J. 1992, ApJ, 398, 394
Arnaud, M., & Rothenflug, R. 1985, A&AS, 60, 425
Avni, Y., & Tananbaum, H. 1986, ApJ, 305, 83
Badnell, N. R. 1991, ApJ, 379, 356
Baldwin, J., Ferland, G., Korista, K., & Verner, D. 1995, ApJ, 455, L119
Baldwin, J. A., Ferland, G. J., Korista, K., Carswell, R., Hamann, F., Phillips, M., Verner, D., Wilkes, B., & Williams, R. E. 1996, ApJ, 461, 664
Baldwin, J., Wampler, J., & Gaskell, C. M. 1989, ApJ, 338, 630
Bautista, M. A., & Pradhan, A. K. 1995, J. Phys. B, 28, L173
Burgess, A. 1965, ApJ, 141, 1588
Ferland, G. J. 1998, in Oak Ridge Symposium on Atomic and Nuclear Astrophysics
Ferland, G. J., Korista, K. T., Verner, D. A., & Dalgarno, A. 1997, ApJ, 481, L115
Ferland, G. J., Korista, K.T., Verner, D. A., Ferguson, J. W., Kingdon, J. B., & Verner, E. M.1998, PASP, in press
Ferland, G. J. 1996, Hazy, a Brief Introduction to Cloudy, University of Kentucky Department of Physics and Astronomy Internal Report
Ferland, G. J., Baldwin, J. A., Korista, K., Hamann, F., Carswell, R., Phillips, M., Wilkes, B., & Williams, R. E. 1996, ApJ, 461, 683
Ferland, G., Binette, L., Contini, M., Harrington, J., Kallman, T., Netzer, H., Pequignot, D., Raymond, J., Rubin, R., Shields, G., Sutherland, R., & Viegas, S. 1995, in The Analysis of Emission Lines, Space Telescope Science institute Symposium Series, eds. R. Williams & M. Livio, Cambridge University Press
Francis, P. J., Hewett, P. C., Foltz, C. B., Chaffee, F. H., Weymann, R. J., & Morris, S. L. 1991, ApJ, 373, 465
Hamann, F., & Ferland, G. 1993, ApJ, 418, 11
Ho, L. C., Filippenko, A. V., & Sargent, W. L. W. 1996, ApJ, 462, 183
Horn, K. 1994, in Reverberation Mapping of the Broad-Line Region in Active Galactic Nuclei, eds. P. M. Gondhalekar, K. Horne, B. M. Peterson (ASP Conference Series Vol. 69), 23
Hummer, D. G., Berrington, K. A., Eissner, W., Pradhan, A. K., Saraph H. E., Tully, J. A. 1993, A&A, 279, 298
Kaastra, J. S., & Mewe, R. 1993, A&AS, 97, 443
Kingdon, J. B., & Ferland, G. J. 1996, ApJS, 106, 205
Korista, K., Baldwin, J., & Ferland, G. 1998, ApJ, in press
Nahar, S. N., & Pradhan, A. K. 1994, J. Phys. B, 27, 429
Nandra, K. 1994, in Reverberation Mapping of the Broad-Line Region in Active Galactic Nuclei, eds. P. M. Gondhalekar, K. Horne, B. M. Peterson (ASP Conference Series Vol. 69), 273

Nussbaumer, H., & Storey, P. J. 1983, A&A, 126, 75
Nussbaumer, H., & Storey, P. J. 1984, A&AS, 56, 293
Nussbaumer, H., & Storey, P. J. 1986, A&AS, 64, 545
Nussbaumer, H., & Storey, P. J. 1987, A&AS, 69, 123
Osmer, P. A., Porter, A. C., & Green, R. F. 1994, ApJ, 436, 678
Osterbrock, D. E. 1989, Astrophysics of Gaseous Nebulae & Active Galactic Nuclei (Mill Valley: University Science Books)
Pequignot, D., Petitjean, P., & Boisson, C. 1991, A&A, 251, 680
Peterson, B.M., 1997, An introduction to active galactic nuclei, (Cambridge, Cambridge University Press)
Reilman, R. F., & Manson, S. T. 1979, ApJS 40, 815, errata 46, 115; 62, 939
Savin, D. W., Bartsch, T., Chen, M., Kahn, S., Liedahl, D., Linkemann, J., Muller, A., Schippers, S., Schmitt, M., Schwalm, D., & Wolf, A., 1997, ApJ 489, L118
Shull, J. M., & Van Steenberg, M. 1982, ApJS, 48, 95
Verner, D. A., & Ferland, G. J. 1996, ApJS, 103, 467
Verner, D. A., & Yakovlev, D. G. 1995, A&AS, 109, 125
Verner, D. A., Barthel, P. D., & Tytler, D. 1994, A&AS, 108, 287
Verner, D. A., Ferland, G. J., Korista, K. T., & Yakovlev, D. G. 1996, ApJ, 465, 487
Worrall, D. M., Giommi, P., Tananbaum, H., & Zamorani, G. 1987, ApJ, 313, 596
Xu, Y., & McCray, R. 1991, ApJ, 375, 190

SOHO: Atomic Physics and the Solar Atmosphere

T.A. Kucera

NASA/Goddard Space Flight Center, Code 682.3, Greenbelt, MD 20771
Space Applications Corp., 901 Follin Ln., Suite 400, Vienna, VA 22180

Abstract. Many aspects of the Sun's corona and wind are studied using data from the ultraviolet spectrum. Accurate atomic parameters are needed to interpret these data correctly, and a good understanding of the behaviors of atoms and ions in plasmas is essential to modeling the Sun's atmosphere. Here I present two examples of studies being carried out using the Solar and Heliospheric Observatory (SOHO) extreme ultraviolet spectrographs. The first of these is the study of flows in the Sun's chromosphere and corona. SOHO has provided new information concerning previous observations of the predominant down-flows in the Sun's lower atmosphere. Accurate measurements of Doppler line shifts have been extended to the corona. It has also been found that the Doppler shifts vary over different parts of the Sun. The second study discussed involves the use of SOHO data to measure elemental abundances in coronal structures know as streamers, giving more information on the "FIP" effect - the observation that there is a relative deficit of elements with high first ionization potentials (FIPs) in the corona and solar wind.

I INTRODUCTION

Ultraviolet spectrographs aboard the Solar and Heliospheric Observatory (SOHO) are allowing us to investigate some of the outstanding puzzles pertaining to the Sun's corona and solar wind. SOHO, a joint project of NASA and the European Space Agency, was launched in December 1995. It is currently in orbit around the Sun at the first Lagrange point, a gravitational balance point between the Sun and Earth. It contains a dozen instruments dedicated to studying the solar interior, atmosphere, and wind [1]. These instruments include three ultraviolet spectrographs which observe the solar atmosphere.

Ultraviolet spectrographs provide an excellent tool for studying the corona. By studying lines from different ions formed over a wide range of temperatures, we can determine the properties of different plasmas at different locations in the corona. Diagnostics based on ultraviolet lines can help us to determine ionic and atomic abundances, temperatures, densities, and velocities in the solar atmosphere. We can also use Doppler techniques to determine flow speeds.

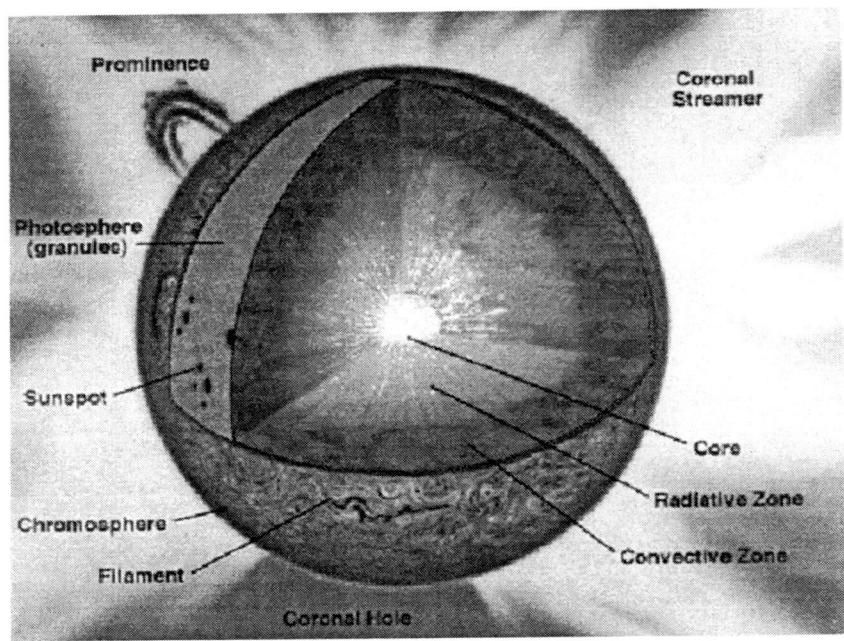

FIGURE 1. Diagram showing different layers of the Sun and solar atmosphere

This paper describes just two of the many investigations of the corona and solar wind being carried out with SOHO data. One of these is the study of flows in the lower solar atmosphere. It has long been observed that at many temperatures the Sun's atmosphere emits red-shifted spectral lines, indicating that the plasma is directed downwards rather than upwards into the solar wind. We are working to quantify and explain this phenomena using measurements of Doppler shifts in ultraviolet lines. It is also known that the distribution of elements in the solar wind is neither constant nor identical to that at the photosphere. Studying the atomic abundances of elements in the corona and wind is essential to developing and testing models of the solar wind. Atomic physics is key to our ability to correctly interpret our data and model the phenomena we study.

A The Corona and Solar Wind

One of SOHO's main focuses is the solar wind. It has been known for some time that there are two kinds of solar wind - the slow speed solar wind, with velocities of 300-500 km/s, and the fast solar wind, which can reach speeds of 800 km/s. It is believed that the high speed solar wind comes from the open field regions marked by the dark coronal holes in X-rays and UV (see Fig. 2). The slow speed wind

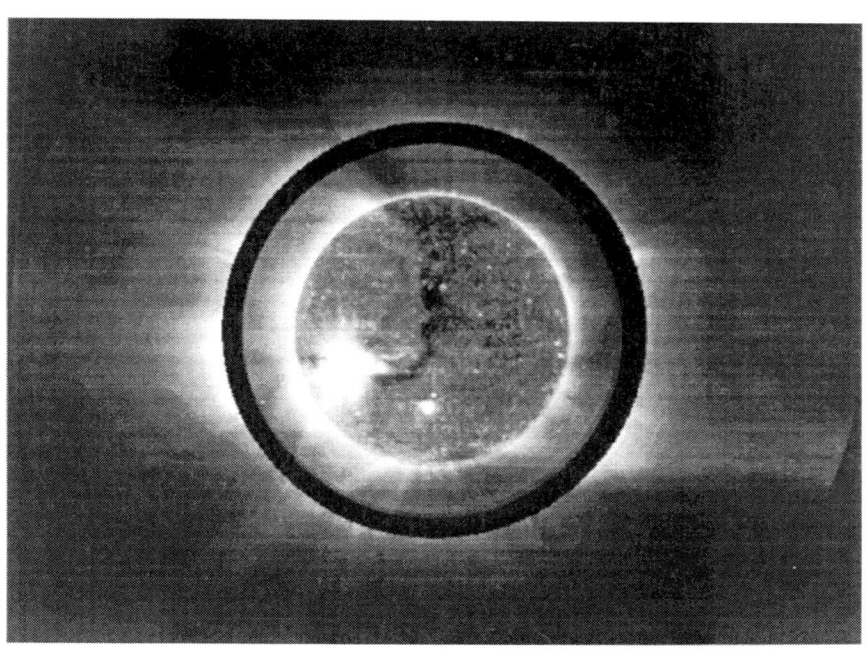

FIGURE 2. Composite of data from the SOHO/Extreme ultraviolet Imaging Telescope (EIT) (center) in a line dominated by Fe XV and the Ultraviolet Coronagraph Spectrometer (edge) in a line of O VI. Note the dark coronal hole extending from the north polar region to the equator. Streamers are visible away from the poles. The largest is centered south of the equator on the left.

comes from the closed field streamer regions which form near the equator and over magnetic active regions.

Although the solar wind is often considered as an extension of the corona, to truly understand it requires that one consider the layers beneath the corona, the ten thousand degree chromosphere and the transition region between it and the million degree corona.

B The SOHO EUV Spectrometers

SOHO contains three extreme ultraviolet (EUV) spectrometers which are used to study the solar atmosphere: the Coronal Diagnostic Spectrometer (CDS), Solar Ultraviolet Measurements of Emitted Radiation (SUMER), and the Ultraviolet Coronagraph Spectrometer (UVCS). For reasons of space this paper discusses only two studies pertaining to UVCS and SUMER.

UVCS [2] is designed chiefly to study the corona from 1.2 to 10 solar radii ($R_\odot$). It can observe in two ultraviolet ranges. One is optimized for measurements in the H I Lyman α line at 1216 Å, but covers the range 1100–1361 Å. The other, designed for observing the O VI lines at 1032 and 1037 Å, includes 937–1126 Å in first order and 469–563 Å in second order. The spectral resolution varies with location and wavelength, but is approximately 0.25 Å for the Lyα channel and 0.16 Å for the O VI channel.

SUMER [3,4] is a tunable spectrograph intended to observe the solar disk and the corona immediately above it. It has a range of 660–1600 Å in first order and 500–805 Å in second order. The spectral resolution is about 0.052 Å, with a bandpass of 45 Å available at any given time.

CDS [5,6] is also intended for studies of the solar disk and low corona. It contains two spectrometers. The Normal Incidence Spectrometer (NIS) observes in the ranges 308–381 and 513–633 Å simultaneously. The Grazing Incidence Spectrometer (GIS) observes in four bands: 151–221, 256–338, 393–493, and 656–785 Å.

SUMER, UVCS, and CDS/NIS all use slits which provide one-dimension of spatial information. Two-dimensional images can be created by taking a series of exposures while shifting the pointing perpendicular to the slit direction. CDS/NIS can also be used with a larger slit to obtain two dimensional images with very low spectral resolution (producing "overlappograms"). CDS/GIS only observes an area of 1 arcsec squared at any one time

II DOWN-FLOWS IN THE SUN'S ATMOSPHERE

It has been known since Skylab [7] that there are systematic red-ward Doppler shifts in ultraviolet spectral lines emitted by the transition region between the chromosphere and the corona (at $\sim 2 \times 10^5$–10^6 K). Given that the corona is moving generally outwards, why is it that the observations show downward flows?

New SUMER observations have allowed us to study this phenomenon at more temperatures, including coronal ones, and to obtain better spatial resolution.

Two studies have recently been performed of average Doppler shifts measured at the center of the solar disk, one by Brekke, Hassler, & Wilhelm [8] and another by Chae, Yun, & Poland [9]. The results of the studies are shown in Figure 3. These graphs show the Doppler shifts measured for different ions plotted as a function of the temperature of peak equilibrium formation of the ion. Assuming that the emitting ions are indeed at those temperatures, this gives us information about the motions of different temperature plasmas in the solar atmosphere. These measurements using SUMER give a more complete picture of the red-shifts as a function of temperature than was available before. In particular, both studies indicate that red-shifts persist into lines formed at coronal temperatures of a million degrees. This was not known before and will require the revision of many models of the red-shift phenomenon.

In addition to the work done with average Doppler shifts, SUMER has also enabled researchers to measure shifts at different locations. Recent studies show that red-shift properties vary with location and that in some areas, primarily coronal holes, blue-shifts are observed in lines produced just below coronal temperatures ($10^{5.8}$ K) [10,11]. This information will have to be incorporated into new models as well.

Of chief importance in interpreting the data is the determination of the measured and rest wavelengths of the lines used in calculating the Doppler shifts. The SUMER instrument does not include an absolute reference source, so the wavelength calibration must be determined using neutral chromospheric lines as references. It is generally assumed that such lines have essentially no Doppler motion. In order to correctly calculate both the dispersions and absolute wavelengths of the spectra, a number of these chromospheric lines are needed in the wavelength band observed. An example of such a band is shown in Fig. 4. In both studies the lines were observed on different parts of the detector to decrease the chance of detector distortion affecting the results.

Chae, Yun, & Poland [9], however, consider more carefully whether the chromospheric lines are truly formed in stationary plasma. They calculate the distribution of electrons with excitation states of the neutral atoms at different temperatures and calculate that some lines generally considered reference lines are emitted at higher temperatures than others. They use the lines formed at the coldest temperatures to determine zero Doppler shift. All other lines, including lines previously assumed to be unshifted, are calibrated from the lowest temperature lines with the result that many of the lines show changes in red-shift of 1-2 km/s.

Another critical issue is the measurement of the rest wavelengths of lines, both the chromospheric calibration lines and the lines for which the Doppler shift is measured. As discussed extensively in papers by Brekke *et al.* [12,8], SUMER data can be used to determine line positions to about 0.004Å (1 km/s). In some cases this is better than the quoted errors on the laboratory determined line positions. In addition, for at least four lines used in the Brekke *et al.* [8] study there are

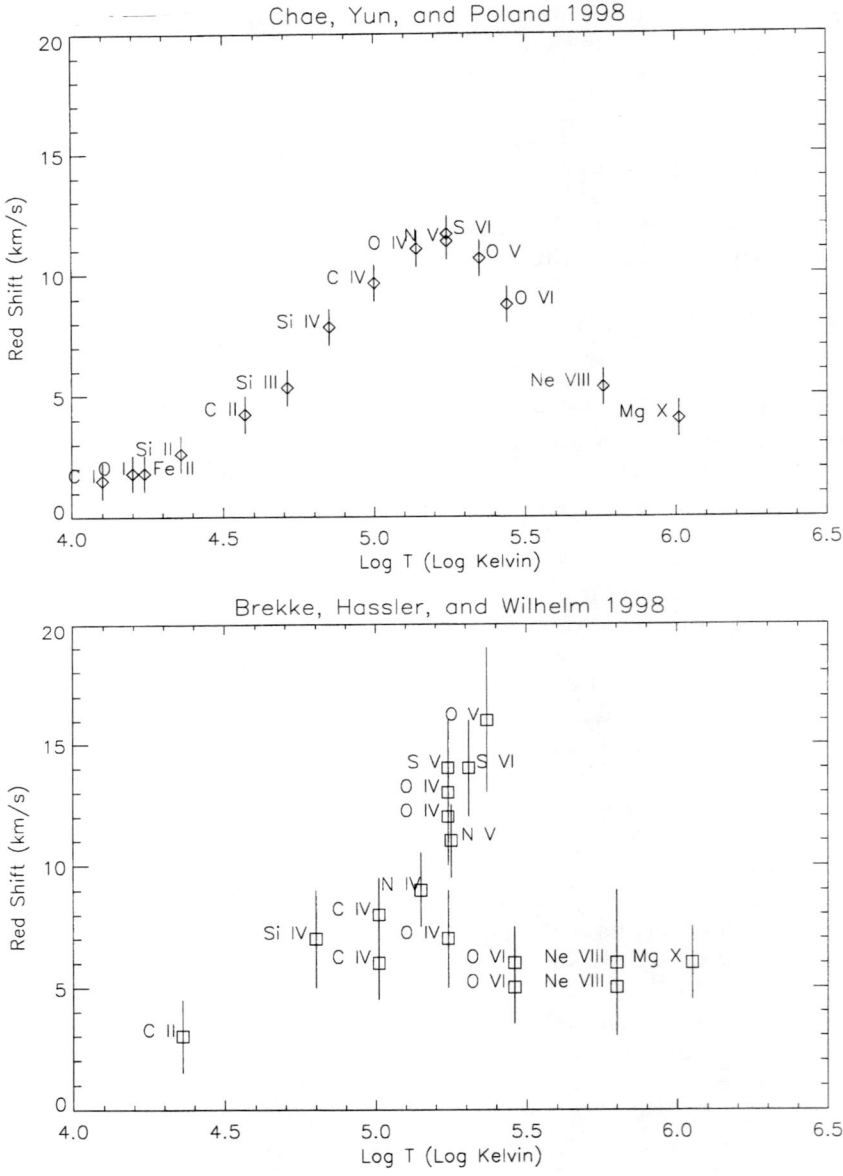

FIGURE 3. Measured red-shifts in the solar atmosphere as a function of temperature, as determined with SOHO/SUMER data by Chae, Yun, & Poland [9] and Brekke, Hassler, & Wilhelm [8]

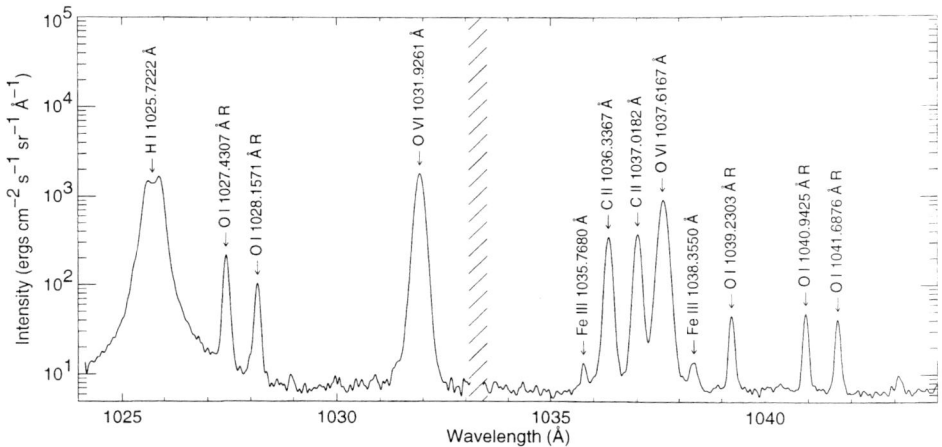

FIGURE 4. A typical quiet Sun spectrum from 1024 – 1044 Å taken with SOHO/SUMER. Laboratory wavelengths are noted above each line and the chromospheric lines used for calibration are marked with an "R". The lines long-ward of 1033 Å (the hatched area) lie on a more sensitive part of the detector. From a paper by Warren et al. [13].

differences corresponding to 4 – 6 km/s between values given by different sources for line rest wavelengths. These are a good fraction of the measured red-shifts, which range from 4 to 16 km/s. Observations of the Sun can be used to asses different laboratory measurements. For instance, if we require that all lines for a particular ion have the same Doppler shift, we can test the validity of the relevant laboratory rest wavelengths.

Essentially, to explain the observations one must develop a model in which the down-flowing plasma emits more radiation than up-flowing plasma in the lines in which red-shifts are observed. In this way the model can account for the observed average red-shifts but still allow an overall up-flow into the solar wind. For instance, in their models of vertically aligned magnetic flux tubes, Chae, Yun, & Poland [14] found that a flux tube containing a down-flow is dominated by the convection associated with that flow and has a gradual temperature gradient. This results in a large total mass over which a particular ion may exist and emit. Tubes containing up-flows, on the other hand, are dominated by conduction and have steep temperature gradients. As a result, any one ion exists over a relatively small distance and there is less emission. This model was developed further [9] by including variations in the cross-sectional area of the flux tube to match the red-shift curves in Fig. 3.

Many other models have been developed to explain the red-shifts, including the nano-flare model of Hansteen [15,16], and the asymmetrically heated loop models of McClymont & Craig [17,18], and Mariska [19]. Commonly, these models make predictions for the red-shift as a function of temperature and predict blue-shifts for lines at coronal temperatures. They will have to be revisited in light of the new data showing red-shifts of coronal lines as well.

In order to adequately construct these models a number of atomic physics issues must be considered. These may include incorporation of ionization fraction calculations, non-Maxwellian velocity distributions, and time dependent non-local thermodynamic equilibrium (non-LTE). Thus far these have been applied to the models to limited extents (for instance [20,21]).

III STREAMERS AND THE SOLAR WIND

It was realized in the 1980s that compared to photospheric abundances there are relatively more elements with low first ionization potentials ($\gtrsim 10$ eV) than with high ones in the solar wind and in the corona (early results were brought together in papers by Meyer [22,23]). This is known as the "FIP" effect. The relative enhancement of low FIP elements to high FIP elements is a factor of 3-5 in the slow-speed solar wind and 1.5-2 in the high-speed wind [24]. The effect varies between different parts of the corona as well [25].

SOHO/UVCS has been able to investigate this in detail. The new data have enabled us to determine absolute abundances of elements for both high and low FIP elements, rather than abundances compared to O or Si, as has often been done in the past.

In a study of streamers with UVCS, Raymond et al. [26] have found that the high FIP elements are depleted (rather than the low FIP elements enhanced) to varying extents in different locations. Abundances were calculated in a quiescent streamer at an altitude of 1.5 $R_\odot$ and in a streamer over an active region at 1.7 $R_\odot$. In the streamers above active regions and in the legs of the equatorial quiescent streamers the high FIP elements are depleted by about a factor of 4. Near the center of quiescent streamers all elements heavier than He are depleted, by a factor of about 5 for the low FIPs and 10 for the high FIPs. These results are shown in Fig. 5. The factor of 4 depletion at the streamer edges is similar to the FIP effect measured in the slow component of the solar wind which is believed to come from streamer regions.

In order to determine an elemental abundance, a number of issues must be addressed (See review by Mason [27]). In the solar corona ion excitation can be due to a combination of collision and radiation. For accurate calculations one must have good models of excitation and ionization states of atoms and ions, including understanding of quasi-stable states, collision, ionization, and recombination rates.

Most models of the FIP effect utilize some sort of diffusion effect in the chromosphere and transition zone (the same area where we observe the down-flows) which affects neutrals and ions differently. A review of several different models is provided by Hénoux [28]. Forces modeled include gravity, temperature and pressure gradients, and electric and magnetic field oriented in various directions.

As an illustration, I describe the model of von Steiger and Geiss [29]. They

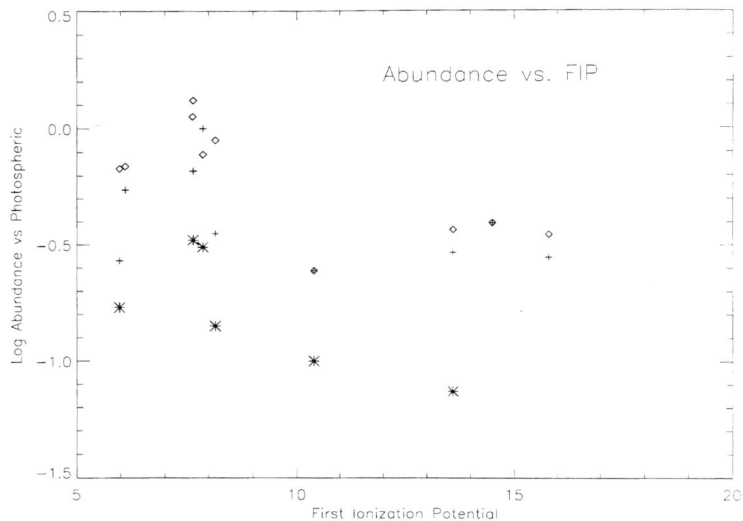

FIGURE 5. Results from Raymond et al. [26] showing the ratio of streamer abundances to photospheric abundances as a function of first ionization potential. The +'s refer to the legs of quiescent streamers, the *'s to quiescent streamer cores, and the ◇'s to active region streamers.

model a vertical magnetic flux tube containing plasma moving upwards. In the area surrounding the tube the flow is downwards. At the base of the tube in the chromosphere the plasma is neutral. As it moves up into hotter regions the atoms ionize via photoionization, with the low FIP elements ionizing to a greater extent than the high FIP ones. During this time, the atoms are diffusing out of the flux tube. However, once an atom ionizes it tends to stay in the flux tube, because in the outer solar atmosphere the magnetic pressure is greater than the gas pressure causing charged particles to follow magnetic field lines. Thus the charged, low FIP elements stay in the flux tube and move upwards into the corona while the neutral, high FIP elements tend to diffuse out of the tube and stay in the lower solar atmosphere. This results in a depletion of high FIP elements in the corona.

This is illustrated in Figure 6. The magnetic field is directed upwards from the solar surface and the flux tube is shown by the heavy dashed lines. At $t = 0$ the plasma being studied is all in the flux tube, but some time later the high FIP elements have diffused out of the tube.

As is discussed by Raymond et al. [26], it may be that the depletion of high but not low FIPs in streamer legs and active region streamers represents the situation at the top of the transition region. Along quiescent streamer legs and in the relatively small active region streamers, the plasma is drawn up into the corona and beyond by the solar wind. In contrast, the cores of quiescent streamers may be sufficiently

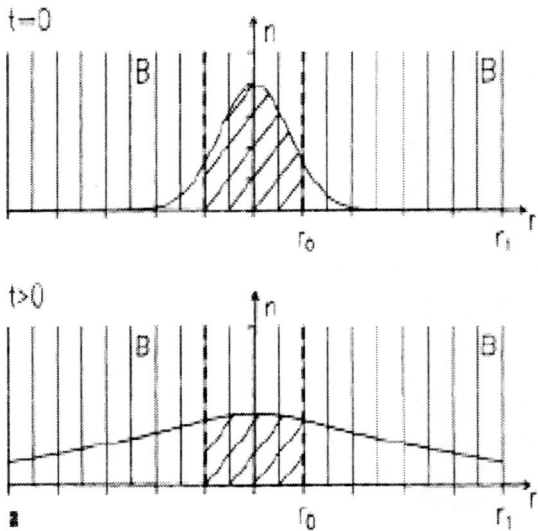

FIGURE 6. Illustration of one of the models of von Steiger and Geiss [29]. The graphs shows the number density of the gas. At time t=0 it is neutral and confined to the flux tube shown by the heavy dashed lines. At the later time, much of the un-ionized gas has diffused out of the flux tube.

quiet that gravitational settling occurs, resulting in a further depletion of all heavier elements regardless of FIP.

IV SUMMARY

The ultraviolet spectrometers aboard SOHO give us new tools to understand the processes at work in the solar atmosphere and wind. Current work is illustrated with two examples of recent work done with SOHO data. Flows in the lower solar atmosphere have been measured by SOHO/SUMER over a greater range of temperatures and with better spatial resolution than possible before. SOHO/UVCS has measured elemental abundances in coronal streamers relative to hydrogen and, again, with unprecedented spatial resolution. Using these data requires accurate atomic parameters including line identifications and rest-wavelengths, ionization, and recombination rates, and accurate modeling of atomic energy states. To adequately model the phenomena observed, researchers must often call on time dependent and non-equilibrium calculations, non-Maxwellian distributions, and other deviations from simple, static, LTE models.

The author would like to acknowledge useful conversations with Drs. Art Poland and Julia Saba. SOHO is a joint project of ESA and NASA.

REFERENCES

1. Domingo, V., Fleck, B., & Poland, A.I., *Solar Physics* **162**, 1 (1995).
2. Kohl, J.L. *et al.*, *Solar Physics* **162**, 313 (1995).
3. Wilhelm, K. *et al.*, *Solar Physics* **162**, 189 (1995).
4. Wilhelm, K. *et al.*, *Solar Physics* **170**, 75 (1997).
5. Harrison, R.A., *Solar Physics* **162**, 233 (1995).
6. Harrison, R.A., *Solar Physics* **170**, 123 (1997).
7. Doschek, G.A., Feldman, U., & Bohlin, J.D., *Astrophys. J. Lett.* **205**, L177 (1976).
8. P. Brekke, D.M. Hassler, & K. Wilhelm, *Solar Physics* in press (1998).
9. J. Chae, H.S. Yun, & A.I. Poland, *Astrophys. J. Supp.* **114**, 151 (1998).
10. Warren, H.P., Mariska, J.T., & Wilhelm, K., *Astrophys. J. Lett.* **490** L187 (1997).
11. Hassler, D.M. *et al.*, in preparation (1998).
12. Brekke, P., & Hassler, D.M., *Laboratory and Astronomical High Resolution Spectra* (eds. A.J. Sauval, R. Blomme, & N. Grevesse), *ASP Conference Series* **81** 589 (1995).
13. Warren, H.P., Mariska, J.T., Wilhelm, K., & Lemaire, P. *Astrophys. J. Lett.* **484** L91 (1997).
14. J. Chae, H.S. Yun, & A.I. Poland, *Astrophys. J.* **480**, 817 (1997).
15. V.A. Hansteen, *Astrophys. J.* **402**, 741 (1993).
16. V.A. Hansteen, P. Maltby, & A. Malagoli, in *Magnetic Reconnection in the Solar Atmosphere* (eds. R.D. Bently & J.T. Mariska), *PASP Conf. Series* 11 (1996).
17. McClymont, A.N., & Craig, J.D, *Astrophys. J.* **312**, 402 (1987).

18. McClymont, A.N., *Astrophys. J. Lett.* **347**, L47 (1989).
19. Mariska, J.T., *Astrophys. J.* **334**, 489 (1988).
20. Spadaro, D., Antiochos, S.K., & Mariska, J.T., *Astrophys. J.* **382**, 338 (1991).
21. Fontenla, J.M., Avrett, E.H., & Loeser, R., *Astrophys. J.* **355**, 700 (1990).
22. Meyer, J.-P., *Astrophys. J. Supp.* **57**, 151 (1985).
23. Meyer, J.-P., *Astrophys. J. Supp.* **57**, 171 (1985).
24. Von Steiger, R., Schweingruber, R.F.W., Geiss, J., & Gloeckler, G., *Advances in Space Research* **15**, (7)3 (1995).
25. Saba, J.L.R., *Advances in Space Research* **15**, (7)13 (1995).
26. Raymond, J.C., et al., *Solar Physics* **175**, 645 (1997).
27. Mason, H.E.,, *Advances in Space Research* **15**, (7)53 (1995).
28. Hénoux, J.-C., *Advances in Space Research* **15**, (7)23 (1995).
29. Von Steiger, R. & Geiss, J., *Astronomy and Astrophys.* **225**, 222 (1989).

Heavy Particle Atomic Collisions in Astrophysics: Beyond H and He Targets

P. C. Stancil, P. S. Krstić, and D. R. Schultz

Physics Division, Oak Ridge National Laboratory, P.O. Box 2008, Oak Ridge, TN 37831-6372
stancil, krstic, schultz@mail.phy.ornl.gov

Abstract. The physical conditions relating to the emission of x-rays from Jovian and cometary atmospheres and to supernova ejecta are briefly described. Emphasis is placed on elucidating the relevance and importance of atomic collision processes, the availability of data, and the outstanding data needs for modeling these environments. Some preliminary theoretical studies of electron capture for important collisions systems, involving molecular and atomic metal targets, are presented.

INTRODUCTION

The amount of neutral atomic hydrogen and helium can be deficient in a surprising variety of astrophysical environments. In many of these situations, low-energy collisions with metal atoms (ions) and molecules can play dominant roles in establishing the excitation and ionization conditions. Unfortunately, as the focus of the atomic collision community has been primarily on H and He targets, the availability, accuracy, and comprehensiveness of the needed metal and molecular target data are wanting. Consequently, the progress in astrophysical modeling is hindered and some proposed explanations of observed phenomena may be subject to revision pending a more complete database.

As the amount of collisional data required for astrophysical applications is enormous, it is impractical to perform close-coupling (CC) investigations for each system, much less to pursue comprehensive experimental evaluations. In fact, a quantal molecular-orbital CC (MOCC) calculation of collisions involving a metal or molecular target could be considered to be near the forefront of state-of-the-art research in theoretical atomic collisions. As an alternative, we are employing a variety of approximate techniques including the quantal decay model (QDM), the classical trajectory Monte Carlo (CTMC) method, and nonadiabatic methods such as the Landau-Zener (LZ) model. A measure of the accuracy can be obtained through the consistency (or inconsistency) of the three approaches. To further gauge the reliability of the results, a program to perform *ab initio* MOCC calculations for select systems is in progress as well as a collaboration with benchmarking experiments at ORNL.

In this article, we describe the physical conditions and collision systems of interest for three different astrophysical environments: Jovian atmospheres, cometary atmospheres, and supernova (SN) ejecta. Some preliminary calculations of applicable total and state-selective electron capture processes are presented.

JOVIAN ATMOSPHERES

Auroral x-ray emission from Jupiter was observed by the *Einstein* satellite [1] and recently by the *Röntgen* satellite (*ROSAT*) [2]. In situ measurements by the *Voyager* spacecrafts detected energetic oxygen and sulfur ions in the Jovian Magnetosphere. Their energy spectra suggest that the oxygen and sulfur ions are of Iogenic origin and their phase space density radial gradient indicates a diffusive flux into Jupiter's atmosphere [3]. The recent *Ulysses* flyby of Jupiter also measured significant quantities of O and S ions [4]. The combined spacecraft measurements indicate an ion energy range of a few keV/u to a few MeV/u. The *Voyager* results led Horanyi et al. [5] to suggest that the auroral x-ray emission was powered by the precipitating ions, the emission being due to radiative decay from highly excited high-charge state O and S ions produced following charge exchange with neutral atmospheric species, primarily H_2. Horanyi et al. [5], and more recently Cravens et al. [6], have constructed numerical models of the ion precipitation including energy deposition, beam ionization balance, and auroral x-ray emission.

The important collisional processes for O ions are listed below. In what follows we describe the atomic data utilized by the Jovian models [5,6], new data or that not presently considered in the Jovian models, and reactions for which data is lacking. For sulfur ions, no data are available. The Jovian models have thus far assumed the cross sections are similar to the cross sections for the oxygen ions. The electron capture reactions

$$O^{q+} + H_2 \rightarrow O^{(q-1)+} + H_2^+, \tag{1}$$
$$\rightarrow O^{(q-1)+} + H + H^+, \tag{2}$$

$$O^{q+} + H_2 \rightarrow O^{(q-2)+} + 2H^+, \tag{3}$$
$$\mathrel{\mathop{\hookrightarrow}} O^{(q-1)+} + e^-, \tag{4}$$

and

$$O^{q+} + H_2 \rightarrow O^{(q-1)+} + 2H^+ + e^-, \tag{5}$$

are the major processes for producing the x-ray emission and along with the projectile ionization (PrI) reaction

$$O^{q+} + H_2 \rightarrow O^{(q+1)+} + H_2 + e^- \tag{6}$$

determine the equilibrium ionization fractions of the precipitating ions [6]. Total single electron capture [SEC, process (1)] cross sections are available from the compilation of Phaneuf, Janev, and Pindzola [7]. As little information is available for state-selective SEC with H_2 targets, the models have adopted two schemes: (i) all captures go to $n = 2$ (80%) and $n = 3$ (20%) or (ii) all available n-states are populated equally. Since the *ROSAT* energy resolution is poor, both schemes can fit the observations reasonably well with nearly equal chi-square statistics [6]. Future observations with NASA's soon to be launched *Advanced X-ray Astrophysics Facility (AXAF)* and the planned Japanese *Astro-E*, however, should place serious constraints on the models as the satellites will have resolutions more than an order of magnitude smaller than *ROSAT*. As such, the need for state-selective SEC data will become more crucial.

Unlike the situation for S ions, some state-selective SEC data for O ions is available. Dijkkamp et al. [8] have measured the cross sections for capture to each of the l levels with $n = 4$ for O^{6+} collisions with H_2 at energies between 0.5 and 7 keV/u, while Beijers et al. [9] measured the $n = 3$ l-distributions from 0.04 to 1 keV/u for O^{3+} collisions with H_2. Meng et al. [10] calculated the n-distributions for O^{8+} + H_2 at 50 and 100 keV/u using the CTMC method for SEC; reaction (5), transfer ionization (TI); reaction (3), double electron capture (DEC); and reaction (4), autoionization (AI). We note that the Jovian models did not consider dissociative electron capture [DiEC, process (2)], DEC, AI, or TI. We are unaware of any other available data for these processes.

There is apparently no cross sections available for projectile ionization (PrI) of O ions. Horanyi et al. and Cravens et al. adopted the C^{q+} and B^{q+}, $q = 2,3$ results of Goffe et al. [11] and extrapolated the trends of the lower charged ions to estimate the higher charge state cross sections.

PrI, or stripping, plays a role in the ion energy loss, but target ionization (TrI) and dissociative TrI (DTrI)

$$O^{q+} + H_2 \rightarrow O^{q+} + H_2^+ + e^-, \qquad (7)$$
$$\rightarrow O^{q+} + H + H^+ + e^-, \qquad (8)$$

are the primary energy loss processes. Cross sections for TrI and DTrI for O^{q+}, $q = 2, 3, 4$, and 5, are available from Shah and Gilbody [12]. The Jovian models extrapolated the cross section behaviors to larger and smaller charge states.

Cross section data are lacking for rovibronic and electronic target excitation (TrE)

$$O^{q+} + H_2 \rightarrow O^{q+} + H_2^*, \qquad (9)$$

and projectile excitation (PrE)

$$O^{q+} + H_2 \rightarrow O^{q+*} + H_2. \qquad (10)$$

The models adopted the arbitrary branching ratios of 0.6 of TrI for TrE and 0.15 of PrI for PrE.

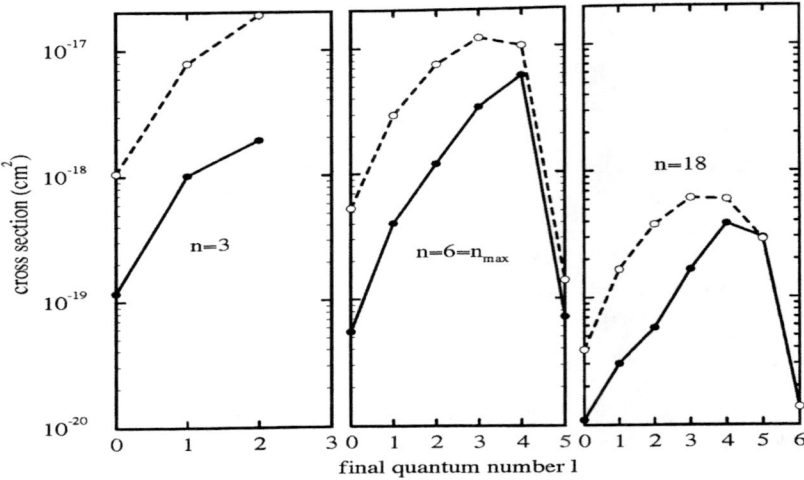

FIGURE 1. n,l distributions for O^{7+} + H_2 collisions at 200 keV/u. Filled circles, SEC. Open circles, TI+AI.

In preparation for future *AXAF* and *Astro-E* observations, we have begun a project to improve the modeling of auroral x-ray emission. We have initiated a program to compute an extensive database of state-selective SEC, DiEC, DEC, AI, and TI cross sections for collisions of intermediate high energy O and S ion collisions with H_2 primarily using the CTMC method. Cross sections are also being computed for H and He targets. While H and He are secondary constituents of the Jovian atmosphere, extensive theoretical and experimental data exists which can be used to test the reliability of the CTMC results. For low to intermediate energies, some MOCC calculations are also being performed.

Figure 1 presents a sample of the CTMC computations, the l-distributions for 200 keV/u collisions of O^{7+} with H_2 for select n values. As $n = 6$ corresponds to the maximum of the n-resolved cross sections, the assumption that all captures go to $n = 2$ and $n = 3$, adopted by the Jovian models, is questionable. Further, the sum of the AI and TI cross sections, processes which were neglected in the models, is greater than the SEC cross sections for most n,l levels. Use of the CTMC data should greatly improve the next generation of Jovian models for the auroral x-ray emission. Measurements of reactions (2) through (8), but particularly for S ions and reactions (2) through (5), would be useful to check the validity of theoretical calculations.

COMETARY ATMOSPHERES

In 1996 an unexpected observation was made by *ROSAT*; Lisse et al. [13] detected strong x-ray emission from comet C/1996 B2 (Hyakutake). Subsequent searches of the *ROSAT* all-sky survey archival data found that four other comets had also been serendipitously observed emitting x-rays [14]. Additionally, x-ray emission has been recently detected from C/1996 Q1 (Tabur) [15], from 6P/d'Arrest with the *Extreme Ultraviolet Explorer* (*EUVE*) [16], and from C/1995 O1 (Hale-Bopp) with *BeppoSAX* [17].

A number of models have been put forth to explain the x-ray emission. Schemes which invoke fluorescent emission and scattering of solar x-rays in the comet's coma [18] or bremsstrahlung [19] and inner-shell excitation of oxygen atoms [20] due to energetic electrons predict x-ray fluxes which are orders of magnitude smaller than observed. Only two models appear to reproduce the observed magnitude of the x-ray flux: scattering of solar x-rays by small dust particles in the coma [21] and electron capture of heavy solar wind ions to highly excited states following collisions with cometary neutrals [22,23]. Mumma et al. [16] have analyzed the observed x-ray brightness offset (i.e., the distance from the comet nucleus along the comet-sun vector), maximum brightness, and x-ray production rates for four comets and concluded that the electron capture model is favored over the dust mechanism. Conversely, Owens et al. [17] argue that the x-ray emissions are consistent with dust scattering of solar x-rays, but inconsistent with the electron capture scenario as the models of Häberli et al. [23] predict that intense narrow lines at energies greater than 0.28 keV which are not observed in their *BeppoSAX* measurements of Hale-Bopp. Further, Owens et al. point out that x-rays were not observed from the dust deficient comet Bradfield 1979X. However, their arguments can be disputed by considering the comet 6P/d'Arrest and the uncertainties in the models of Häberli et al. The dust and gas production rates in the coma of 6P/d'Arrest are 30% smaller and three times larger, respectively than observed for Bradfield 1979X, though an x-ray flux one-tenth that of Hale-Bopp was measured by the *EUVE* [24] suggesting no correlation of x-ray emission with dust. While the models of Häberli et al. include an elaborate three-dimensional magnetohydrodynamic simulation of the cometary plasma environment, their treatment of the atomic collision processes, as admitted by the authors, is perhaps too simplistic, owing to the lack of available collision data. The primary process for producing the x-rays is the SEC reaction

$$A^{q+} + X \rightarrow A^{(q-1)+} + X^+ \tag{11}$$

where A is a heavy solar wind ion and X a cometary neutral. Häberli et al. considered only A = C, O, and Ne and X = H$_2$O, OH, O, H, and CO. Since SEC data are only available for H targets, they adopted the H total SEC cross sections of Phaneuf et al. for X = O and H$_2$ target cross sections for X = H$_2$O, OH, and CO. A constant cross section of 4×10^{-15} cm^2 was utilized for all Ne ions with all neutral targets. For n, l distributions, they assumed that the n distribution was maximum for $n \propto q^{3/4}$ and for intermediate n levels, that the capture takes place to high l.

The x-ray emission models due to electron capture can be improved if cross sections for the actual neutrals are utilized. Neutral production rates due to sublimation from the surface of the comet nucleus as inferred from numerous measurements (e.g. Ref. [25]) suggest that the coma is composed of $\gtrsim 75\%$ H_2O and its photodissociation products OH, O, and H; $\sim 15\%$ CO; and $\sim 5\%$ CO_2. Traces of complex hydrocarbons have also been observed: CH_4 ($\sim 1\%$), C_2H_2 ($\sim 0.5\%$), and C_2H_6 ($\sim 0.5\%$). In addition to C, O, and Ne ions, the solar wind also contains significant quantities of N, Si, S, and Fe ions.

The quantity of charge transfer data for heavy ion collisions with cometary neutrals is sparse. We briefly summarize what is currently available. For the dominant neutral species, H_2O, total SEC due to He^{2+} (4-113 keV/u) [26] and O^+ (0.01-7 eV/u) [27] have been measured. We are not aware of any data for OH targets, but total and state-selective SEC impact-parameter calculations have been performed for O^+ (6-600 eV/u), O^{2+} (20 eV/u), S^+ (16-300 eV/u), S^{2+} (0.3-300 eV/u), and S^{3+} (20 eV/u) with O [28]. Numerous data are available for H targets [7,29]. Total SEC measurements for CO targets are available for H^+ (5-150 keV/u) [30,31], He^+ (1-88 keV/u) [32], He^{2+} (0.3-113 keV/u) [26,33], and Ar^+ (0.002-0.15 eV/u) [34]. Thermal SEC rate coefficient measurements for N^{2+} [35], O^{2+} [36], Ne^{2+} [37], and Kr^{2+} [37] for CO gas have also been performed. For CO_2 targets, total SEC measured cross sections are available for H^+ (5-150 keV/u) [31], He^+ (1-88 keV/u) [32], and He^{2+} (4-113 keV/u) [26] and thermal SEC rate coefficients for Ne^{2+} [37], Ar^{2+} [38], and Kr^{2+} [37]. Some data are available for the trace constituent target CH_4 with H^+ (0.05-150 keV/u) [31,39], He^+ (1-88 keV/u) [32], He^{2+} (4-113 keV/u) [26], C^{4+} (0.7-300 eV/u) [40], Ne^{2+} [37], Ar^{2+} [38], and Kr^{2+} [37]; C_2H_2 with H^+ (0.1-7 keV/u) [41]; and C_2H_6 with C^{4+} (0.7-300 eV/u) [40]. While the low charge state ions will not produce x-rays, the cross sectional data may be useful for testing future theoretical calculations or investigating ejected electron induced processes and the overall ionization structure.

To ascertain whether charge transfer is a viable mechanism for the production of cometary x-rays, detailed total and state-selective SEC as well as DEC, DiEC, TI, and AI cross sections are needed primarily for H_2O, but also for OH, O, CO, and CO_2. We are currently in the process of extending the CTMC and MOCC calculations for H_2 targets discussed in the previous section to the cometary neutrals. Measurements of these collision systems would be highly desirable.

SUPERNOVA EJECTA

Type Ia Supernovae

The ionization balance in the ejecta of a SN is strongly affected by charge transfer reactions. However, since the abundances of hydrogen and helium are negligible in the core regions due to inefficient mixing of the elements, the important reactions involve metal atoms and metal ions, systems which have received little attention in the atomic collision community. For example, the iron core of a Type Ia SN is composed primarily of Fe, Co, and Ni with charge states of $q = 1 - 3$ and neutrals. Liu, Jeffery, and Schultz [42] have found that the iron ionization structure is controlled by SEC

$$Fe^{q+} + Fe \rightarrow Fe^{(q-1)+} + Fe^+ \qquad (12)$$

and electron impact ionization. Similar processes determine the Co and Ni ionization structures. The SEC rate coefficients utilized in this model were estimated from QDM and CTMC calculations assuming that DEC would be negligible [43]. The only guidance as to the reliability of this assumption are SEC and DEC measurements of Mg^{2+} collisions with Mg and Zn [44]. While these measurements are limited to energies $\gtrsim 50$ eV/u, they suggest, however, that for smaller energies DEC will dominate. This is due to the resonant or quasiresonant conditions for DEC with both targets. To test the validity of our methods, we show in Figure 2 cross sections for SEC in Mg^{2+} collisions with Mg computed with QDM, CTMC, and LZ [43] in comparison with the measurements. Two types of LZ calculations are presented; one using the Olson-Salop [45] empirical (model) radial coupling and the other utilizing a radial coupling deduced from an *ab initio* computation of the Mg_2^{2+} potentials. The QDM and CTMC results overestimate the low energy SEC cross section as they assume a quasi-resonant situation exists when in actuality only two exit channels exist which are exoergic by 7.4 and 3 eV. The LZ method, which accounts for non-resonant interactions, gives good agreement with the experiment if an accurate estimate of the coupling can be obtained. On the other hand, LZ calculations which apply empirical couplings, constructed for H and He targets, to metal targets may be inappropriate [46]. The *ab initio* LZ calculations further suggest that SEC will be unimportant for the Mg^{2+} + Mg system [43].

However, compared to the simple $3s^2$ configuration of Mg, the iron-group elements of interest for Type Ia SNe have complicated valence structures ($3d^64s^2$ for Fe) as well as large angular momenta and spin. This results in many low-lying atomic excited states and a multitude of molecular states for the interacting iron-group pair. As an example, consider the molecular states of Fe_2^{2+} which correlate to the ground states of Fe and Fe^{2+}. The Wigner-Witmer rules predict that there are a total of 15 molecular states from Σ_g^+ to Γ_g for each of five multiplicities $2S + 1 = 1, 3, 5, 7, 9$, for a total of 75 molecular states. A large number of these should be easily capable of coupling to a number of lower-lying exoergic channels correlating to $Fe^+ + Fe^{+*}$. It would appear that for the iron-group elements, the quasi-resonance conditions would be satisfied and that SEC would be at least as important as DEC in the low energy regime. A study is currently underway to compute *ab initio* potential curves and perform MOCC calculations for some of these iron-group systems to verify if the quasi-resonance condition exists.

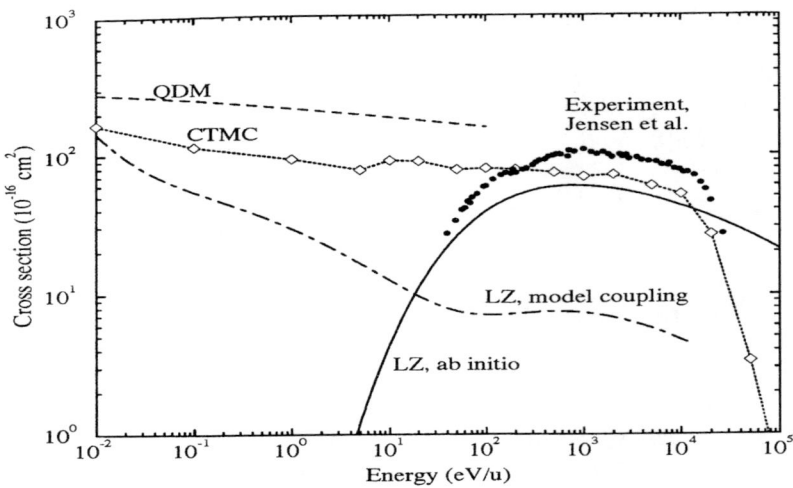

FIGURE 2. SEC for $Mg^{2+} + Mg$.

Type II Supernovae

In 1987 molecular emission from a SN was detected for the first time in the type II object SN1987a [47,48]. The emissions originate from vibrational bands of CO, SiO, and possibly CS [49]. Emission features near 2.3 μm in the low resolution spectra of SN1995ad are also thought to be due to the first overtone band of CO [50]. Since the discovery of molecular emission from SN1987a, a number of groups have attempted to model the emission of CO (see Ref. [51] for a review) with fair success. As the density is below the critical value, the vibrational populations will depart from local thermodynamic equilibrium (LTE). The model of Liu et al. [52] assumes that the vibrational levels are populated by electron impact excitation

$$e^- + CO(v'') \rightarrow e^- + CO(v') \qquad (13)$$

for which only the $v'' = 0$ to $v' = 1$ cross section has recently been measured by Allan [53] and calculated by Morgan [54]. This is thought to be a good approximation as the ejecta is ionized with an electron fractional abundance as high as 10^{-2}. To estimate the rate coefficients for $\Delta v = 1$ transitions for other vibrational levels, Liu et al. [52] assumed these were proportional to the absorption oscillator strengths. As no experimental or theoretical electron impact data for SiO are available, Liu and Dalgarno [55] estimated the excitation rate coefficients using the Born approximation and the available Einstein A-coefficients. Liu [57] made the same approximation for CS excitation.

However, the most abundant species in the zone of the ejecta where the molecules are formed is atomic oxygen suggesting that excitation due to O impacts

$$O + CO(v'') \to O + CO(v') \tag{14}$$

may play a role. Adopting the vibrational excitation rate coefficient of CO due to H collisions from Hollenbach and McKee [58], but scaled by the O-CO collision system reduced mass suggests that the rates for electron and O impact excitation of CO are comparable. Therefore, to improve models of molecular emission from Type II SNe, experimental and theoretical data are needed for electron and O impact excitation of CO, SiO, and CS.

CONCLUSIONS

Three astrophysical environments have been described in which atomic collisions with metal and molecular targets are important. Much of the necessary data to model these environments are unavailable. The major processes of interest are total and state-selective electron capture due to multiply charged ion collisions with H_2, H_2O, OH, O, CO, and CO_2 and collisions of iron-group ions with iron-group neutrals. Other processes of interest include projectile and target ionization and excitation due to oxygen and sulfur ion collisions with H_2 and vibrational excitation of CO, SiO, and CS due to O collisions.

ACKNOWLEDGMENTS

This work was supported by the U S DOE Office of Fusion Energy Sciences and Office of Basic Energy Sciences at Oak Ridge National Laboratory which is managed by Lockheed Martin Energy Research Corp under Contract DE-AC05-96OR22464. We thank W. Liu and D. Jeffery for helpful discussions.

REFERENCES

1. Metzger, A. E., et al., *J. Geophys. Res.* **88**, 7731 (1983).
2. Waite, J. H., Jr. et al., *J. Geophys. Res.* **99**, 14,799 (1994).
3. Gehrels, N. and Stone, E. C., *J. Geophys. Res.* **88**, 5537 (1983).
4. Lanzerotti, L. J., et al., *Science* **257**, 1518 (1992).
5. Horanyi, M., Cravens, T. E., and Waite, J. H., Jr., *J. Geophys. Res.* **93**, 7251 (1988).
6. Cravens, T. E., Howell, E., Waite, J. H., Jr., and Gladstone, G. R., *J. Geophys. Res.* **100**, 17,153 (1995).
7. Phaneuf, R. A., Janev, R. K., and Pindzola, M. S., *Atomic Data for Fusion*, vol. 5, ORNL-6090, Washington, D.C.: U. S. Dept. of Energy, 1987.
8. Dijkkamp, D., et al., *J. Phys. B* **18**, 737 (1985).
9. Beijers, J. P. M., Hoekstra, R., and Morgenstern, R., *J. Phys. B* **29**, 1397 (1996).

10. Meng, L., Reinhold, C. O., and Olson, R. E., *Phys. Rev.* A **42**, 5286 (1990).
11. Goffe, T. V., Shah, M. B., and Gilbody, H. B., *J. Phys.* B **12** 3763 (1979).
12. Shah, M. B. and Gilbody, H. B., *J. Phys.* B **14**, 2831 (1981).
13. Lisse, C. M., et al., *Science* **274**, 205 (1996).
14. Dennerl, K., Englhauser, J., and Trümper, J., *Science* **277**, 1625 (1997).
15. Dennerl, K., Englhauser, J., Trümper, J., and Lisse, C., *IAU Circular* **6495** (1996).
16. Mumma, M. J., Krasnopolsky, V. A., and Abbott, M. J., *Astrophys. J.* **491**, L125 (1997).
17. Owens, A., et al., *Astrophys. J.* **493**, L47 (1998).
18. Krasnopolsky, V. A., *Icarus* **128**, 268 (1997).
19. Northrop, T. G., *Icarus* **128**, 480 (1997).
20. Bingham, R., et al., *Science* **275**, 49 (1997).
21. Wickramasinghe, N. C. and Hoyle, F., *Astrophys. Space Sci.* **239** (1996).
22. Cravens, T. E., *Geophys. Res. Lett.* **24**, 105 (1997).
23. Häberli, R. M., et al., *Science* **276**, 939 (1997).
24. Abbot, M., Krasnopolsky, V. A., and Mumma, M. J., *IAU Circ.* **6667** (1997).
25. Mumma, M. J. et al., *Science* **272**, 1310 (1996).
26. Rudd, M. E., Goffe, T. V., and Itoh, A., *Phys. Rev* A **32**, 2128 (1985).
27. Li, X., Huang, Y.-L., Flesch, G. D., and Ng, C. Y., *J. Chem. Phys.* **102**, 5100 (1995).
28. McGrath, M. A. and Johnson, R. E., *J. Geophys. Res.* **94**, 2677 (1989).
29. Olson, R. E. and Schultz, D. R., *Phys. Scripta* **28**, 71 (1989).
30. Browning, R. and Gilbody, H. B., *J. Phys.* B **1**, 1149 (1968).
31. Rudd, M. E., DuBois, R. D., Toburen, L. H., Ratcliffe, C. A., and Goffe, T. V., *Phys. Rev* A **28**, 3244 (1983).
32. Rudd, M. E., Goffe, T. V., Itoh, A., and DuBois, R. D., *Phys. Rev* A **32**, 829 (1985).
33. Itoh, A. et al., *J. Phys. Soc. Japan* **64**, 3255 (1995).
34. Kobayashi, N., *J. Phys. Soc. Japan* **36**, 259 (1974).
35. Fang, Z. and Kwong, V. H. S., *Phys. Rev.* A **55**, 440 (1997).
36. — **51**, 1321 (1995).
37. Adams, N. G. and Smith, D., *Int. J. Mass. Spectrom. Ion Phys.* **35**, 335 (1980).
38. Smith, D., Grief, D., and Adams, N. G., *Int. J. Mass. Spectrom. Ion Phys.* **30**, 271 (1979).
39. Kimura, M., Li, Y., Hirsch, G., and Buenker, R. J., *Phys. Rev.* A **52**, 1196 (1995).
40. Soejima, K., Okuno, K., and Kaneko, Y., *Organic Mass. Spectrom.* **28**, 344 (1993).
41. Kimura, M., Li, Y., Hirsch, G., and Buenker, R. J., **54**, 5019 (1996).
42. Liu, W., Jeffery, D. J., and Schultz, D. R., *Astrophys. J.* **494**, 812 (1998).
43. Krstić, P. S., Stancil, P. C., and Schultz, D. R., in *Second Oak Ridge Symposium on Atomic and Nuclear Astrophysics.* Edited by A. Mezzacappa. Bristol: Institute of Physics, 1998, in press.
44. Jensen, B. and Pedersen, E. H., *J. Phys.* B **23**, 1501 (1990).
45. Olson, R. E. and Salop, A., *Phys. Rev.* A **14**, 579 (1976).
46. Swartz, D. A., *Astrophys. J.* **428**, 267 (1994).
47. Spyromilio, J., et al., *Science* **334**, 327 (1988).
48. Rank, D. M., et al., *Nature* **331**, 505 (1988).
49. Meikle, W. P. S., et al., *Mon. Not. Roy. Astron. Soc.* **238**, 193 (1989).

50. Spyromilio, J. and Leibundgut, B., *Mon. Not. Roy. Astron. Soc.* **283**, 89 (1996).
51. Dalgarno, A., Stancil, P. C., and Lepp, S, *Astrophys. Space Sci.* **251**, 375 (1997).
52. Liu, W., Dalgarno, A., and Lepp, S., *Astrophys. J.* **396**, 679 (1992).
53. Allan, M., *J. Electron. Spectrosc. Rel. Phenomena* **48**, 219 (1989).
54. Morgan, L. A., *J. Phys. B* **24**, 4649 (1991).
55. Liu, W. and Dalgarno, A., *Astrophys. J.* **428**, 679 (1994).
56. Liu, W. and Dalgarno, A., *Astrophys. J.* **471**, 480 (1996).
57. Liu, W., *Astrophys. J.* in press (1998).
58. Hollenbach, D. and McKee, C. F., *Astrophys. J* **342**, 306 (1989).

SPECTROSCOPY AND DIAGNOSTICS I

Plasma Spectroscopy of Pulsed Power Driven Z-Pinch Plasmas

J. Davis, R.W. Clark, J.L. Giuliani, Jr., and J.W. Thornhill

Radiation Hydrodynamics Branch, Plasma Physics Division
Naval Research Laboratory, Washington, D.C. 20375

Over the years there has been a sustained interest in and fascination with Z-pinch plasmas. Whether the interest is in radiation source development, fusion plasmas, and hohlraum physics or basic research there exits an extensive bibliography of literature presenting a variety of claims regarding the performance of Z-pinch plasmas. In this paper we will study the dynamics of Z-pinch plasma loads and focus primarily on the production of soft x-ray radiation. Many of the issues and revelations are common to Z-pinch plasmas in general but the attention here will be focused on the loads radiative performance. For the cases investigated here the load is heated by a multi-terawatt pulsed power accelerator. A radiation hydrodynamic model self-consistently driven by a circuit describes the dynamic evolution of the massive Al multiwire load. Predictions are made for the K- and L-shell soft x-ray emission as a function of the ionization dynamic model. The ionization dynamic models are represented by 1) a time-dependent NonEquilibrium (NEQ) model, 2) a Collisional Radiative Equilibrium (CRE) model and 3) a Local Thermodynamic Equilibrium (LTE) model. For all three scenarios the radiation is treated 1) in the free streaming optically thin approximation where the plasma is treated as a volume emitter and 2) in the optically thick regime where the opacity for the lines and continuum is calculated online and transported through the plasma. Each simulation is carried out independently to determine the sensitivity of the implosion dynamics to the ionization and radiation model, i.e., how it affects the radiative yield and emission spectra. Results will be presented for the L- and K-shell radiation yields and emission spectra representing the radiation intensity as a function of photon energy from 10 eV to 10 keV. The predictions from the various ionization models will be compared and contrasted with experimental results wherever possible.

+ This work was supported by DSWA.

INTRODUCTION

The theoretical description and numerical simulation of Pulsed power driven Z-pinch material loads in a variety of configurations and geometries is based on models that span the gamut from simple slug models to complicated multidimensional scenarios. Each model is then utilized to predict, postdict, and analyze some property of the plasma as it evolves from the initial cold start phase to the final collapsed phase characterized by a short burst of x-rays. The accuracy and reliability of the model is assessed by how well it predicts the dynamic evolution and plasma properties in comparison with the observations. In some instances simple slug models with two level atom ionization dynamic models have been employed to estimate the implosion time and the magnitude of the emitted K-shell x-ray pulse. At the other extreme there are 2- and 3-D RMHD (Radiation MagnetoHydroDynamic) scenarios that include a portfolio of subroutines for the equation of state, non-LTE ionization physics, transport properties, etc. These multidimensional models attempt to describe the entire plasma morphology and focus on specific issues such as the onset and growth of plasma instabilities and how they affect the integrity of the implosion and the emitted x-ray pulse.

Model validation is generally accomplished by comparing theoretically generated databases with experimentally observed results for selected plasma parameters. These quantities often include the magnitude of the radiative yield, the x-ray pulse length and shape, x-ray pinhole photos of the assembling plasma at selected wavelengths, and emission spectra. Most other plasma parameters are inferred from unfolding the experimental results. In order to demonstrate the uniqueness and sensitivity of the emission spectrum to variations in the model we have performed a series of numerical simulations of an aluminum wire array driven by a circuit similar to the Saturn accelerator. The simulations were performed using a 1-D radiation MHD model to illustrate how the emission features and radiation signatures are influenced as the atomic and opacity models are modified. The use of a full 1-D model is sufficient to illustrate the sensitivity of the emission history to the choice of atomic and opacity model. For example, in the case of the multiple wire Al array loads (having 40 or more individual wires) investigated on the Saturn accelerator, the implosion appeared to be less influenced by azimuthal asymmetries and, to first order, adequately represented by a 1-D implosion model until the final phase of stagnation.[1] Also, comparing and contrasting the synthetically generated spectrum with the experimentally observed spectrum has the potential to validate the reliability of the simulation and help identify its strengths and weaknesses and identify any shortcomings the model possesses.

MODEL

The morphology of z-pinch plasmas generally exhibits three (and sometimes four) distinct and sometimes overlapping stages. The first stage is the initial heating and expansion phase. In the second stage the JxB force drives the implosion. The third stage is characterized by the plasma reassembling on axis converting kinetic energy of run-in to thermal energy. During this third stage the newly formed hot dense plasma emits an intense kilovolt x-ray pulse. On occasion, a final and fourth stage may develop when the current is large enough to maintain plasma confinement, and additional heating and compression occurs accompanied by additional x-ray production.

The numerical simulations presented here to characterize the dynamics of a radially imploding Z-pinch plasma are based on a one-dimensional two temperature multi-zone non-LTE-radiation-magnetohydrodynamic model, DZAPP, with a transmission line circuit model representing the driving generator.[2] In a cylindrical coordinate system (r,z,t) the standard equations of continuity,

$$\frac{\partial \rho}{\partial t} + \frac{1}{r}\frac{\partial}{\partial r}(r\rho u) = 0, \qquad (1)$$

total momentum,

$$\frac{\partial}{\partial t}(\rho u) + \frac{1}{r}\frac{\partial}{\partial r}(r\rho u^2) + \frac{\partial (p_i + p_e)}{\partial r} + \frac{1}{r}\frac{\partial}{\partial r}(rQ_{vis}) = -\frac{1}{c}J_z B_\phi, \qquad (2)$$

ion internal energy,

$$\frac{\partial}{\partial t}(\rho \varepsilon_i) + \frac{1}{r}\frac{\partial}{\partial r}(r\rho \varepsilon_i u) + p_i \frac{1}{r}\frac{\partial}{\partial r}(ru) - \frac{1}{r}\frac{\partial}{\partial r}(r\kappa_i \frac{\partial T_i}{\partial r}) + Q_{vis}\frac{\partial u}{\partial r} = Q_{ie} \qquad (3a)$$

and electron internal energy,

$$\frac{\partial}{\partial t}(\rho \varepsilon_e) + \frac{1}{r}\frac{\partial}{\partial r}(r\rho \varepsilon_e u) + p_e \frac{1}{r}\frac{\partial}{\partial r}(ru) - \frac{1}{r}\frac{\partial}{\partial r}(r\kappa_e \frac{\partial T_e}{\partial r}) = -Q_{ie} + \eta J_z^2 - \Lambda \qquad (3b)$$

are integrated over each radial zone and then followed in a Lagrangian fashion. Here ρ is the mass density, u is the radial velocity, p_i the ion pressure, p_e the electron pressure,

Q_{vis} the rr component of a tensor artificial viscosity for smoothing shocks, J_z the axial current density in the pinch, B_ϕ the azimuthal magnetic field, $\varepsilon_{i(e)}$ the specific internal energy per unit mass of the ions(electrons), T_e the electron temperature, T_i the ion temperature, κ_e the electron thermal conductivity, κ_I the ion thermal conductivity, η the electrical resistivity, and Λ the radiation cooling (negative) or heating (positive) rate. The transport coefficients $\kappa_{i,e}$, η, and the ion-electron equilibration rate Q_{ie} are taken from Braginskii.[3] The numerical or artificial viscosity Q_{vis} for Lagrangian type codes is fully discussed by Schulz.[4] In the present paper we use $Q_{vis}=\beta\rho\Delta u C_s$, where Δu is the velocity difference across a zone and C_s is the local sound speed. Q_{vis} is zero if the zone is expanding. A value of 10 for the β parameter is used to soften the Z-pinch implosion and prevent radiative collapse.[5] Such a viscous enhancement is found to improve the agreement between experiment and simulation for the observed stagnation densities and radii.[6] Admittedly, the large viscous term is used to mimic the effect of presently unknown pinch physics leading to a softened implosion. The current density and magnetic field are related through Ampere's law

$$J_z = \frac{c}{4\pi} \frac{1}{r} \frac{\partial}{\partial r}(r B_\phi).$$

(4)

The magnetic field is determined by an induction equation by combining a generalized Ohm's law with Faraday's law:

$$\frac{\partial B_\phi}{\partial t} + \frac{\partial}{\partial r}(u B_\phi) = \frac{\partial}{\partial r}(c\eta J_z).$$

(5)

The ion internal energy includes the ion thermal energy as well as the atomic excitation + ionization specific energy (ε_X):

$$\varepsilon_i = \frac{3}{2}\frac{k_B T_i}{m_i} + \varepsilon_x$$

(6a)

The electron internal energy is represented by

$$\varepsilon_e = \frac{3}{2}<Z>\frac{k_B T_e}{m_i}$$

(6b)

where n_i is the ion number density, m_i is the ion mass, k_B is Boltzmann's constant and $<Z>$ is the mean charge per ion.

The transmission line circuit model is similar to the BERTHA code of Hinshelwood.[7] The model follows the forward and backward propagating voltage waves assuming only transverse electromagnetic modes are present. Circuit elements have associated transit times Δt such that a line element of impedance Z has an resistance Z, an inductance ZΔt or a capacitance Δt/Z. Wave propagation across junctions between line elements is solved through reflection and transmission coefficients derived from Kirchoff's laws. The junction at the dynamic load is treated implicitly to ensure strict conservation of electrical + plasma energy.

The ionization dynamics and radiation transport are time split from the magnetohydrodynamics, i.e., the ionization (and excitation) specific energy ε_X and charge state Z are held constant during a hydro timestep. As the excitation/ionization dynamics is updated within a zone, subject to the conservation of internal energy in that zone, the atomic level populations are determined by a set of rate equations

$$\frac{\partial f_{li}}{\partial t} = \sum_m (R_{mli} \, f_{mi} - R_{lmi} \, f_{li}),$$

(7)

where f_{li} is the fractional population of level l in the ith ionic species, and R_{mli} is the net rate describing the transition in the ith species from the initial level m to the final level l. The net rate includes collisional ionization, excitation, de-excitation, 3-body and radiative recombination, photo-ionization, radiation pumping and radiative decay. For the time dependent simulations the time derivative is maintained in eq. (7) and is neglected for equilibrium calculations. Details on the solution techniques and rate coefficients are summarized in Duston, et. al.[8]

The aluminum model contains all the ground states and a manifold of excited states distributed throughout the various ionization stages. Only the strongest lines, i.e., lines with the largest oscillator strengths, were included in the simulation. Finally, except where noted, the majority of numerical simulations were carried out using an atomic model that contained a full K-shell excited state manifold with some 50 spectral lines while the L and M shells contained all ground states and an excited state manifold with about 150 selected spectral lines, respectively. This assumption represents a reasonable compromise between the atomic model and computational constraints, particularly when the medium is opaque and radiation needs to be transported through the plasma.

Radiation emission, including opacity effects, from the plasma is dependent upon the local atomic-level populations, for not only are photons created by radiative recombination and spontaneous decay, such photons also can lead to population redistribution. Thus the ionization dynamics and radiation transport are a strongly coupled interactive system and must be solved together as demonstrated in ref. 8. The radiation transport of the bound-bound and bound-free transitions is carried out using the probability of escape formalism described by Apruzese.[9] Multifrequency transport is performed for the free-free radiation. An additional benefit of performing detailed ionization modeling is the diagnostic capabilities that emerge directly from the calculation. Since spectral features are maintained throughout the calculation, they are easily reproduced in a form suitable for comparison with experimental data.

RESULTS

The current profile that results from driving a Saturnlike generator circuit coupled to a changing load inductance is shown on Fig. 1. The peak short circuit current is about 10 mega-amperes. The load in this case is represented by ninety, 1.3 µm diameter aluminum wires with a total mass of 615 µgms. The load is 2 cm long and the initial array diameter is 1.72 cm.

A snapshot view of the evolution of some of the hydrodynamic parameters from the DZAPP model is shown on Fig. 2 as a function of radius [(x) in cm] and time [in µsec] with a time dependent collisional-radiative ionization dynamic model (referred to as NEQ). The electron and ion temperatures are represented as E-Temp and I-Temp, respectively, and are measured in eV. The density is in gm/cm^3 and the velocity in cm/µsec. The plasma is expanding for positive values and is compressing for negative values of velocity, respectively. The "gray-scale" adjacent and to the right of each insert on the Figure quantifies the degree of shading. The "first" bounce occurs around 80 ns. The interpretation of the observations beyond the first bounce is clouded by the possibility that structure evolved and dominated the subsequent morphology vitiating the 1-D approach. The structure may have significantly influenced the emission and x-ray pulselength voiding any meaningful comparison with the observational data beyond the first bounce. Therefore, the simulation was terminated after the first bounce but was allowed to expand out to 2.5 times the minimum radius. At peak compression the ion temperature is about 160 keV and the electron temperature rises to about 15 keV in a highly localized region. The density at this time is about 0.51 gm/cm^3. Several other values are identified on each of the inserts for convenience.

To illustrate the consequences resulting from the application of each ionization dynamic model an emission spectra was synthesized for each model. The comparison of synthetic emission spectra with the observed spectra is often used as one of the benchmarks to assess the model and evaluate how well the numerical simulation described the implosion. In the past, analysis of a few selected spectral lines originating in the K-shell along with the continuum slope were unfolded to provide a measure of internal plasma conditions and evaluate the generators' energy-load coupling efficiency and its radiative performance. We are unaware of any attempts to create a temperature and density profile from the unfolded experimental data in order to re-run the models, generate new synthetic spectra, and compare with the original predictions and observations. All too often the observed spectra is not compared with the synthetic one because the radiation physics modeling has been compromised to accommodate other aspects of the implosion modeling. It is ironic that our information and understanding of the implosion history comes from the measured observed "light", i.e., emitted radiation, but that the radiation modeling contained in many simulation models often remains crude and primitive and is treated in a superficial "window dressed" fashion.

The ionization dynamic models used here to quantify and construct an emission spectra are represented by NEQ, CRE (Collision Radiative Equilibrium), and LTE (Local Thermodynamic Equilibrium) for optically thick and thin simulations. All the simulations are self-contained in the sense that for each ionization model a full hydrodynamic simulation was performed rather than the post processing of a single hydrodynamic simulation. This produced differences in the implosion times and minimum radius. All the simulations were integrated in time out to a final radius about 2.5 times the minimum implosion radius and include the first bounce and subsequent re-expansion out to the terminated radius. The total time integrated spectra produced from the NEQ model will serve as our benchmark simulation and is shown on Fig. 3. Some selected spectral lines are identified along with the transition energy (keV), radiated power (MW/cm), and radiated energy (kJ). The helium- and hydrogen-like resonance lines account for about 50 % of the radiated K-shell energy. The calculated radiated energy emitted in the L- and K-shells is presented in Table 1.

RADIATIVE YIELDS

Ionization Model	Rad Transfer	Rad Yield (kJ) L-Shell	Rad Yield (kJ) K-Shell
NEQ	THICK	148.05	37.49
CRE	THICK	146.55	34.51
LTE	THICK	122.25	26.83
NEQ	THIN	112.80	35.54
CRE	THIN	126.30	44.16
LTE	THIN	138.00	39.10

TABLE 1

The THIN simulations are presented here for completeness because this approximation is employed in many simulations to take into account cooling by radiative losses. As the plasma evolves the opacity (which is a function of plasma conditions, pathlength, and oscillator strength) influences both the population of excited states and consequently the emitted radiation. A comparison of the magnitude of the L- and K-shell radiative yield shows that the CRE simulations are in reasonably good agreement with the NEQ calculations. The Table also illustrates that the LTE assumption, even with opacity, predicts an L-shell yield reduced by about 20% and a K-shell yield lower by about 30% than the NEQ model. The apparently good agreement with the CRE THICK result is not surprising since for aluminum the rates are fast enough for the populations to adjust themselves over hydrodynamic time cycles. As the density increases during peak compression the NEQ and CRE L-shell excited state populations begin to approach LTE. Therefore, the radiative energetics for the NEQ and CRE ionization models do not differ greatly in magnitude from each other over the entire plasma. However, for reasons of spectral fidelity and reproducibility it is necessary to provide information on the distribution of radiative energy as a function of photon energy. That can only come from the emission spectrum.

The experimental observations for the 90 multiple wire Al loads produced a total yield between 175 to 350 kJ and a K-shell yield between 35 and 51 kJ. As noted above, direct comparison with the experimental observations is moot because of the uncertainty in structural integrity during the final stages of the implosion. Just prior to the first bounce the plasma showed signs of breaking up and producing intense bright spots superimposed on the bulk background plasma. It may very well be the bright spots that are the source of differences between the experimental results and model predictions. The simulations predict a conversion of kinetic energy to total radiation of about 80% and to K-shell radiation of about 20% which is in agreement with earlier conversion efficiencies. Although the magnitude of the yield is of the same order as the observations the distribution of radiated energy, particularly in the K-shell, may be more revealing of any spectral structural differences.

The total emission spectra are shown on Figs. 3-8 for the cases represented in Table 1. Fig. 3 represents the NEQ result and is considered the benchmark simulation. Figs. 4-8 reproduce the spectrum corresponding to the NEQ, CRE and LTE thick and thin models for comparison with Fig. 3. At first glance the NEQ, CRE, and LTE thick appear identical. However, careful analysis will reveal distinct variations in specific lines and in the underlying continuum. Each Figure contains a listing of K-shell transitions, radiated power and radiated energy making it convenient to identifying individual lines. The overall appearance as well as individual line features of the THIN simulations are remarkably similar to each other since the L-shell populations are, for the most part, tending toward LTE during the final stages of compression. However, there are observable differences in the THIN and THICK simulations that can be identified easily. The trend toward LTE is clearly indicated in the departure coefficients, which tend toward unity in regions of higher density at peak compression. (The departure coefficients will be presented elsewhere in order to stay within the page limitations imposed in these Proceedings.)

Since the L-shell is so rich and densely populated in spectral lines, the individual spectral differences become blurred when making comparisons between the different model simulations. However, the differences are substantial. On the other hand, the K-shell lines are not as densely populated and individual features can be easily identified and better observed. The details of these calculations are seen on the

Figures as part of the spectrum and as an insert. The K-shell inserts exhibit profound differences that can be used as an effective footprint in validating code results.

In summary, the intent of this investigation was to 1) produce and characterize the emission spectra from a massively imploding multiwire Al array driven by a 10 MA generator, 2) compare and contrast synthetically generated spectra for three different but standard models of the ionization dynamics, and 3) use the synthetic spectra as a tool for validating the accuracy and reliability of the numerical simulation model. We have addressed and commented on all 3 issues and have concluded that it is feasible to use synthetically generated spectra to benchmark and validate numerical simulations. The key to benchmarking and validating models and numerical simulations is in performing well-defined experiments with good diagnostics. In addition, the emission spectrum is an indispensable tool when unfolding and analyzing conditions within the plasma as it evolves.

ACKNOWLEDGEMENTS

The authors would like to thank Drs. Apruzese, Terry and Whitney for interesting and lively discussions and for making available the results of their analysis of multiple wire experiments prior to publication. This work was supported by DSWA.

REFERENCES

1. Sanford, G.O. Allshouse, et. al., Phys Rev Lett 77, 5063 (1996).
   K. Whitney, J. Thornhill, et. al., Phys Rev E 56, 1 (1997).
   T. Sanford, T. Nash, et. al., Rev Sci Instrum 68, 852 (1997).
2. J. Davis, J. Giuliani, Jr., M. Mulbrandon, Phys Plasmas 2, 1766 (1995).
3. S. Braginskii, *Review of Plasma Physics*, Ed. By M. A. Leontovich [Consultants Bureau, New York, 1, 205 (1965)].
4. W. Schultz, *Methods in Computational Physics*, Ed. By B. Alder, S. Fernbach, and M. Rotenberg (Academic Press 3, 1 (1964)).
5. J. Shearer, Phys Fluids 19, 1426 (1976).
6. J. Thornhill, K. Whitney, C. Deeney, P. LePell, Phys Plasmas 1, 321 (1994).
7. D. Hinshelwood, Naval Research Lab Memo Rept No. 5185 (1983). Also, Technical Information Service Document No. 135024.
8. D. Duston, R. Clark, J. Davis, J. Apruzese, Phys Rev A27, 1441 (1983).
9. J. Apruzese, J Quant Spect & Rad Transfer, 25, 419 (1981).

Saturn Equivalent Circuit

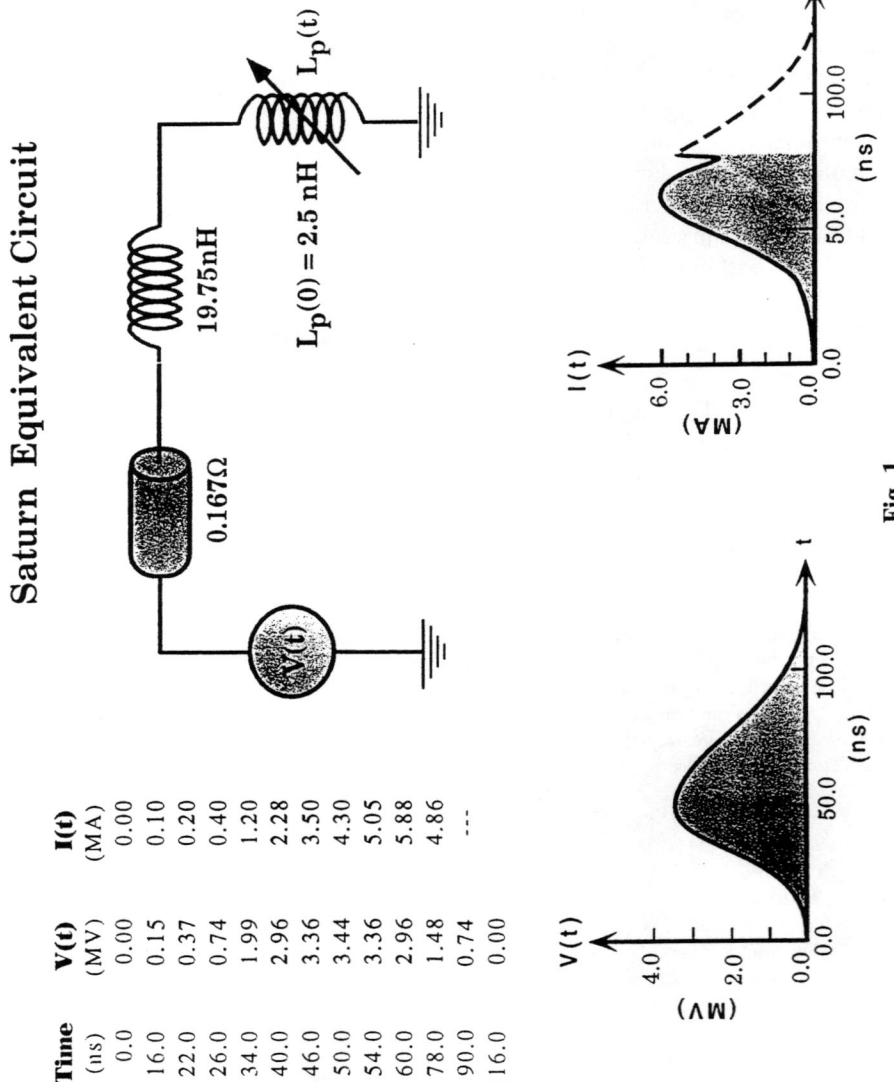

Time (ns)	V(t) (MV)	I(t) (MA)
0.0	0.00	0.00
16.0	0.15	0.10
22.0	0.37	0.20
26.0	0.74	0.40
34.0	1.99	1.20
40.0	2.96	2.28
46.0	3.36	3.50
50.0	3.44	4.30
54.0	3.36	5.05
60.0	2.96	5.88
78.0	1.48	4.86
90.0	0.74	---
116.0	0.00	

Fig. 1

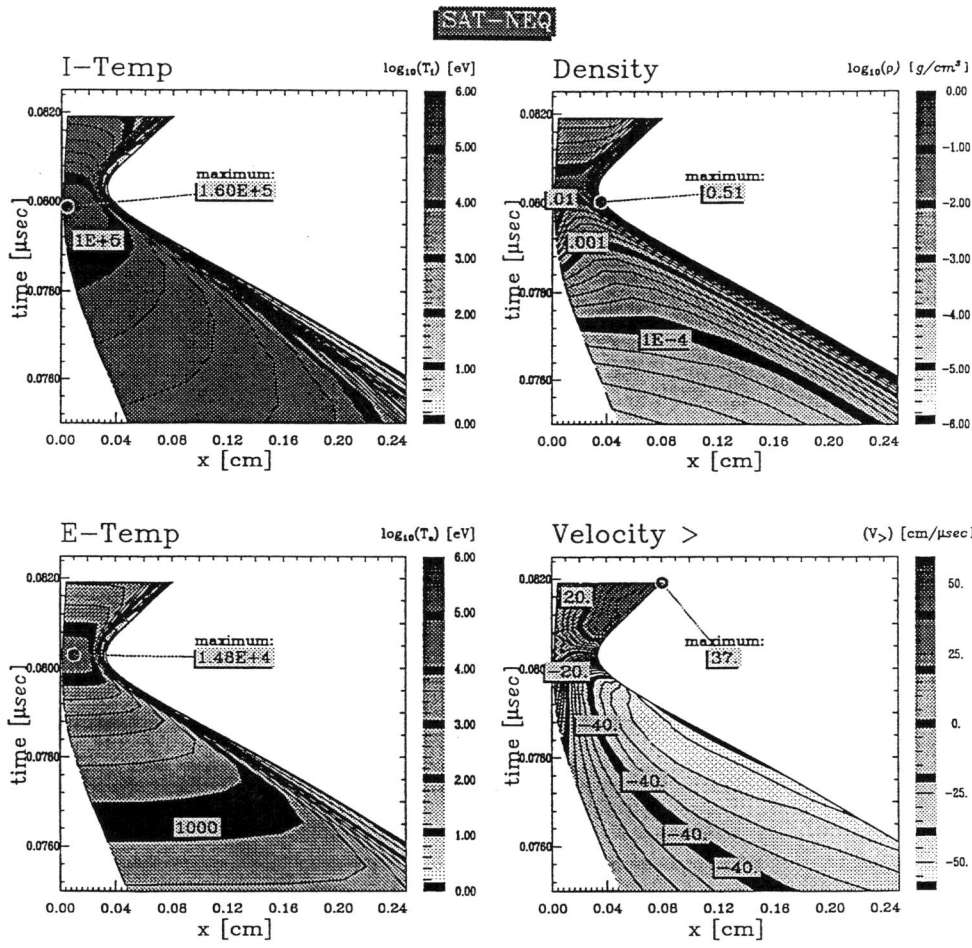

Fig. 2

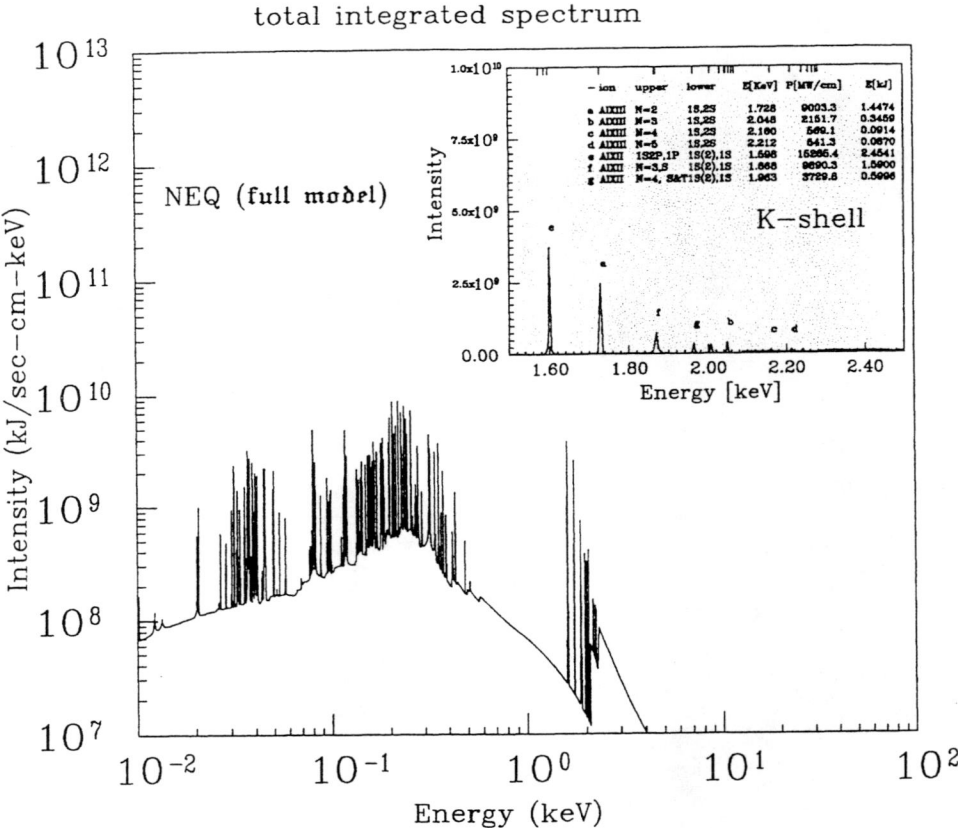

Fig. 3

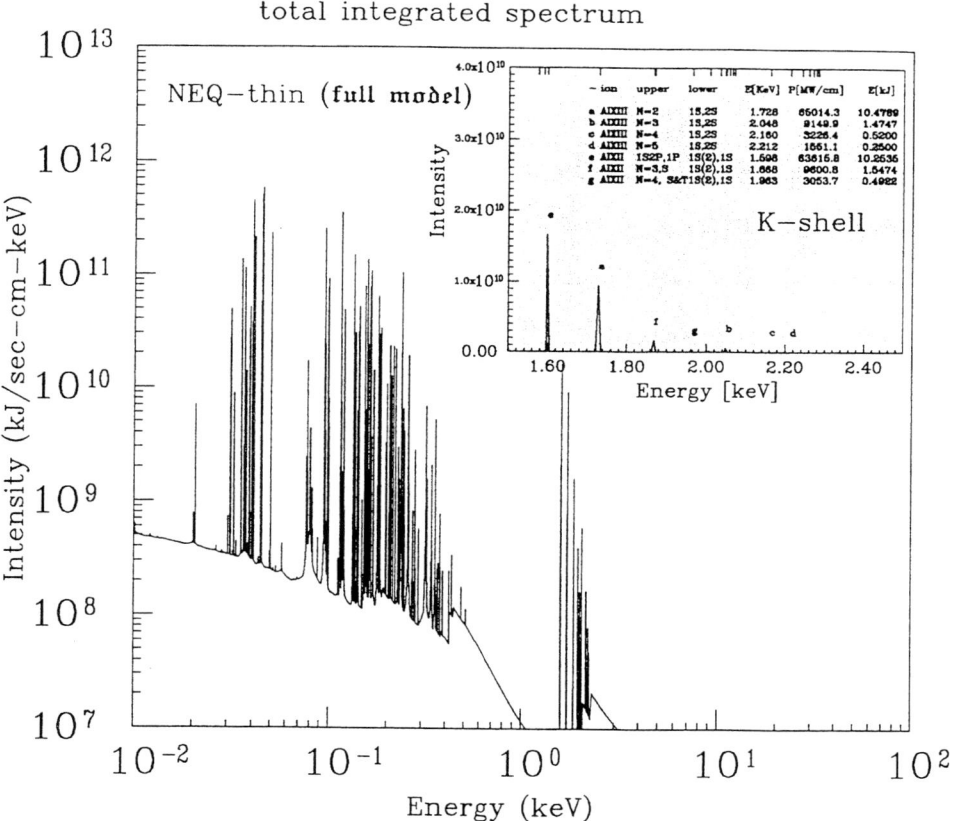

Fig. 4

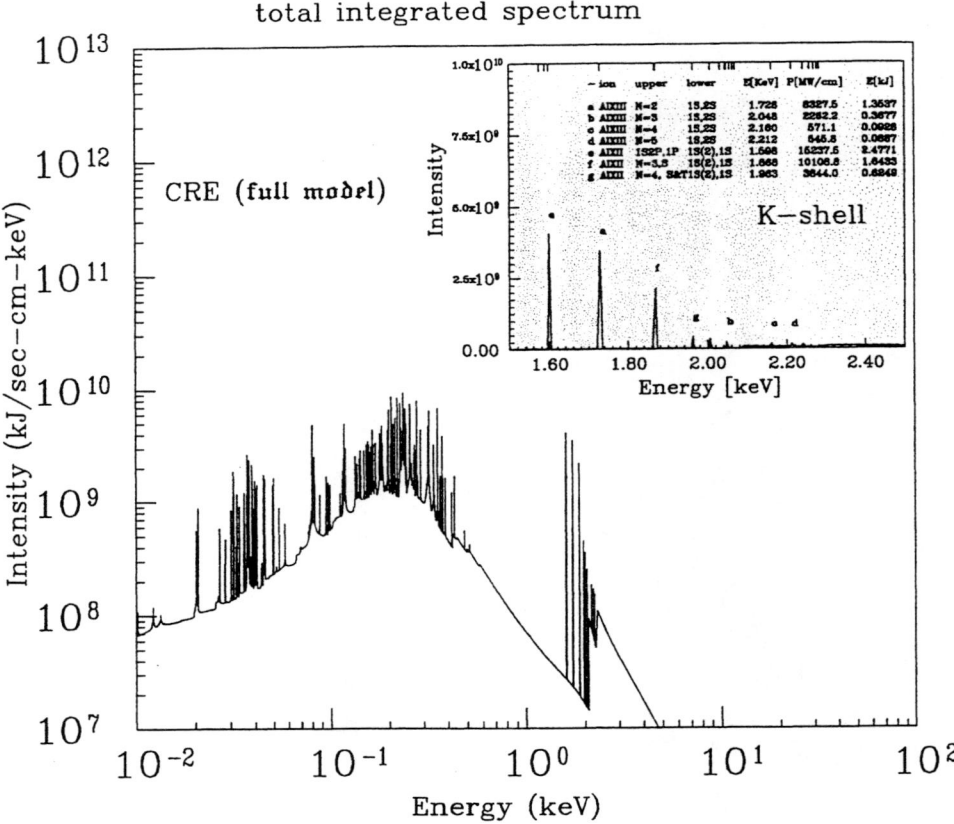

Fig. 5

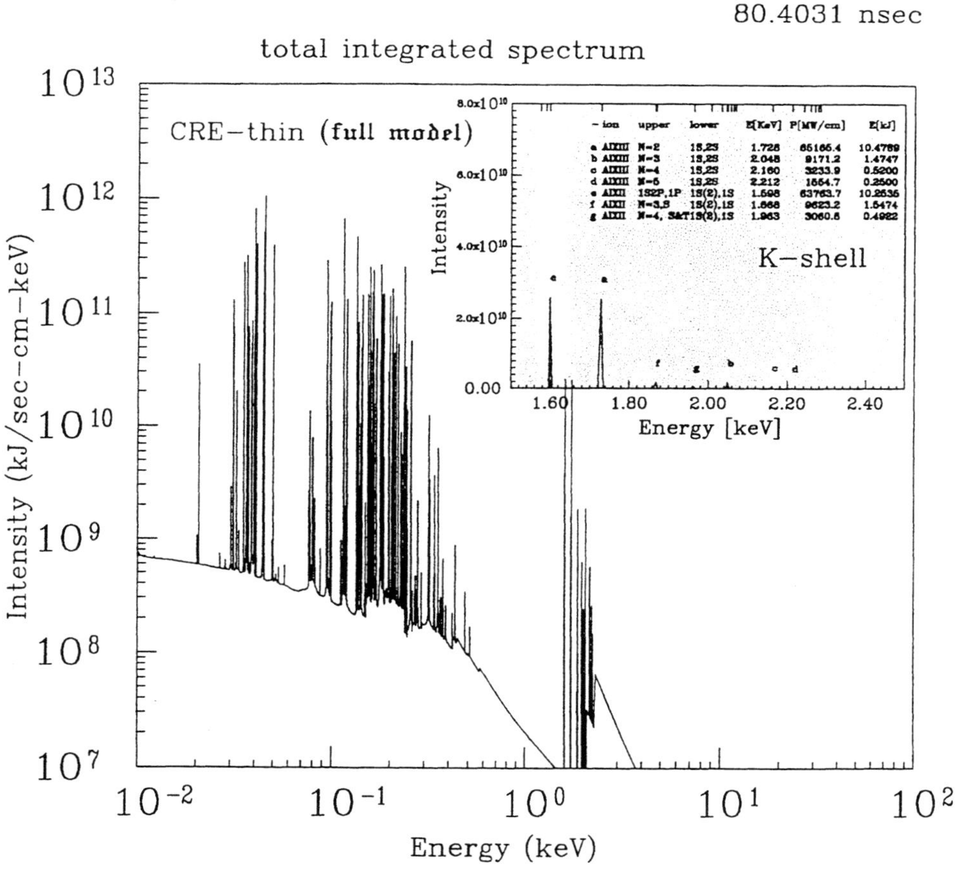

80.4031 nsec

Fig. 6

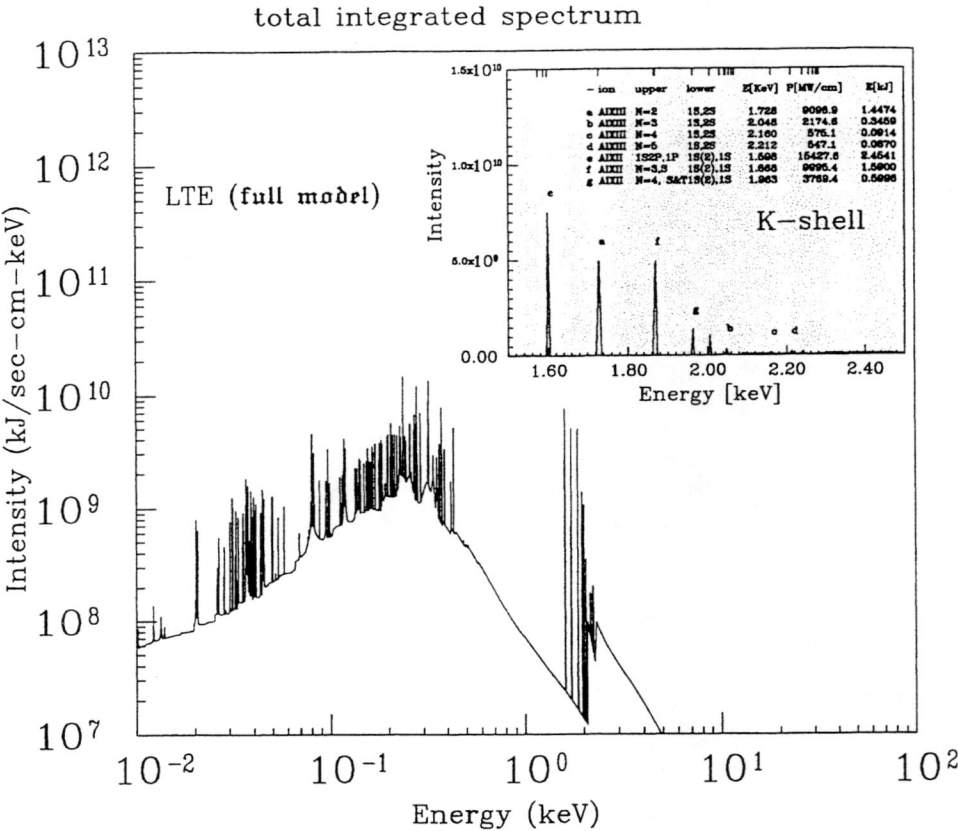

Fig. 7

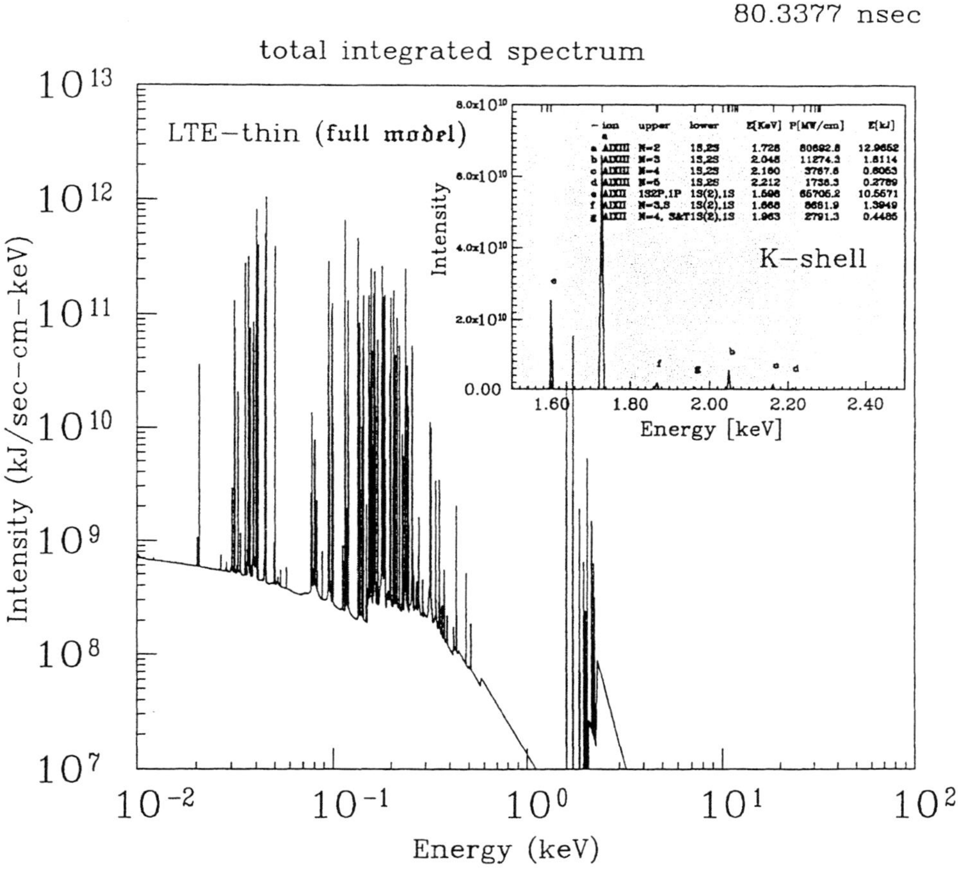
80.3377 nsec
total integrated spectrum
LTE−thin (full model)
Intensity (kJ/sec-cm-keV)
Energy (keV)
10^{13}
10^{12}
10^{11}
10^{10}
10^{9}
10^{8}
10^{7}
10^{-2}
10^{-1}
10^{0}
10^{1}
10^{2}
K−shell
Intensity
Energy [keV]
8.0x10^10
6.0x10^10
4.0x10^10
2.0x10^10
0.00
1.60
1.80
2.00
2.20
2.40
– ion upper lower E[KeV] P[MW/cm] E[kJ]
a AlXIII N=2 1S,2S 1.728 80692.8 12.9652
b AlXIII N=3 1S,2S 2.048 11274.3 1.8114
c AlXIII N=4 1S,2S 2.160 3767.8 0.6063
d AlXIII N=5 1S,2S 2.212 1738.3 0.2789
e AlXII 1S2P,1P 1S(2),1S 1.598 65705.2 10.5671
f AlXII N=3,S 1S(2),1S 1.868 8681.9 1.3949
g AlXII N=4, S&T 1S(2),1S 1.963 2791.3 0.4485

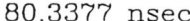

Fig. 8

Recent Advances in Spectroscopy of Strongly Correlated Plasmas

E.Leboucher-Dalimier, P.Sauvan, P.Gauthier, P.Angelo, H.Derfoul, S.Alexiou, A.Poquerusse, T.Ceccotti, A.Calisti*

*Physique Atomique dans les Plasmas Denses, LULI, CNRS
Université Paris VI, 4 Place Jussieu, 75252 Paris Cedex 05, France
and Ecole Polytechnique, 91128, Palaiseau Cedex, France.*

**Physique des Interactions Ioniques et moléculaires, Université de Provence, Centre de St Jérôme, 13397, Marseille, France*

in collaboration with S.J.Rose (RAL Didcot), E.Förster (XROP Jena), A.Devdariani (University of St Petersburg), R.W.Lee (LLNL)

Abstract - The Quasimolecular Model using a Two Centre basis to describe the electronic emitting structure gives an alternative treatment of line broadening in dense and hot plasmas. Two codes are developed : IDEFIX for the radiative properties, QMSPECTRA (postprocessed to the first one) for the spectral line shapes. The observability of dense plasma effects (PPS, asymmetries and satellite features) in spectroscopic measurements is analysed within the proposed model and taking care of the eventual integrations over density gradients.

INTRODUCTION

The effects of high material densities on atomic processes and the spectral line shapes characterizing dense and hot plasmas are investigated in this paper. The interest of this subject is obvious for the study of strongly correlated plasmas relevant to ICF and astrophysics (i.e.the centre of the sun) ; moreover high energy density plasmas give rise to new basic problems to be solved for understanding the transition between plasma physics and solid state physics. The alternative model proposed in the present work uses a quasimolecular approach which is justified. This model allows the exhibition of ion-ion correlation effects in the radiative properties of dense and hot plasmas and as a consequence in the emitted spectral lines. These effects can lead to new diagnostics useful for the drive of implosion experiments ; also, they may have an interest for the study of the energy transfer from stars and that is why new experiments have been designed for the exhibition and the measure of dense plasma effects.

THE TWO CENTRE MODEL : AN ALTERNATIVE APPROACH

The plasmas of interest are highly correlated ($N_e > 10^{23} cm^{-3}$) : the mean interionic spacing $R_i = (3/4\pi N_i)^{1/3}$ is comparable to the orbital extent of the outer orbitals $r_n = n^2 a_0 / Z$. Because of the resulting overlapping of neighboring ion orbitals, we can use a N centre basis. This N centre emitting structure is then confined with free electrons in a bounded volume. But as the temperature is high ($T_e > 300 eV$) the plasmas are moderately coupled : i.e. the ionic coupling parameter Γ is estimated at about 1. Therefore *the N centre states are not established* (i.e. as in a molecule bound by a chemical bond). The formation of bound molecules is plainly not possible for the plasma conditions probed here because the ion kinetic energy is much higher than any binding energy. But nevertheless this representation is justified because the radiative transitions are instantaneous on the ionic motion time scale and in this way the Franck-Condon principle is reliable.

Thanks to Molecular Dynamics (MD) simulations it can be shown that the *Two Centre model* is an alternative approach for moderately coupled plasmas. To support this assumption we have first demonstrated that the Nearest Neighbor (NN) interaction is a good approximation for the microfield distribution. MD simulations made for argon in Ar-filled implosion conditions ($N_e = 5.10^{24} cm^{-3}$, $T_e = 800 eV$) show that this approximation gives the exact microfield (all neighbor ions involved) within 20% (see Fig.1).

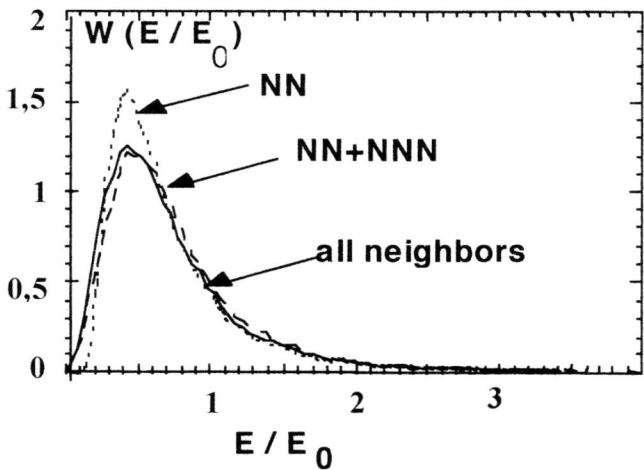

FIGURE 1. Molecular Dynamics microfield distributions for argon plasma from the nearest neighbor contribution only, from the nearest and next nearest neighbor only and from all neighbors. E_o is the microfield for the mean interionic separation.

Secondly MD simulations for a F^{8+}-F^{9+} transient molecule reveal a very small change of the interionic distance R(t) on the ionic motion time scale ; so that the two centre formation is highly probable during the time of interest (ω_{pi}^{-1} the inverse of the ionic

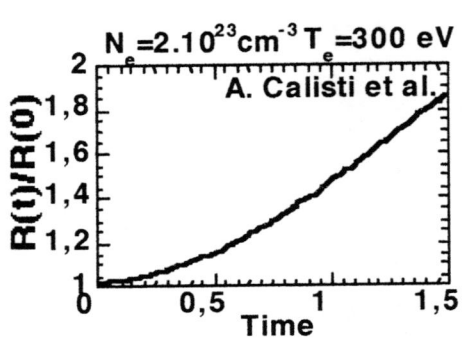
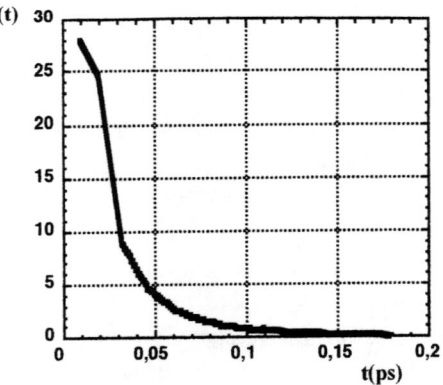

FIGURE 2. Interionic distance evolution with time (ω_{pi}^{-1} time unit) for F^{8+}-F^{9+} transient molecule.

FIGURE 3. F^{8+}-F^{9+} quasimolecule lifetime distribution for $N_e=10^{23}$cm^{-3} and $T_e=300$eV

plasma frequency, estimated at about $3\ 10^{-1}$ ps) (see Fig 2). More precisely, the lifetime of the molecule can be estimated at 30 fs (see Fig 3). As a consequence we chose a transient dicenter model to describe the emitting structure. The probability distributions needed for the emission were computed also by MD simulations using an effective electron screening. We present in Fig 4 Nearest Neighbor NN probability densities P(R) in a fluorine plasma for different electron densities ranging from 10^{22}cm^{-3} to 10^{25} cm^{-3} and the same temperature 300 eV. They give the radii of interest and we remark that the radius connected with the maximum decreases strongly with the density. It has been shown too that it increases with Z (1).

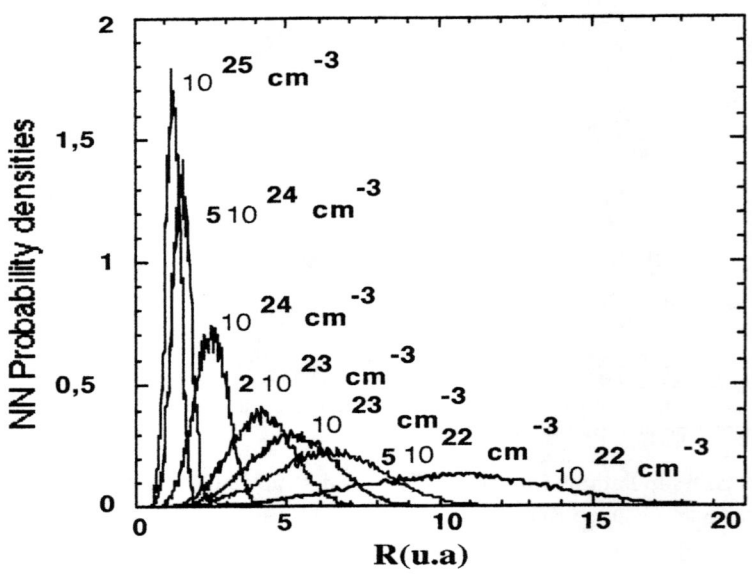

FIGURE 4. Probability densities for a fluorine plasma (Te=300eV)

THE ELECTRONIC STRUCTURE

The emitting structure, two NN ions (with one or two bound electrons) embedded in a non uniform electron gas, is confined in a volume satisfying the quasi-neutrality condition. Using a self-consistent treatment to solve the coupling between the Schrödinger and the Poisson equations, we determine with the code IDEFIX (2,3) the adiabatic energies, the transition energies and the oscillator strengths. The model involves an exact treatment of multipolar interactions to all orders and the outputs for the previous quantities are parameted with the interionic spacing R. This code runs for H-like and He-like states and for a large range of Z and plasma conditions (N_e, T_e).
The main points asserting the efficacity of the code are the following :

i) *the Two Centre model is adapted to exhibit at short interionic spacing the splitting of the transition energies, the possible extrema in the transition energies and the possible strong variation of the oscillator strengths.* Figure 5 shows for the F^{8+}-F^{9+} system the splitting of the transition energies due to the NN interaction of the n=3 energy level for a mean electron density $N_e=2.10^{23}$ cm^{-3} and an electron temperature $T_e=300$ eV.

ii) *the self consistent treatment is efficient for solving the shielding of the NN interaction by plasma free electrons.* Two major effects in the transition energies at large internuclear separations result from the free electron screening : a reduction of the level splitting and a Plasma Polarisation Shift (PPS). These two effects can be seen from the comparison between Fig.5a (without screening) and Fig.5b (with screening). The importance of the screening has also been demonstrated (2) on the ground state energy of a H-like fluorine ion (Fig 6) leading to a depression of the Ionisation Potential.

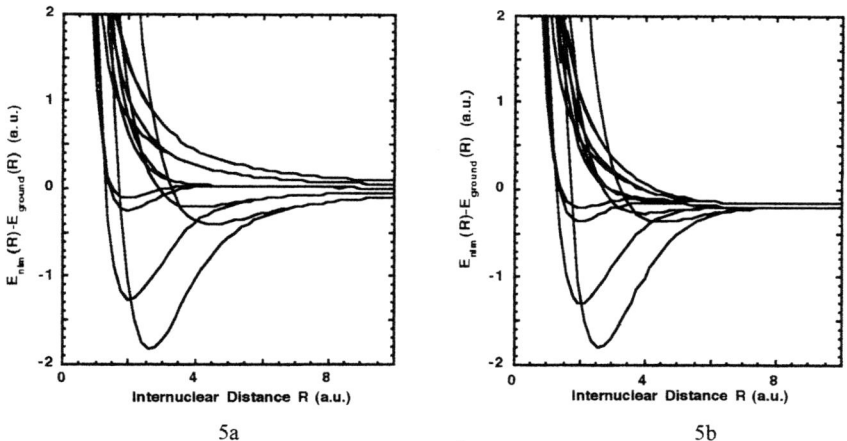

5a   5b

Figure 5. Variation of the transition energies of an F^{8+} ion, as a function of the distance from its nearest neighbor. Electron screening is considered in 5b ($N_e=2.10^{23}$ cm^{-3}) but not in 5a. The calculations were performed for the same electron temperature $T_e=300$eV.

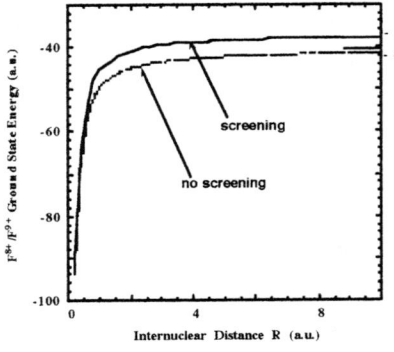

Figure 6. Variation of the ground state energy of a fluorine ion as a function of its distance from its NN in the plasma for $N_e=10^{23}$, $T_e=300$eV (solid line) and $N_e=0$ (dotted line). In the case of the finite electron density plasma, the F^{8+} ground state energy predicted by the Ion Sphere Model is progressively recovered as R increases.

THE LINE BROADENING WITHIN THE TWO CENTRE MODEL

In this section, dense plasma effects on spectral line shapes (photoabsorption K edge and emitted spectra) are analysed. The line broadening described here uses at first a quasistatic approach for the emitting or absorbing ionic configuration. Thus, the spectral features analysed involve averages over all configurations weighted by the NN probability P(R). The electronic structure computed from the code IDEFIX is given as an input to the line broadening.

The photoabsorption K edge position

The K edge position requires an average of the ground state energy (E_k) over NN distances :

$$I(N_e,T_e) = \frac{\int_0^\infty dR P(R,N_e,T_e) E_k(R,N_e,T_e)}{\int_0^\infty dR P(R,N_e,T_e)} \qquad (1)$$

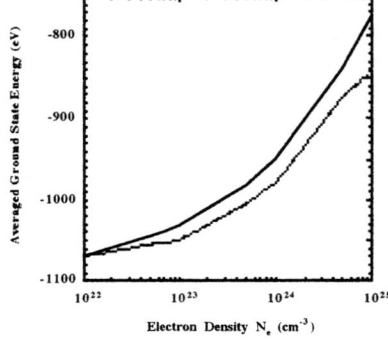

Figure 7. Averaged ground-state energy of H-like fluorine as a function of the plasma electron density. Quasimolecular computations (dotted line) are compared with results from the Ion Sphere Model (solid line). Calculations have been performed with $T_e=300$eV.

In figure 7 it is given for H-like fluorine as a function of the density (2). For densities lower than 5.10^{22} cm^{-3}, results from the two-centre model and from the ion sphere model (4) come together. For higher densities there is a discrepancy. This effect could be of importance and should be considered in the opacity calculations (5).

The line shape in very dense plasmas : the Quasi-static approach

The Stark-broadened line shapes are computed in QMSPECTRA code by averaging the electron energy levels $E_{i,f}(R)$:

$$I(\omega) \propto \sum_{if} \int dR\, P(R)\, f_{if}(R)\, \delta\!\left(\Delta E_{i,f}(R) - \omega\right) \qquad (2)$$

For given plasma parameters (N_e, T_e), $\Delta E_{if}(R)$ represents the transition energy between sublevels i and f, and f_{if} is related to the oscillator strength.
Electron collisional broadening of the various sublevels (not taken into account in formula (2)) was included in our code in the usual impact approximation using second order perturbation theory (6).
In the interionic distance window open by the NN probability density P(R), the transition energies are widely splitted and shifted and the oscillator strengths vary strongly.
These are arguments for the *shifts, asymmetries and satellite-like features* exhibited in dense plasmas.
The following results show *the importance of using a molecular basis*. We present in Fig. 8 Flyβ line profiles at 300eV for two densities (a) 10^{22}cm^{-3} and (b) 2.10^{23}cm^{-3}. The profiles are computed with the standard code Pim Pam Poum (PIIM Marseille), the dotted lines, and with the Two Centre model QMSPECTRA, the solid lines. While at low densities (curve (a)) the spectra come together, at high densities there is a discrepancy (curve (b)).

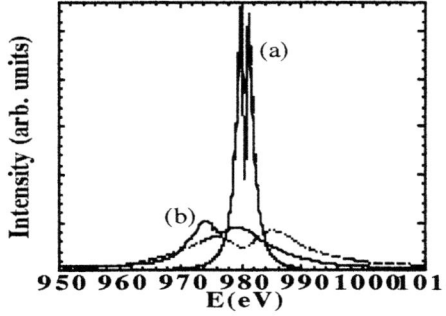

Figure 8. Flyβ line shapes at T_e=300eV for N_e=10^{22}cm^{-3} (curve (a)) and N_e=2.10^{23}cm^{-3} (curve (b)) : Standard Pim Pam Poum simulations (dotted lines) and QMSPECTRA simulations (solid lines).

The comparison between the two approaches is relevant as long as the quasi-static approximation is used for ions and the only NN microfied distribution is taken into account in the Pim Pam Poum simulations.

Calculations performed at various densities show *the sensitivity of the line shapes to electron screening*. In figure 9 we present Flyβ line profiles at T_e = 300eV for $N_e = 5.10^{22}$cm^{-3} (curves (a) and (b)) and $N_e = 2.10^{23}$cm^{-3} (curves (c) and (d)). Curves (a) and (d) include the electron screening whereas this influence has not been included in curves (b) and (c). Plasma screening leads to a reduction of the Stark effect (i.e. a reduction of the Full Width at Half Maximum and a reduction of the splitting of the components) and a global Plasma Polarization Shift (PPS) of the line in the direction of the lower energies.

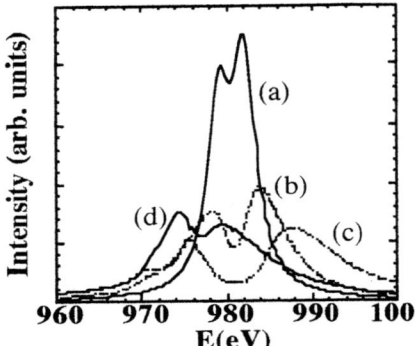

Figure 9. Flyβ line profiles for $N_e=5.10^{22}$cm^{-3} ((a) and (b)) and $N_e=2.10^{23}$cm^{-3} ((c) and (d)) (T_e=300eV for all cases). Calculations have been performed with QMSPECTRA. For profiles (b) and (c) electron screening effects have not been included.

At high densities the line profiles may exhibit satellite-like features. Figure 10 gives schematically reasons for such exhibitions. Three situations are analysed thanks to drafts for the transition energies ΔE and the oscillator strengths f , both plotted versus the interionic distance R in the NN probability window (P(R)).

The first example given in Fig.10(a) shows that the *splitting at short R of transitions degenerated at large R* can lead to a satellite (here a blue satellite) when the oscillator strengths are comparable in the window open by P(R). Secondly a satellite-like feature may correspond to the *excitation at short R of a transition forbidden at large R*. In Fig.10(b) the transition b is splitted from the transition a and gets an important oscillator strength in the probability window P(R). The last draft (Fig.10(c)) shows the possibility to get a satellite when the transition energy *ΔE exhibits an extremum* simultaneously with an oscillator strength of importance in the NN probability window.

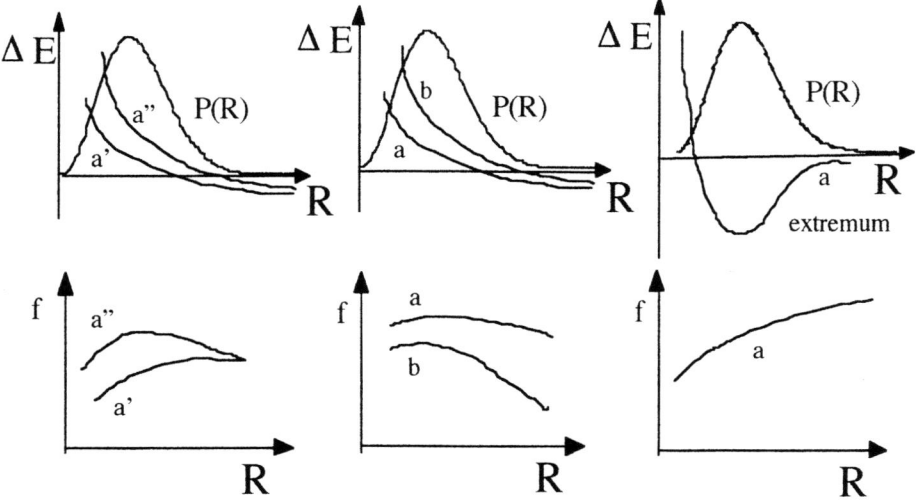

Figure 10. Satellite-like features exhibitions.
(a) Blue satellite due to the splitting of transitions
(b) Excitation of a forbidden transition b in the blue wing of the transition a
(c) Red satellite due to an extremum in ΔE.

In fact the exhibition of satellite-like features is questionable because these satellites may not survive different smoothings originated from theoretical or experimental grounds :
i) the *electron collisional broadening* increasing with the density and then smoothing the satellites
ii) the *ion dynamics* correction originated within our model in the transient molecule formation. This effect will be discussed in the next sub-section
iii) the *space/time integrations* over different plasma conditions (N_e, T_e). This effect will depend on the satellite origines. It has ben checked (2,3) that satellites due to ΔE extrema survive these integrations to a certain extent, because the interionic distance corresponding to an extremum is not sensitive to the density. Only the evolution of the NN probability with the density determines the allowed density boundaries. In the last section devoted to the analysis of experimental results we will see that the space/time integrations may smooth or generate satellite-like features according to the strength of density gradients.

The line shape in very dense plasmas : the dynamics approach.

This problem had been first tested analytically with simplifying assumptions for dense plasma (7,8). In this work the ion dynamics is treated selfconsistently with the plasma. This approach is the second step for the QMSPECTRA code.

As explained before, the molecule formation is basically transient. The resulting dynamics correction is not negligible in the vicinity of profile singularities ω_S (satellite-like features) giving a new reference for the time of interest $(\omega - \omega_S)^{-1}$.

Let us consider an ion moving on a classical trajectory depending on the potential energy $E_p(R_1R_2)$ due to the NN fixed ion and a non uniform free electron gas.

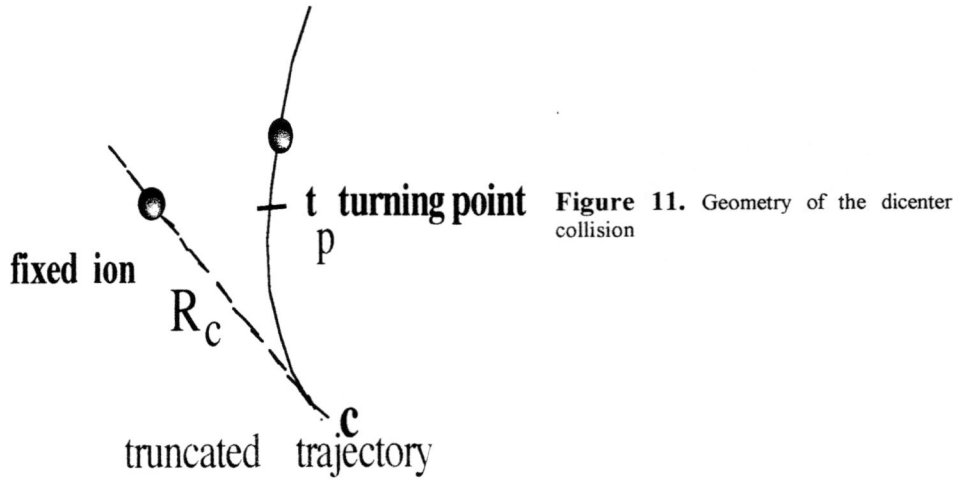

Figure 11. Geometry of the dicenter collision

This trajectory is truncated because of the cut off R_c corresponding to the interionic distance of the molecule in the confined quasineutral volume. The line profile within this approach is given by :

$$I(\omega) \propto \sum_{if} \int d\vec{R}_c d\vec{v}_{R_c} \, P(\vec{R}_c, \vec{v}_c) \left| \int_{R_{tp}}^{R_c} \frac{dR}{v} D_{if}(R) \cos\left(\int_{R_{tp}}^{R} \omega - \Delta E_{if}(R') \frac{dR'}{v} \right) \right|^2 \quad (3)$$

It depends on the history of the collision controlled through the interionic distance R and the radial velocity v. R_{tp} corresponds to the turning point and v_c to the radial velocity for the cut off. The ensemble average involves the conditional probability $P(\vec{R}_c, \vec{v}_c)$ computed by Molecular Dynamics. The transition energies $\Delta E_{if}(R)$, the dipole moments $D_{if}(R)$ and the potential energy $E_p(R,R_c)$ are given by IDEFIX. Electron collisional broadening, not expressed in (3), has also been included in our calculations.

The enhancement of satellite-like featues due to extrema in $\Delta E_{if}(R)$ with the dynamics correction is a highlight result. In figure 12 for Flyβ a comparison between the two versions (quasistatic / dynamics) of the Two Centre model (QMSPECTRA) is given for high density conditions ($N_e = 4.10^{23}$ cm^{-3}, $T_e = 300$eV). The dynamics correction smoothes the central part of the profile but enhances the satellites. These satellites involve interionic distances shorter than 3 a.u. (see Fig.5) which can be achieved for all trajectories such as $R_{tp} < 3$ a.u. In the quasistatic approach the observability of the satellites is only conditionned by the more restrictive NN probability window $P(\vec{R}_c, \vec{v}_c = 0)$.

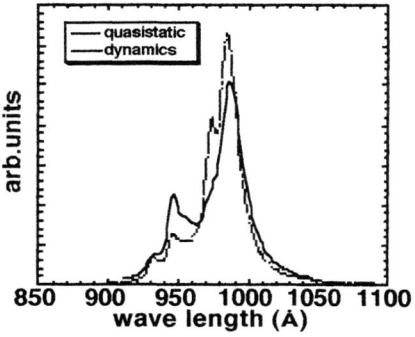

Figure 12. Comparison between quasistatic and dynamics Flyβ profiles.

The profiles computed with the dynamics version of QMSPECTRA have been compared to standard profiles issued from the ion dynamics version of Pim Pam Poum. As seen in figure 13 for Flyβ ($N_e = 2.10^{23}$ cm^{-3}, $T_e = 300$eV), our model gives the same near wings shape as PPP and the molecular satellites in addition.

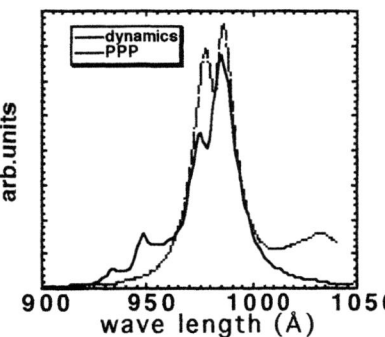

Figure 13 Comparison between Flyβ profiles computed with Pim Pam Poum (the dotted line) and the dynamics version of QMSPECTRA (the solid line). For the comparison the PPS shift (not taken into account in Pim Pam Poum) has been eliminated in our computation.

The dynamics correction is sensitive to the temperature. Figure 14 gives the evolution of Flyβ with the temperature. The satellites are enhanced when the temperature is increasing.

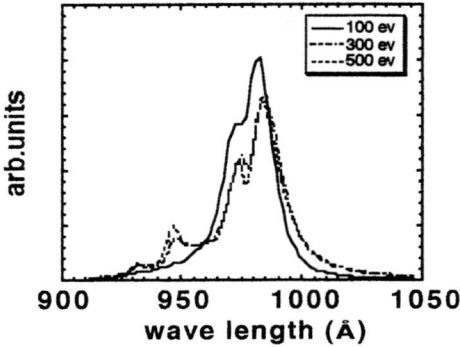

Figure 14 Flyβ profiles computed with the dynamics version of QMSPECTRA for different temperatures. Satellite are not exhibited at 100eV.

THE TWO CENTRE MODEL FOR THE ANALYSIS OF EXPERIMENTAL RESULTS

In this section some experimental results obtained for high density conditions are discussed and it is shown that the NN approximation is relevant to reproduce asymmetries and satellite-like features.

Ar He α in implosion experiments

The time-resolved (60ps) spectra analysed here are merging from implosions of Argon - filled plastic shell targets (experiments made at Rochester (9)). It is shown that ArHeα exhibits a blue satellite feature at the peak compression. A theoretical ArHeα profile computed by a standard model using an atomic basis (Woltz et al(10)) is not able to explain the experimental results. On the contrary, the Two Centre model, in its quasistatic version, has been shown to be efficient to reproduce the asymmetry and the satellite feature. This blue satellite is due to the splitting of the resonant component ($^1S - {}^1P, 1s^2 - 1s2p$) increasing rapidly as the NN distance decreases (Fig.15).

For the comparison, electronic and opacity broadenings have been included. As the blue satellite is very sensitive to the density (here $6.10^{24} cm^{-3}$), it potentially provides a density diagnostic. The possible overlapping of the excited states orbitals taken into account in the Two Centre model explaining the anomalous splitting of components may indicate the initiation of a plasma-solid phase transition.

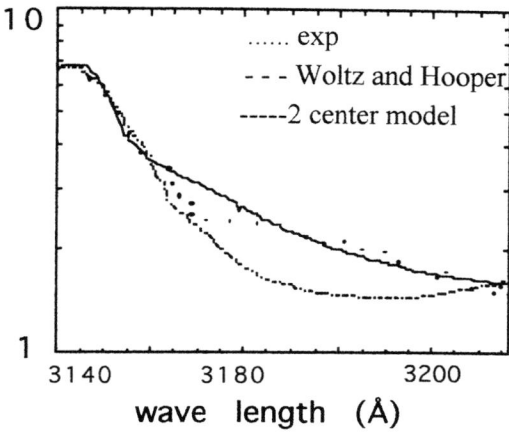

Figure 15 Ar $1s^2$ - $1s2p$ high energy wing at the peak compression. The Two Centre model (QMSPECTRA) is in agreement with the experimental data. There is a discrepancy with the standard model.

Flyβ in colliding foil experiments

This study deals with experiments made at LULI using the colliding foil technique. Two intense lasers (4.10^{14}Wcm^{-2} each beam) irradiate two thin Teflon foils about 100µm separated. The technique has been expertised for the generation of a dense plasma (1). Figure 16 gives an X ray 2D imaging of the collision allowing the knowledge of the plasma geometry.

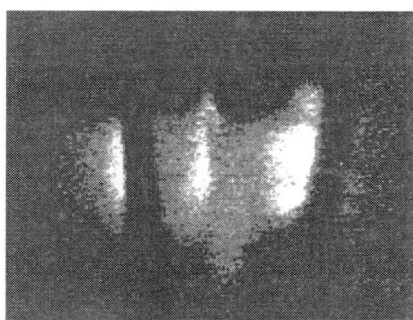

Figure 16 X-ray 2D imaging of two 5µm CF_2 foils initially separated by 100µm. The laser beam intensities are 4.10^{14}Wcm^{-2}.

A TlAP crystal coupled to a X Streak camera recorded the Flyβ spectrum merging from the most emissive plasma at the peak density. The time resolved (50ps) spectrum exhibits an asymmetrical profile with two satellite - like features.

We present in figure 17 a synthetic theoretical profile which fits the experimental data. This profile uses the Two Centre model (QMSPECTRA) and takes account for the electronic broadening and the spatial integration (25μm) over strong density and emissivity gradients. These gradients are determined previously from a 1D1/2 hydro code (FILM).

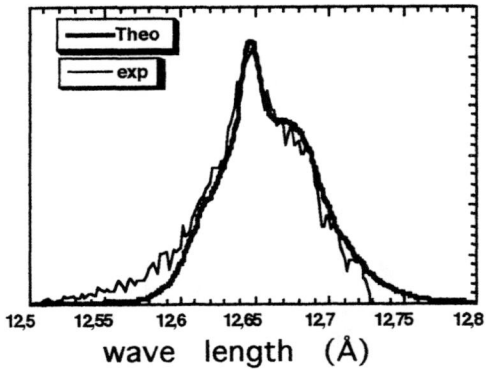

Figure 17 Flyβ time resolved spectrum from a colliding foil experiment. A forbidden transition and a PPS induced asymmetry are clearly exhibited.

Two dense plasma effects are reproduced :
(i) a quadrupolar transition 3d-1s excited on the blue side
(ii) a red satellite feature attributed to PPS (Plasma Polarization Shift) induced effects on several transitions combined to density integrations.

CONCLUSION

An alternative treatment of line broadening is provided with the Two Centre model when the NN approximation is relevant. This model gives rise to tractable dense plasma effects : PPS, asymmetries and satellite features exhibitions which can be used as diagnostics.
It is shown that ion dynamics corrections may be of importance at very high density for satellites due to extrema in the energy transitions. It is emphasized that a comparison between experimental and computed profiles must be taken carefully if space/time integrations are involved.
It is suggested that the model should be extended in the future to take account for nonadiabatic effects occuring for level crossings.

REFERENCES

1. Leboucher-Dalimier, E. et al., J.Quant.Spectrosc.Radiat.Transfer **58**, 721 (1997)
2. Gauthier, P., et al., Phys. Rev. E **58**, n°1 (1998)
3. Gauthier, P., et al., J.Quant.Spectrosc.Radiat.Transfer **58**, 597 (1997)
4. Nguyen, H., et al., Phys. Rev. A **33**, 1279 (1986)
5. Rose, S. J, et al., J. Phys. B **31**, L127 (1998)
6. H. R. Griem, *Spectro line broadening by plasma*, Academic Press, NY (1974)
7. Devdariani, A., et al., *LULI Report*, NTIS PB97- 170963 (1997)
8. Sauvan, P., et al., *LULI Report*, NTIS PB97- 170963 (1997)
9. Hooper, C. F., et al., Phys. Rev. Lett. **63**, 267 (1989)
10. Woltz, I. A., et al., Phys. Rev. A **38**, 4766 (1989)

INTERACTION OF INTENSE LASER PULSES WITH NEUTRAL GASES AND PREFORMED PLASMAS.

A.J. Mackinnon, M. Borghesi, A. Iwase, M.W.Jones, and O. Willi.
The Blackett Laboratory, Imperial College of Science, Technology and Medicine, London SW7 2BZ, U.K.

Abstract. The interaction of a high intensity laser pulse with a neutral gas or preformed plasma has been studied over a wide range of target and laser conditions. It was found that the propagation of 2ps laser pulses ($\lambda = 1.054\mu m$, P = 5-10TW, I ~ 5×10^{14} - 1×10^{18} Wcm^{-2}) in neutral gases with atomic densities greater than 0.001 of critical was strongly influenced by ionisation induced refraction. Preformed density channels were effective in overcoming refraction but the channel length was found to be limited by ionization induced defocusing of the prepulse.

I. Introduction

The propagation of intense laser pulses through gases and preformed plasmas is an important area of study for many topical applications of laser produced plasmas. Including such schemes as: the laser particle accelerator[1], recombination x-ray laser[2], harmonic generation[3] and the fast ignitor scheme[4]. These applications generally require an ultra-short high power laser pulse to interact with a neutral gas or plasma at high density (with $n_e > 0.001 n_c$ where n_e and n_c are the electron and critical density respectively) and high irradiance (10^{14} - 10^{19}Wcm^{-2}) over distances ranging from a few mm to several cm. At extremely high laser irradiance (>10^{18}Wcm^{-2}) self-channelling due to relativistic and ponderomotive effects can greatly enhance the propagation length [5]. However when lower irradiance laser pulses (10^{14} – 10^{17} Wcm^{-2}) are focused into an initially neutral gas the interaction length can be severely limited by ionization induced defocusing to values much less than the diffraction limited focal depth[5]. In these cases it is possible to extend the interaction length by focusing a low energy prepulse into the gas to form a preformed density channel [6-9]. To date most experiments using this scheme have been carried out in the low density regime where the length of the guiding channel is limited by the depth of focus of the preforming laser. In fact guiding over 20x Rayleigh lengths ($L_R = \pi\omega_0^2/\lambda$, where ω_0 is the laser spot size of a Gaussian beam of wavelength λ) has been observed in these circumstances when an axicon was used to maximise the length of the plasma channel[7]. Applying this result to higher plasma densities is not straightforward for short (1-10ps) prepulses because as the density increases the length of the channel will be limited by ionization defocusing of the preforming pulse [10-11] not diffraction. There is presently little or no information on the effect of ionization defocusing on the formation of preformed density channels using short prepulses. In addition there are very few direct measurements of how the second pulse interacts with and modifies the channel. This paper will present the results of a series of experimental measurements of the interaction of intense laser pulses with gases and preformed density channels.

II. Experimental arrangement

The experiments were performed at the Rutherford Appleton Laboratory using the VULCAN Nd:glass laser[12]. A $1\mu m$, 1-3ps, laser pulse was focused onto the edge of a neon gas jet with atomic density $>1\times10^{19}$cm^{-3} at a vacuum intensity of $0.5-1\times10^{18}$Wcm^{-2}. An F/5 off axis parabola focused a 1-3ps pulse onto the gas vacuum boundary with a typical focal spot of 30μm full width at half maximum. The vacuum intensity was $1-3\times10^{18}$Wcm^{-2} with a 200ps pedestal at an intensity of 10^{13}Wcm^{-2}. The target (which has been described in detail elsewhere [11,13]) consisted of a solenoid pulsed gas jet with a 1mm diameter cylindrical nozzle. In this letter we describe the results of interactions with neon or helium gas with a peak neutral gas density over the range 3×10^{19}-1.0×10^{20}cm^{-3}. The plasma channels were formed by focusing a co-linear prepulse (with temporal and spatial characteristics identical to those of the main pulse) into the gas jet. During the experiment the energy of the pre-pulse was varied from 0.1 to 1J (1% or 10% of the main beam energy respectively) and the relative delay between the two pulses was varied from 500ps to 2.3ns. A schematic of the experimental arrangement is shown in Fig.1.

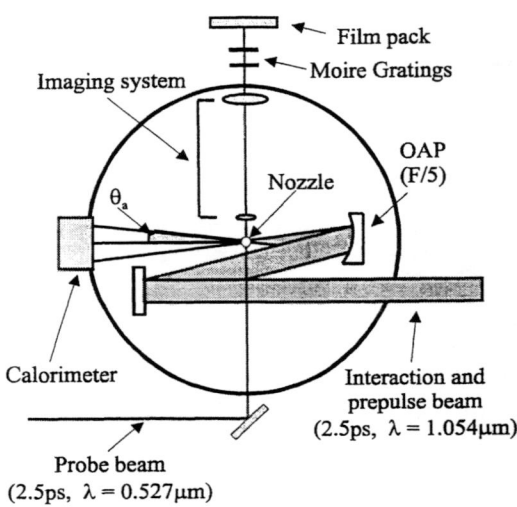

Figure 1: Experimental arrangement. The main pulse (E = 20J, P > 5TW) and prepulse (E=0.1 or 1J) are colinear and are both focused by the same off axis parabola (OAP) into the gas flow out of the nozzle. The calorimeter measures the laser energy transmitted within θ_a (the cone angle of focusing optics, $\theta_a = 5.5°$).

The plasma was diagnosed with a temporally independent probe pulse, split off from the main uncompressed heating beam. The probe pulse was compressed to 3 ps with a pair of gratings and frequency doubled to 0.527μm in a KDP crystal. The plasma was imaged onto the film plane by a collimated telescope, with a magnification of 50x. A Moiré deflectometer (comprising of a pair of 20 lines per mm Ronchi gratings placed 10cm apart, near the image plane of the telescope) provided time resolved measurements of the electron density profiles [14], at discrete intervals up to 2.5ns before or after the interaction pulse. The spatial resolution along the fringe direction was limited by diffractive effects between the gratings to 25μm [15]. Finally the laser energy transmitted through the target within a half angle of 5.5° was measured with an absorbing glass calorimeter.

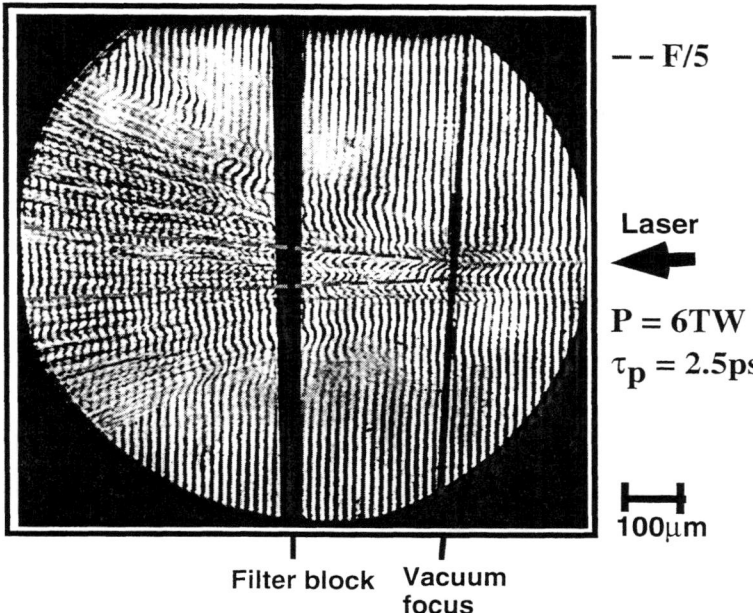

Figure 2: Moiré deflectogram taken 6ps after a 6TW laser pulse was focused into neon gas at $n_a = 3 \times 10^{19} \text{cm}^{-3}$. The laser is incident from the right and the gas flow is the vertical direction. The dashed lines mark the cone angle of the focusing optics, while the solid lines mark the radial boundary of the plasma. It is clear from this figure that as the laser propagates through the gas it suffers from strong ionisation defocusing, which leads to plasma formation far outside the cone angle of the focusing optics.

III. IONIZATION DEFOCUSING

Fig 2 shows a Moiré deflectogram taken 6ps after the interaction of a 6TW laser pulse with neon gas at a neutral density of $3 \times 10^{19} \text{cm}^{-3}$. For these conditions the Rayleigh range, $L_R = \pi \omega_0^2 / \lambda$, $= 350 \mu m$ (where ω_0 is the laser spot size and $\lambda = 1.054 \mu m$), the defocusing length [10] was $L_d = \lambda(n_c/2n_e) = 15 \mu m$ (assuming, for simplicity, singly ionized neon). Under such circumstances ionization induced refraction would be expected to strongly influence the laser propagation. It can be seen from Fig.2 that this was indeed the case - ionization induced density gradients are produced as soon as the laser begins to interact with the gas (i.e. the shifting fringes at the far right of the field of view). These gradients cause the beam to defocus and break up, resulting in a plasma whose radius increases rapidly along the propagation direction[11]. In addition the energy transmitted into the 5.5° acceptance angle of the calorimeter through the plasma in these circumstances was only 2.5%. For these conditions the absorbed fraction was calculated to be around 5% or less, so the most probable cause of the extremely low transmission within the calorimeter acceptance angle was due to refraction. If for example the confined beam expanded at an angle θ_R, then only the fraction $(\theta_a/\theta_R)^2$ would be intercepted within the acceptance angle θ_a of the calorimeter. From Fig.2 the plasma expands at an angle $\theta_R \sim 30°$ which gives $T \sim 2\%$ for $\theta_a \sim 5.5°$. This expansion angle was also consistent with the order of magnitude estimate of the refraction angle caused by the experimentally measured density gradients. These observations showed conclusively that the interaction was indeed dominated by ionization defocusing as predicted by simple theory (a more detailed description of these and other measurements will be published elsewhere [19]).

IV. INTERACTIONS WITH PREFORMED DENSITY CHANNELS

The feasibility of using a preformed density channel to overcome defocusing in such conditions was investigated by focusing a prepulse into the gas ahead of the main pulse and studying the subsequent interaction with Moiré deflectometry. Information on the degree of confinement afforded by the channel was obtained by varying the following experimental parameters: (a) the relative delay between the two pulses, (b) the target gas material and density and (c) the energy content of the prepulse. A deflectogram, taken 6ps after the interaction of a 5TW pulse with a preformed channel is shown in Fig 3(a). This preformed channel was formed by focusing a 0.1J pre-pulse into a neon gas jet (with a neutral density of $3 \times 10^{19} \text{cm}^{-3}$), 2.3ns ahead of the main pulse. The walls of the density channel can clearly be seen in this figure as the two well defined regions of high-density plasma (shown by strong fringe shifts) running parallel to the laser propagation axis at radii of 100μm. The strong filamentary structures within the channel (to the right of the filter block), are the result of optical ionization by the interaction beam. It is important to note that there is no evidence of plasma formation anywhere outside the channel. Consequently at the channel output (800μm downstream of the vacuum focus) the pulse is confined within a radius of 100μm. This is a significant improvement over the neutral gas interaction, where there was plasma formation out to a radius >350μm. The laser intensity within the channel is thus approximately an order of magnitude greater than with the neutral gas interaction.

V. EFFECT OF VARYING THE PREPULSE - MAIN PULSE DELAY

The degree of confinement was found to depend strongly on the relative timing between the two pulses. This is clear from the deflectogram shown in Fig.3(b), taken 6ps after a 5TW pulse was focused into a channel with a prepulse–main delay of 500ps. It is apparent from this figure that along the length of the channel the walls are not as well developed as in Fig. 3(a). Although strong fringe shifts can be seen close to the propagation axis there is also significant plasma formation outside the channel. There is clearly less confinement at 500ps compared to 2.3ns. The reason for this difference is that the shock wave structure at the channel walls was not properly formed at 500ps, allowing significant leakage of the main pulse. Hydro-code simulations, which predict a stronger shock structure at 2.3ns compared to 500ps, support this interpretation.

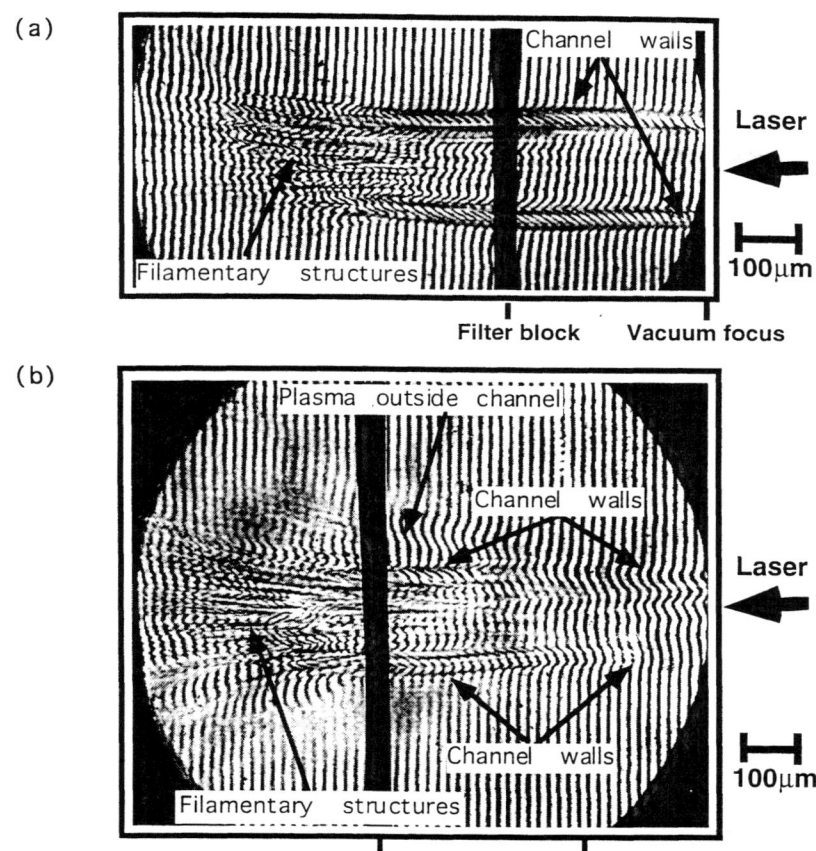

Figure 3: (a) Deflectogram taken 6ps after the interaction of a 5TW beam with a preformed density channel. The channel in this case was formed by focusing a 2.5ps duration prepulse with an energy of 0.1J into neon gas (at $n_a = 3 \times 10^{19} cm^{-3}$) 2.3ns before the incidence of the main (5TW) pulse. (b) Interaction with channel after prepulse-main delay of only 500ps. Direct comparison of (a) and (b) demonstrated that the main pulse was more effectively confined in the 2.3ns channel compared to the 500ps one. This is explained by the stronger shock structure of the channel at 2.3ns compared to 500ps.

Further evidence of increased levels of pulse confinement at longer delays was also provided by the transmitted energy measurements. The fraction of laser energy transmitted within 5.5° was found to increase from 2% to 10% as the delay increased from 500ps to 2.3ns. Although this increase in transmitted energy was relatively small it demonstrated that the properties of the channel wall, not just pre-ionization of the gas, were the most important factor leading to confinement of the pulse.

VI. THE STRUCTURE OF THE CHANNEL

Further insight into the channel forming process was obtained by examining the 2.3ns channel a few ps before the arrival of the interaction pulse, as shown in Fig.4. The deflectogram can be spilt into two distinct parts; (a) a uniform region clearly showing the strong shock wave structure that confined the beam and (b) a non uniform region, where the plasma is turbulent and the fringe shifts show a peak on axis, implying an axial density maximum. An electron density profile in the uniform portion of the guide is shown in Fig.5. The profile is characteristic of a strong shock wave [16] with an axial density minimum increasing to a sharp maximum at the channel walls. This represents a refractive index profile that is capable of confining an intense laser[6-9]. A simple but instructive estimate of the largest ray angle confined by the uniform part of the channel can be obtained from Snell's law applied to the density profile. Assuming a discontinuity in refractive index at the shock front, the critical angle, $\theta_c = \cos^{-1}(\mu_w/\mu_c)$, where μ_w and μ_c are the refractive indices at the channel wall and centre respectively. For an average density of $5 \times 10^{19} cm^{-3}$ at the channel wall and $1.0 \times 10^{19} cm^{-3}$ on axis, $\theta_c \sim 10\text{-}15°$, which is consistent with the experimental observations.

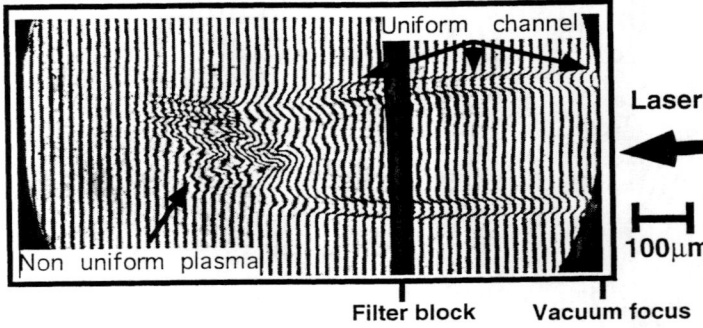

Figure 4: Moiré deflectogram taken 6ps before the arrival of the main pulse. This channel was formed by focusing a short prepulse (τ_p = 2.5ps) into neon gas (at $n_a = 3 \times 10^{19}$ cm^{-3}), 2.3ns before the main pulse.

VII. Important aspects of the main pulse - channel interaction

Another important and novel feature of this data was the significant increase in the fringe shifts at the walls when the main beam interacted with the 2.3ns channel (this can be seen by comparing Figs 3(a) and 4). A quantitative measurement of this effect was obtained by using helium at a neutral density of $3 \times 10^{19} cm^{-3}$ as the target gas. Fig.6 shows density profiles taken before and after focusing the interaction pulse into a channel formed 2.5ns after the 0.1J prepulse. It can be seen that the electron density at the channel walls increased by a factor of 2.5 when the interaction beam is focused into the channel. For a 4TW pulse confined within a channel of radius of 100µm an assumption of a top hat laser spatial profile gives an order of magnitude estimate of the mean irradiance within the channel as $\sim 1\times 10^{16}$Wcm^{-2}. This intensity is near threshold for doubly ionized ($Z^*=2$) helium and the density increase at the channel walls was therefore consistent with optical ionization increasing $Z^*=0.8$ to 2. This observation is important because it conclusively shows that the main pulse interacted with the channel walls.

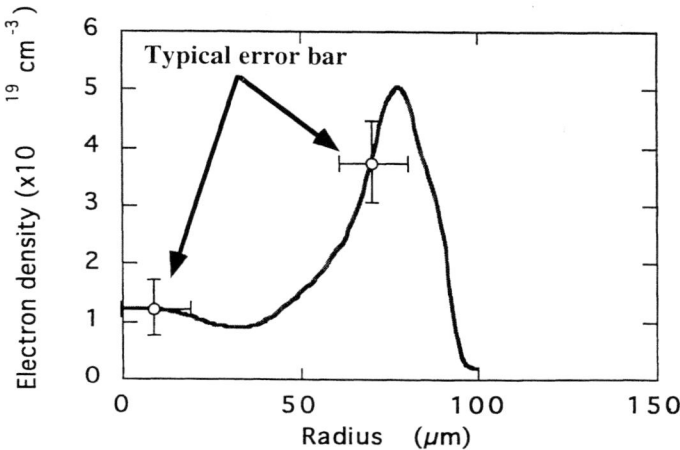

Figure 5: Measured density profile in the uniform part of the channel (shown in Fig 4) in neon gas. The profile was able to confine ray angles up to 10-15°.

Further independent confirmation of the level of pulse confinement was also obtained in a separate experiment using 0.35ps KrF laser pulses. In this experiment a laser pulse was focused at a vacuum irradiance of 1×10^{17}Wcm^{-2} into an identical gas nozzle with a peak neutral density around $0.01 n_c$. The output of the channel was imaged with a quartz F/2.5 lens into a CCD camera with a magnification of 10x and spatial resolution of 5 µm. This diagnostic also clearly showed that the interaction pulse was confined within the channel, leading to an irradiance five times greater than the neutral density case. These results, will be discussed in a forthcoming longer article[19].

VIII. INFLUENCE OF IONIZATION DEFOCUSING ON THE CHANNEL LENGTH

The non-uniform plasma at the end of the channel in Fig.4 was due to the finite depth of focus of the preforming laser pulse. Confirmation of this interpretation was obtained by the observations that an increase in the prepulse energy from 0.1 to 1J resulted in a channel that was uniform over the length of the nozzle (~1mm) and the transmitted energy increased from 10 to 25%. This represents an order of magnitude increase in transmission from the 2% value obtained in the neutral gas.

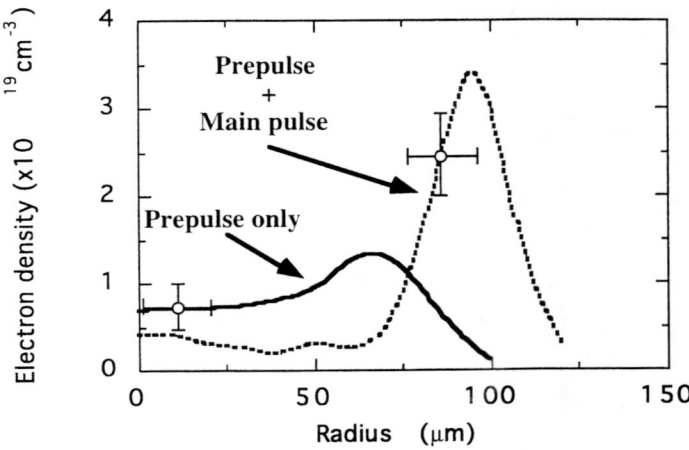

Figure 6: Measured increase in density at the channel walls in helium due to the arrival of the main pulse.

The most probable cause of this relatively low transmission within the calorimeter acceptance angle was due to residual refraction occurring within the channel. If for example the confined beam expanded at an angle θ_c, then only the fraction $(\theta_a/\theta_c)^2$ would be intercepted within the acceptance angle θ_a of the calorimeter. For these channels $\theta_c \sim 12°$ which gives $T \sim 21\%$ (for $\theta_a=5.5°$), consistent with the experimental measurement.

The mechanism limiting the channel length was examined using the blast wave model of Zel'dovich and Raizer[15]. In cylindrical symmetry this model gives the channel radius at time t, r(t) as a function of: the energy deposited per unit length E_d, the gas density ρ, and t, through the similarity relation $r(t) \propto (E_d/\rho)^{1/4} t^{1/2}$. It has been shown in a previous publication that this model described the channel expansion well for t > 100ps, but a 1D hydro-code was required to simulate the complete temporal evolution of the channel [18]. In this case the asymptotic expansion radius can be directly related to the original laser energy deposited within the centre of the channel. With this technique the energy deposited in the uniform part of the channel was estimated. Around the region of best focus the channel expanded from 15 to 100μm in 2.3ns and the simulations gave $T_e \sim 300$-400eV. This agreed well with the value of $T_e=380$eV, predicted by the absorption model in these circumstances (a 0.1J prepulse with an intensity of 1.5×10^{16}Wcm^{-2} and neon with a neutral density of 1×10^{19}cm^{-3})[17].

In contrast at the other end of the channel the radius was only 25μm after 2.3ns and a shock wave had not developed at all. The prepulse intensity at this position, due to the cone angle of the focusing optics, was around 3×10^{14}Wcm^{-2} resulting in ionization stage only marginally higher than threshold. For these circumstances the absorption model predicted $T_e \sim 1$-5eV and the plasma would not undergo any significant radial expansion in the 2.3ns before the main pulse arrived, in agreement with the experimental observation.

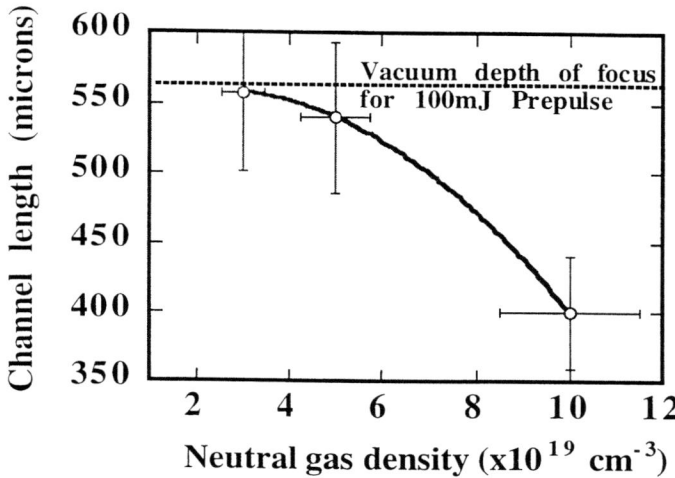

Figure 7: Measured reduction in channel length due to increase in neutral gas density of neutral gas. As the gas density increased ionization defocusing limited the useful length of the channel.

The cool, unexpanded plasma region at the end of the channel formed a barrier to the interaction pulse and limited the effective confinement length of the channel. This interpretation also explained why an increase in the neutral density led to a reduction in channel length, as this was consistent with enhanced levels of ionization defocusing of the pre-pulse at the higher neutral density. The variation of channel length as a function of gas jet density is plotted Fig.7. As the neutral gas density increased from 3×10^{19} to $1 \times 10^{20} cm^{-3}$ a 50% reduction in channel length from 550 to 350μm was observed. This showed that ionization defocusing presented a fundamental limitation on the use of short laser pulses to form plasma waveguides in refracting plasmas.

IX. CONCLUSIONS

This work has presented direct observations of the interaction of an intense picosecond pulse with a plasma channel preformed in a gas jet target. At low densities the first clear evidence of the interaction between a confined pulse and the channel wall was observed. This data also demonstrated definite pulse confinement by the channel leading to a laser irradiance that was significantly increased compared to an interaction with the neutral gas. At higher densities, however, the channel length was not limited by the depth of focus of the optical system but by ionization defocusing of the prepulse. A series of short pulses or a single longer duration prepulse (~100-500ps) could be used to overcome this limitation [9]. However more work is required in this area since a single long duration prepulse would help to overcome ionization defocusing but it may possibly lead to additional heating which would not be beneficial for some applications[2].

X. ACKNOWLEDGEMENTS

The authors would like to acknowledge useful discussions with Dr R.J.Taylor. The authors acknowledge the contributions made by the staff of the Central Laser Facility especially F. Walsh, C. Danson and D. Pepler. The authors are also very grateful to G.J. Pert for the use of his laser absorption code. This work was funded by EPSRC/MoD grants.

XI. REFERENCES

1 P.Sprangle and E.Esarey, Phys.Fluids B **4**, 2241 (1992).
2 D.C.Eder, P.Amendt and S.C.Wilks, Phys. Rev. A. **45**, 6761 (1992).
3 X. Li et al., Phys. Rev. A. **39** 5751 (1989), A. L'Huillier K. J.Schafer, K.C. Kulander, J.Phys. B.:At. Mol.Opt. Phys. **24** 3315 (1991).
4 M.Tabak et al, Phys.Plasmas **1**, 1621 (1994).
5. M. Borghesi et al., Phys.Rev.Lett. **78**, 879 (1997); S.Y. Chen et al., Phys.Rev.Lett. **80**, 2610 (1998); J. Fuchs et al., Phys.Rev.Lett. **80**, 1658 (1998); K Krushelnick et al., Phys.Rev.Lett. **78**, 4047 (1997).
6. S.A.Mani, J.E.Eninger and J.Wallace, Nuclear Fusion **15**, 371 (1975); C.S.Liu and
   V.K.Triptati, Phys.Plasmas **1**, 3100 (1994); R.Rankin, et al., Opt. Lett. **16**, 835 (1991).
7 C.G.Durfee III and H.M.Milchberg, Phys.Rev.Lett. **71**, 2409 (1993).
8 L.C.Johnson and T.K.Chu, Phys.Rev.Lett. **32**, 517 (1974); T.R.Clark and H.M.Milchberg, Phys.Rev.Lett. **78**, 2373 (1997).
9. V.Malka et al, Phys.Rev.Lett. **79**, 2979 (1997).
10 S.Rae, Opt.Comm. **97**, 25 (1993); W.Leemans et al, Phys.Rev.A. **46**, 1091 (1992); P.Monot et al, J.Opt Soc. Am. B 1579 (1992); C.D.Dekker, D.C.Eder and R.A.London, Phys.Plasmas **3**, 414. (1996).
11 A.J.MacKinnon et al, Phys.Rev.Lett. **76**, 1473 (1996).
12 C.Danson et al , Opt.Comm. **103**, 392 (1993).
13 Y.M.Li and R.Fedosojevs Meas.Sci.Tech. **5**, 1197 (1994).
14 O.Kafri, Opt.Lett. **5**, 555 (1980).
15 E.Keren and O.Kafri, J.Opt Soc. Am. A. **2**, 111 (1985).
16 Ya.B.Zel'dovich and Yu.P.Raizer, *Physics of Shock Waves and High Temperature Hydrodynamic Phenomena* (Academic Press New York, 1966) Vol. I. p.97.
17 G.J.Pert, Phys.Rev.E. **51**, 4778 (1995).
18 M.Dunne et al, Phys.Rev.Lett. **72**, 1024 (1994).
19 A.J.Mackinnon et al, to be submitted to Phys.Rev.E; A.J.Mackinnon, University of London thesis (unpublished) (1996).

IONIZATION
AND
RECOMBINATION I

Electron-Impact Ionization and Recombination of Atomic Ions Studied at Storage Rings

A. Müller*, T. Bartsch*, C. Brandau*, G. Gwinner[†], A. Hoffknecht*,
C. Kozhuharov[§], A. A. Saghiri[†], S. Schippers*, M. Schmitt[†], A. Wolf[†]

*Institut für Kernphysik, Strahlenzentrum, Universität Giessen, D-35392 Giessen, Germany

[†]Max-Planck-Institut für Kernphysik, D-69117 Heidelberg, Germany

[§]Gesellschaft für Schwerionenforschung (GSI), D-64291 Darmstadt, Germany

Abstract. The combination of accelerator technology and merged-beams techniques in storage ring experiments has opened up new experimental opportunities for electron-ion collision studies. In particular, charge changing electron-ion interactions at energies reaching from keV down to μeV with meV (and lower) electron beam temperatures have become accessible with ions of almost all charge states between H^- and U^{92+}. Examples of recombination and ionization measurements are presented and specific issues such as interference effects, high-resolution spectroscopy and rate enhancements in the presence of external fields are discussed.

INTRODUCTION

Charge-changing electron-ion collisions are important fundamental processes in all kinds of plasmas. They determine the charge-state balance of the ions and, hence, also the spectrum of electromagnetic radiation emitted by the plasma. Understanding and diagnosing the state of a plasma, whether of astrophysical origin or man made, relies on information about cross sections and rate coefficients for electron-ion interactions.

Colliding beams of electrons and ions [1] have been used since almost forty years now for studies of electron-ion collision processes. By far the most of the available data were obtained with small-scale equipment, i.e. with an ion source on an electrostatic potential of typically several kV, providing beams of slow ions, and with an (intersecting) electron beam of eV to keV energies in combination with the necessary equipment to characterize the beams and their overlap and to accomplish signal recovery.

By the combination of the well known merged-beams approach with ion accelerator technology a new era of electron-ion collision studies began little over a decade ago [2]. Nine heavy ion storage rings equipped with electron cooling devices have become available world wide [3]. The majority of these devices are concentrated in Europe. They provide unique new possibilities to study electron-ion collisions with high precision in a wide parameter range. By merging bright, cooled ion beams with very cold intense and well characterized electron beams of the cooling device, energies in the electron-ion center-of-mass frame from 0 to several keV are accessible. Energy spreads as low as about 10 meV have been experimentally verified at low energies. Intense ion beams (up to 1 mA of completely stripped uranium) can presently be made available, depending on the choice of accelerator facility. Four of the rings (TSR in Heidelberg, a schematic of which is shown in Figure 1, ESR in

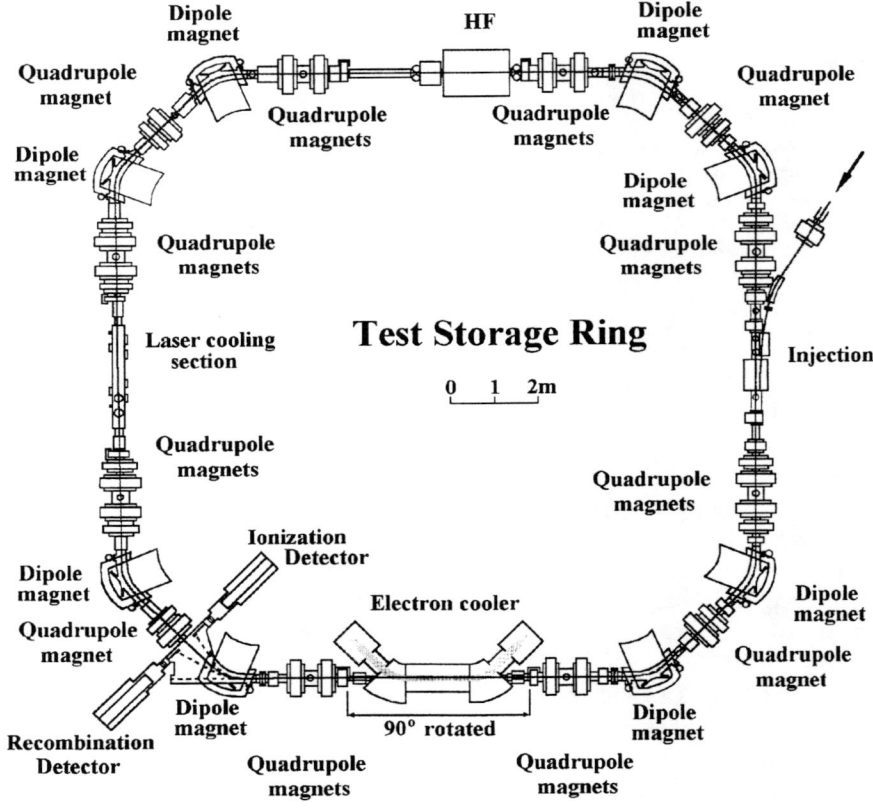

FIGURE 1. Schematic of the heavy-ion storage ring TSR in Heidelberg.

Darmstadt, ASTRID in Aarhus and CRYRING in Stockholm) have devoted most of their beamtime to atomic physics. They provide complementing opportunities for experiments. The authors have taken advantage of these opportunities at several different storage rings to study charge-changing electron-ion collisions covering ionization and recombination of many ions ranging from Li^+ to U^{89+}.

Objects of research have been manifold. A large fraction of the experiments have addressed issues of fundamental nature. Hence, measurements were carried out for ions with only few core electrons where theoretical techniques can be sensitively tested. These studies involve recombination of completely stripped ions with electrons, interactions with ions carrying 1, 2 or 3 core electrons and collisions of ions with only one active electron outside closed shells such as Fe^{15+}. However, the applied aspects of electron-ion collisions for plasma and accelerator physics also led to efforts to study systems with more complex electronic structures. Among these are fluorine-like Se^{25+} and Fe^{17+}. While the first of these ions is important in connection with the attempts to develop short-wave-length lasers, the latter is relevant to the understanding of the radiation spectrum from astrophysical plasmas. Recently, an experimental program has been launched at the TSR aiming at collision cross sections and rate coefficients of the astrophysically abundant Fe^{q+} ions with $q=15,16,...,23$. Measurements are available now also for many-electron ions such as Au^{50+}. There, the goal of experiments is the understanding of the observation of unexpected high recombination rates of such ions at low relative energies, i.e. under electron-cooling conditions. Recombination losses of ions during the preparation of intense ion beams in storage rings has become a serious issue in accelerator physics. The recombination studies carried out so far indicate effects of the strongly coupled cold dense plasmas envisaged during ion cooling in the electron beam of the cooling device.

The following major issues have been addressed in electron-ion collision studies: (I) accurate rates and cross sections for ionization and recombination; (II) spectroscopy of doubly excited states by the observation of resonances; (III) field effects on recombination rates; (IV) clarification of observed recombination rate enhancements at low energies; (V) interference phenomena; and (VI) indirect ionization mechanisms for ions of intermediate high charge states. Examples of recent measurements related to these topics are presented below.

BASIC CONSIDERATIONS

The processes

The electron collision processes that have been studied so far with atomic ions A^{q+} at storage rings are:
(photo)recombination

$$e + A^{q+} \to\ ...\ \to A^{(q-1)+} + \text{photon(s)}, \tag{1}$$

and net single ionization (including electron detachment from negative ions)

$$e + A^{q+} \to \ ... \ \to A^{(q+1)+} + 2e. \tag{2}$$

Both observation channels are characterized by the presence of direct and indirect mechanisms. The indirect reactions proceed via intermediate excited states indicated by the dots in Equations 1 and 2. Connections between ionization and recombination can be established by the observation of resonant intermediate excited states. Such states are populated by radiationless (dielectronic) capture (DC)

$$e + A^{q+} \to \left[A^{(q-1)+}\right]^{**}, \tag{3}$$

of the incident electron by the ion A^{q+} with simultaneous excitation of a core electron. In the subsequent relaxation process the ion can stabilize its new charge state by photoemission, but it can also emit one or two electrons. In the case of radiative stabilization the process termed dielectronic recombination (DR) is completed while for two-electron emission

$$\left[A^{(q-1)+}\right]^{**} \to A^{(q+1)+} + 2e \tag{4}$$

the final product is an ionized ion. In both cases the intermediate excited state gives rise to a resonance in the cross section with the resonance energy and the resonance width determined by the intermediate electron-ion complex. With their high energy resolution, storage ring experiments are particularly well suited for the detection and detailed measurement of such resonance phenomena.

Experimental features

For the measurement of rates and cross sections ions of the desired species are injected into the ring from an accelerator. High beam currents up to the mA range can usually be accumulated and the ions are cooled by the interaction with the electron beam of the cooling device. So far, this beam has to serve both for the cooling of ion beams and as an electron target for collision experiments. By the implementation of an electron target in addition to the cooler, future experiments will profit from easier beam handling, still better energy resolution and a wider range of accessible electron-ion center-of-mass energies.

When the ion beam is sufficiently cold the energy of the electron beam is detuned from "cooling" and accordingly, the relative energy between the ion and electron beams is increased from zero. In the collisions with electrons of the cooler, stored ions change their original charge state and, by that, drop out of the ring and can be detected without disturbing the circulating ion beam. Suitable particle detectors are mounted both at the outer and the inner side of the ring, typically behind the first beam bending magnet downbeam from the electron cooling device (see Figure 1) such as to facilitate collection of recombined and ionized product ions, respectively. Counting rates R_{exp} are recorded as a function of the electron energy which is rapidly scanned over a certain range usually in small steps with intermittend beam cooling after each voltage step.

Normalized collision rates α are determined from the relation

$$\alpha(E_{rel}) = \frac{R_{exp}\,\gamma^2\,v_i\,q\,e}{I_i\,\ell_{eff}\,n_e\,\varepsilon}. \tag{5}$$

Detection efficiencies ε of the particle detectors are usually close to 1, I_i is the measured electrical current of the circulating ion beam, v_i the ion velocity, $q\,e$ the ion charge, ℓ_{eff} the interaction length of the two beams, n_e the electron density, and $\gamma = \gamma(v_i)$ the relativistic Lorentz factor for the ion beam. The rates α basically depend on the velocity spread within the electron beam and the collision cross section σ:

$$\alpha(v_{rel}) = <\sigma\,v_{rel}> = \int \sigma(v)\,v\,f(v_{rel},\vec{v})\,d^3v\ . \tag{6}$$

For the particular case of merged electron and ion beams in storage cooler rings, two velocity coordinates are commonly used to describe the electron-velocity distribution in the rest frame of the ions: $v_\|$, the velocity component in electron-beam direction, and $v_\perp$, the velocity component perpendicular to the electron-beam direction. The energy (or velocity) spreads are therefore characterized by two corresponding temperatures $T_\|$ for the longitudinal and $T_\perp$ for the transverse direction. In the accelerated electron beam, these temperatures are quite different with $T_\| \ll T_\perp$, so that $f(\vec{v})$ is highly anisotropic. Its mathematical form is given by

$$f(v_{rel},\vec{v}) = \frac{m_e}{2\pi kT_\perp}\exp\left(-\frac{m_e v_\perp^2}{2kT_\perp}\right)\sqrt{\frac{m_e}{2\pi kT_\|}}\exp\left(-\frac{m_e(v_\| - v_{rel})^2}{2kT_\|}\right), \tag{7}$$

where m_e denotes the electron rest mass. The quantity v_{rel} in this formula is the average longitudinal center-of-mass velocity

$$v_{rel} = \frac{|\,v_{e,\|} - v_{i,\|}\,|}{1 + \frac{v_{i,\|}\,v_{e,\|}}{c^2}}, \tag{8}$$

where c is the vacuum speed of light, and where $v_{e,\|}$ and $v_{i,\|}$ are the longitudinal velocity components of the electron and ion beams in the laboratory frame, respectively. They are determined from

$$v_{e,\|} = c\sqrt{1 - [1 + \frac{E_e}{m_e c^2}]^{-2}} \tag{9}$$

and

$$v_{i,\|} = c\sqrt{1 - [1 + \frac{E_i}{m_i c^2}]^{-2}}\ . \tag{10}$$

The ion rest mass is represented by m_i. The energies E_e and E_i are determined by electron and ion acceleration voltages, respectively. The relative velocity v_{rel}, as defined by Eq. (8), can be different from the velocity v_{cm} in the electron-ion center-of-mass frame. This is especially true for low energies, where E_{cm} comes close to $kT_\perp$ and $kT_\parallel$. Since only E_{rel} is directly accessible to a measurement, experimental data are usually displayed as a function of the relative energy

$$E_{rel} = (\gamma_{rel} - 1) m_e c^2 \tag{11}$$

with

$$\gamma_{rel} = [1 - (v_{rel}/c)^2]^{-1/2}. \tag{12}$$

When $E_{rel} \gg kT_\perp, kT_\parallel$ the velocity v_{rel} approaches v_{cm} and hence Eq. (6) can be rewritten as

$$\alpha(v_{rel}) \approx v_{rel} \int \sigma(v) f(v_{rel}, \vec{v}) d^3v = v_{rel} \sigma_{app} \tag{13}$$

The apparent cross section σ_{app} results from the convolution of the real cross section σ with the experimental velocity (or energy) distribution. This quantity makes physical sense when the width of the distribution function is much smaller than v_{rel} (or E_{rel}, respectively). In this case apparent cross sections

$$\sigma_{app} = \alpha(v_{rel}) / v_{rel} \tag{14}$$

are used instead of rate coefficients α.

Particularly in recombination experiments, the measured rate coefficients and cross sections are modified by field ionization of Rydberg states in the beam bending magnets. For applications in plasma physics, this truncation of the Rydberg state distribution has to be considered. Statistical uncertainties of the measured data can in most cases be reduced to insignificance, total systematic uncertainties are typically of the order of ±20%.

An important issue in most electron-ion collision experiments is the energy resolution. Since the first application of the merged-beams techniques to collision studies involving multiply charged ions enormous progress has been achieved in reducing electron beam temperatures. While in the pioneering experiments the electron beam was characterized by $kT_\perp \approx 5\text{eV}$ and $kT_\parallel \approx 60\text{meV}$ (see e.g. [4]), the first storage ring cooler used in single pass merged-beams experiments provided $kT_\perp \approx 0.15\text{eV}$ and $kT_\parallel \approx 2\text{meV}$ [5]. The race for lower temperatures is going on and the present record is held with the cooler of CRYRING where evidence has been found for $kT_\perp \approx 1\text{meV}$ and $kT_\parallel \approx 0.06\text{meV}$ [6]. The cooler of the TSR is only slightly behind and a new electron target development in Heidelberg is promising to push the frontier to yet lower temperatures. The consequence of these efforts becomes obvious in Figure 2 where the results of three measurements of DR of C^{3+} ions are displayed. Identical scales have been applied to each display box. The differences in energy resolution need no further comment, nor the accessibility of very low relative energies. Truncation of the Rydberg state population is different in the three experiments and so is the influence of external electromagnetic fields. Hence, the areas under the different curves cannot be directly compared.

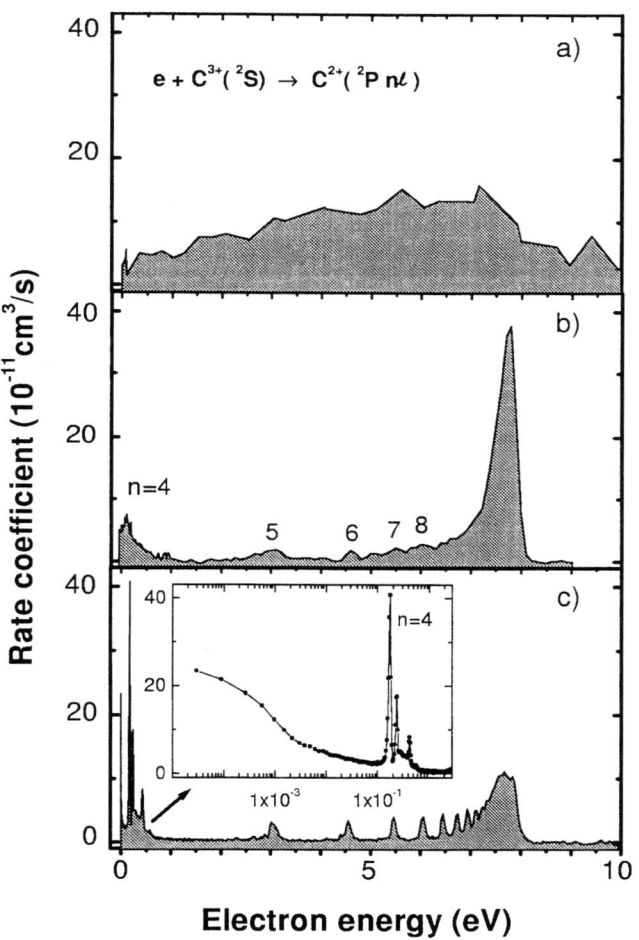

FIGURE 2. Comparison of three measurements of DR of C^{3+} ions. The resonance features are due to transitions $e + C^{3+}(1s^2 2s) \rightarrow C^{2+}(1s^2 2pn\ell) \rightarrow C^{2+}(1s^2 2s^2)$ + photons. In the experiments different external electromagnetic fields and different levels of field ionization were involved. From top to bottom the data are from references [4,5,7] and show the development of energy resolution in such experiments over a time span of little over ten years. The TSR data under c) are preliminary.

RECENT RESULTS

Electron-impact net single ionization

Removal of an electron from an ion by electron impact would ideally be carried out at ion energies where collisions with the residual gas do not play a role. While this demand leads to low ion energies, the efficient production of beams of highly charged ions by stripping in a foil requires high ion energies. Also, the merged-beams approach, needed e.g. to access the threshold for electron detachment from negative ions at only 0.75eV, calls for high ion energy. In this respect colliding beams experiments with highly charged ions have to compromise and hence, the experimental conditions at storage rings for ionization are very much less favorable than for recombination studies. Huge backgrounds from electron stripping in the residual gas limit the gain of statistics in a given beamtime. Electron-impact ionization may provide only a few percent of the total counting rate of ionized and stripped ions. Nevertheless, it has been possible to measure fine details in the ionization cross sections of Li-like ions Si^{11+} and Cl^{14+} [11], as well as of Na-like

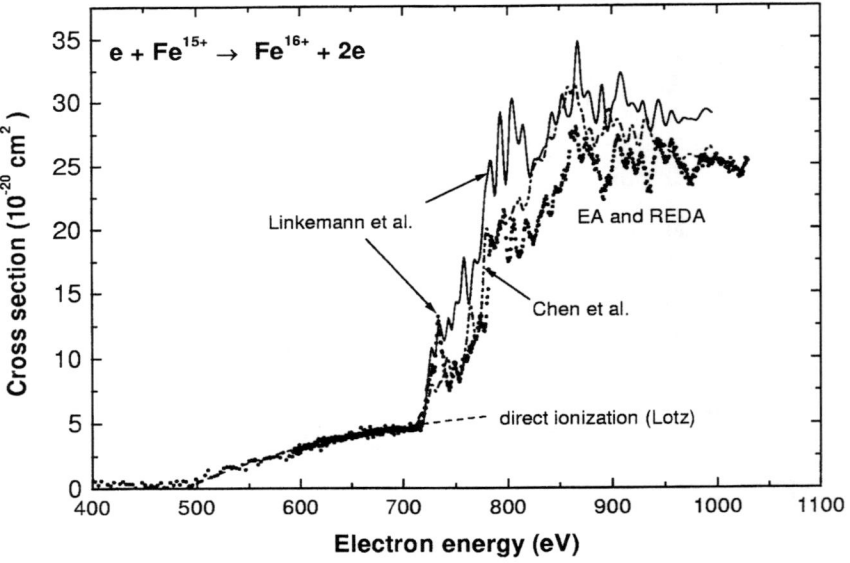

FIGURE 3. Cross sections for electron-impact ionization of Fe^{15+} ions. The acronyms EA and REDA stand for "excitation-autoionization" and "resonant-excitation-double-autoionization". The experimental data [8] are indicated by solid dots. The solid and dotted lines are the results of extensive theoretical calculations [8,9]. Direct ionization was estimated from the Lotz formula [10].

ions Cl^{6+}, Fe^{15+}, and Se^{23+} [12]. Although the lower-charge Na-like ions are in the feasibility range of modern ion sources, delivering sufficiently intense beams at the favorable low energies, the storage rings offer a decisive advantage. In the hot plasmas of ion sources producing highly charged ions, usually large amounts of ions in excited states are generated. Many of these states live sufficiently long to reach the electron-ion interaction region and thus cause severe normalization problems. An additional complication had to be faced in previous ionization experiments with Na-like ions: the parent beams contain metastable autoionizing states which produce huge background signals when decaying in the interaction path. In a storage ring experiment, however, storage of the ions for only a few seconds make the metastable states completely go away and then the measurement can start without that background.

An example for the results obtained at the TSR in Heidelberg is shown in Figure 3. The experimental cross section for electron-impact net single ionization of Fe^{15+} ions is displayed together with two theoretical calculations. Above 750eV the cross section is dominated by indirect processes (termed EA and REDA, see caption of Figure 3). Particularly remarkable is the wealth of resonance features which are due to DC and subsequent emission of two electrons (see Equations 3 and 4). While theory reproduces the overall size and structure of the measured cross section, the fine details observed in the experiments are not matched by the calculations.

The experiments clearly show an increase of the importance of indirect relative to direct ionization mechanisms. While the contributions of EA and REDA are about the same as for the direct process in Cl^{6+}, the ratio of relative strengths moves towards 10 as one goes along the isoelectronic sequence to Se^{23+}.

Electron-ion recombination

Applications in astrophysics

Many of the plasma collision-data needs can be met by the application of heavy ion storage rings. Particularly recombination measurements with ions in intermediately high charge states are relatively straightforward. Most ions present in the solar corona or in a fusion plasma can easily be studied for example at the TSR.

The expected high quality of X-ray spectra from astrophysical objects to be obtained in future satellite experiments has led to a research program at the TSR in which recombination and ionization of the cosmically abundant ions Fe^{q+} with $q=15,16,...,23$ is studied. Recently, the measurement of the recombination spectrum of fluorine-like Fe^{17+} ions provided a surprise for the astrophysics community [13]. Different from all previous expectations the experiment showed the importance of M1 excitations of the $^2P_{3/2}$ core of Fe^{17+} ions in DR. Hence, in the cold plasma of X-ray driven interstellar nebulae, the recombination rate can be much higher (by two orders of magnitude) than previous calculations. The experimental results from the TSR are shown in Figure 4. Experimental data are also available at this

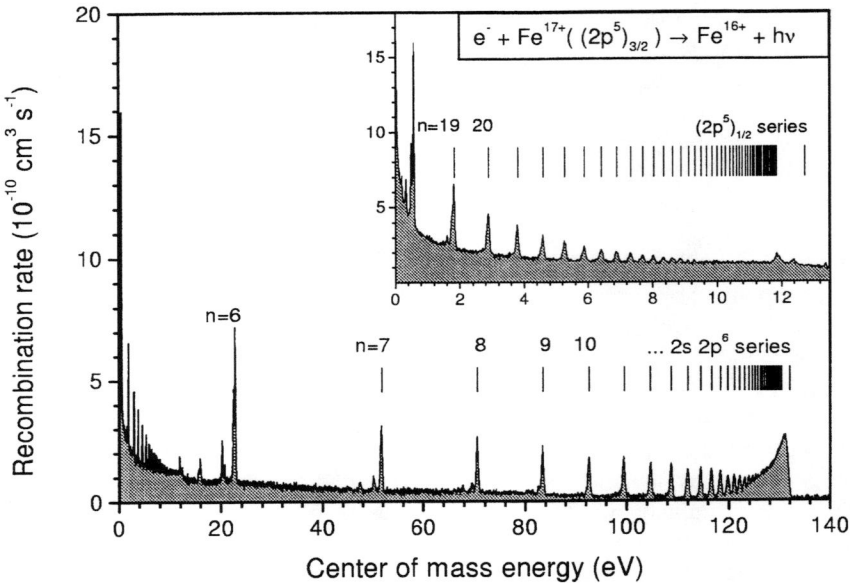

FIGURE 4. Recombination rates of Fe^{17+} ions measured [13] at the TSR.

time for Fe^{15+} [8,12], measurements with Fe^{18+} and Fe^{19+} ions have recently been successfully completed and are presently analyzed.

Interference effects

In particular cases, direct (radiative) and dielectronic recombination (RR and DR) can by no means be distinguished by experiments, because initial and final states of the interaction process including the photons may be identical. This situation can lead to interference of the amplitudes for both pathways of the recombination process. Unified photorecombination theory predicts distortions of the DR resonance profiles as a result of such interference [16] and also for overlapping resonances of equal symmetry [17]. An experimental observation of interference of RR and DR has been reported previously for differential cross section measurements carried out at an ion trap [18].

Recent predictions by Gorczyca et al. [14] suggested easy observability of interference effects in the recombination of argon-like Sc^{3+} ions with the lowest excitation channels associated with 3p → 3d and 3p → 4s core transitions. In experiments at the TSR an effort was made to measure recombination of these ions and to find interference effects [15]. No clear-cut evidence for the predicted effects could be found so far. The data for Sc^{3+} (see Figure 5) measured with the handicap

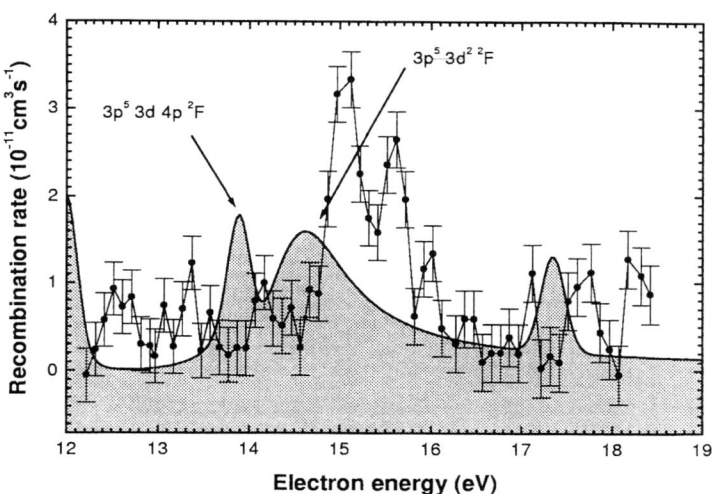

FIGURE 5. Recombination rates of Sc^{3+} ions. The experimental data [15] were obtained at the TSR. They are shown with their statistical uncertainties. The solid line is based on a calculation by Gorczyca et al. [14] convoluted with the experimental electron energy distribution which is characterized in this case by $kT_\parallel = 0.5$meV and $kT_\perp = 10$meV.

of huge backgrounds from residual-gas collisions clearly show that the theoretical predictions have to be revised. A similar conclusion has to be drawn from a second measurement with (isoelectronic) Ti^{4+} ions which yielded much better statistics as compared to the Sc^{3+} run.

Influences of external fields

While electric fields were experimentally demonstrated to influence the size of DR cross sections for singly charged ions [19] there were discrepancies between theory and experiment and inconsistencies in the understanding of these effects particularly also for multiply charged ions. First quantitative measurements covering a wide range of controlled electric and magnetic fields have been carried out for Si^{11+} ions [20] showing that further detailed studies of field effects on DR will be necessary. Remaining discrepancies with state-of-the-art theory sparked the prediction of a new mechanism of DR rate enhancement in the presence of an additional magnetic field perpendicular to the electric field [21,22]. Such magnetic fields have been involved in all but one previous DR measurements. Qualitatively, the assumption of additional m-mixing by magnetic fields on top of the ℓ-mixing caused by electric fields and the resulting enhancement of DR cross sections is in accordance with the experimental observations. A clear-cut experimental proof of magnetic field effects in crossed E- and B-fields, however, remains to be found.

Metastables and lifetimes

Storage rings have been extensively used for lifetime measurements of excited atomic, molecular and nuclear states [3]. The observation of DR starting from excited long-lived parent states has provided high precision data for transition probabilities in few-electron atomic ions [23]. The technique used in such experiments promises to be widely applicable. Experiments were carried out at the TSR to determine the lifetime of metastable 2^3S helium-like Li^+ [24]. The preliminary result of the analysis is (47.3 ± 2.9) s.

An important further result of the experiments with beams of Li^+ ions was the measurement of recombination rate coefficients for both the $1s^2$ ground state and the $1s2s\,^3S$ metastable state. Although the He-like structure of the parent ion would suggest easy theoretical access to the problem, theory had to include more details and involve a more complex treatment than had been anticipated. Nevertheless, the agreement between the TSR experiment and advanced calculations using R-Matrix techniques is not yet fully satisfying. At the low charge state of the ion involved here the electron-electron interaction plays a relatively important role in comparison with the situation in a highly charged heavy ion where the electron-nucleus interaction dominates.

Enhanced rates at low energies

New and unexpected effects on the recombination of ions were found at very low electron-ion collision energies [25–27]. Recombination rate enhancements far beyond the expectations for RR lead to enormous intensity losses during electron cooling of ion beams [28], with beam lifetimes as low as 2 seconds. Again, measurements at storage rings provide a suitable experimental approach to the understanding of this exciting and important phenomenon. Measurements on the sequence of Au^{q+} ions with $q=49,50,51$ were carried out at the TSR [29]. Measurements at a high-density electron target with Au^{25+} showed the record high enhancement factor of 365 of the measured recombination rate above theoretical expectations for RR [30]. As in the TSR experiments with the more highly charged gold ions this huge enhancement is probably due to DR resonances at very low energies. An increase of the electron density in the Au^{25+} experiment from $3.3 \cdot 10^8$ cm^{-3} to $3.7 \cdot 10^9$ cm^{-3} had no significant effect on the measured recombination rate [30]. Measurements of similar nature with variation of experimental parameters were also carried out with Cl^{17+}, U^{89+} and Au^{76+} ions. Recombination rate enhancements are observed in all three cases although no DR resonances are expected for these ions at low energies (and are impossible for the completely stripped Cl^{17+} ions). The results of that work are still being analyzed. Negligible effects of electron density variations in accordance with previous observations at CRYRING [31] seem to contradict the expectations based on molecular dynamics calculations [32] simulating the plasma conditions in the electron cooler.

An important new experimental observation is the influence of the magnetic guiding field of the electron beam in the interaction region. The analysis of the Au^{25+} measurements revealed a clear and strong increase of the recombination rate coefficient at zero energy when the magnetic field was increased. Recently a much better controlled measurement on magnetic field effects was performed with F^{6+} ions at the TSR in Heidelberg. By the observation of resonance line shapes of DR well below 1eV, the experiment carried an inherent thermometer for the determination of the electron beam temperature. The preliminary results of this experiment also indicate an enhancement of the recombination rate at zero energy when the magnetic field is increased.

Spectroscopy of high-Z ions

The development of electron beams with increasingly better energy resolution in connection with the bright cold ion beams available in the rings has paved the way towards a precision spectroscopy of singly and multiply excited states of ions ranging from singly charged low-Z to very highly charged high-Z ions, where Z is the atomic number. With the resolution obtained in previous experiments, QED effects on level energies could already be sensitively probed [33]. A new step towards the

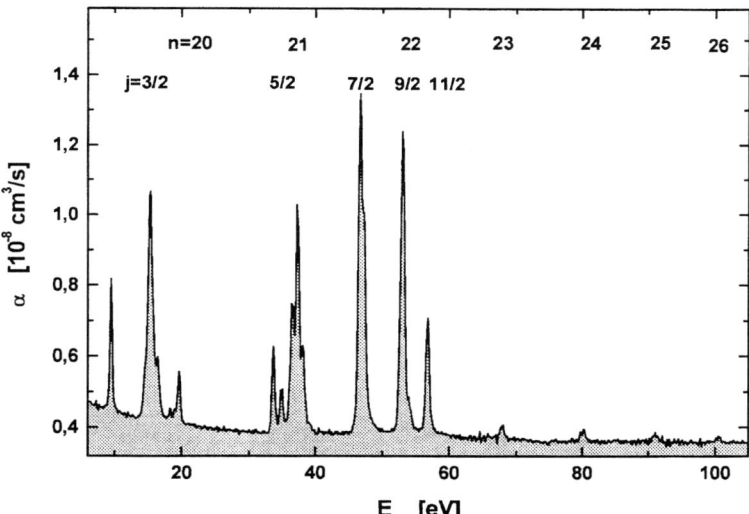

FIGURE 6. Recombination rates of Au^{76+} ions measured [34] at the ESR. Both the energy and rate scales are preliminary, background has not been subtracted yet. The resonances arise from two series of Rydberg states associated with excited $2p_{1/2}$ and $2p_{3/2}$ cores in DR processes $e + \mathrm{Au}^{76+}(1s^22s) \to \mathrm{Au}^{75+}(1s^22pn\ell_j) \to \mathrm{Au}^{75+}(1s^22s^2) +$ photons.

best possible precision of energy determinations of excited states in very highly charged ions and of Lamb shift measurements based on the observation of DR resonances appears to be possible now [34]. Few-electron systems studied recently in our ESR collaboration comprise lithium-like Au^{76+}, Pb^{79+} and U^{89+} ions. An example for the quality of the results obtained at the ESR is presented in Figure 6.

SUMMARY AND OUTLOOK

Collisions of ions with free electrons lead to complex interactions. Due to their fundamental character they provide an ideal testing ground for our understanding of atomic structures, transition probabilities and collision dynamics. In particular, DR can be employed for a collisional spectroscopy of multiply excited states. Extrapolation of Rydberg resonance energies to their series limit can even provide precise excitation energies for singly excited states and thus opens new pathways towards high precision tests of QED calculations of energy levels in very highly charged few-electron systems. The physics of electron-ion recombination also involves unique interesting phenomena, such as interference of different recombination channels or different, but overlapping resonances of equal symmetry. Effects of external fields on DR provide an intriguing field of research. Surprising results are obtained at the very low relative energies which can now be accessed by the merged beams technique in storage rings using the cold electron beams of the cooling devices. At least for energies up to about 1 keV the storage cooler rings provide the method of choice for studies of electron-ion recombination. These studies will be further extended in the near future. Electron-impact ionization studies at storage rings are more difficult. Only very selected ion species will be studied using rings. The extension of accessible electron-ion center-of-mass energies is required for that. A big step in the potential and the quality of measurements will be possible when additional, dedicated electron targets in the rings will become available.

ACKNOWLEDGEMENTS

The authors thank the staff at the different storage ring facilities for their invaluable help. Fruitful interaction with numerous colleagues in theoretical physics has been stimulating and illuminating for the present research program. Support by the Gesellschaft für Schwerionenforschung (GSI), Darmstadt, by the Max-Planck-Institut für Kernphysik, Heidelberg, by the German Ministry of Education, Science, Research and Technology (BMBF), and by the Human Capital and Mobility Programme of the European Community is gratefully acknowledged.

REFERENCES

1. Brouillard F. (ed.), *Atomic Processes in Electron-Ion and Ion-Ion Collisions*, Plenum, New York and London, 1986.
2. Graham W. G., Fritsch, W., Hahn, Y., and Tanis, J. H. (eds.), *Recombination of Atomic Ions*, NATO ASI Series B: Physics Vol. 296, Plenum, New York, 1992.
3. Müller, A. and Wolf, A., *Accelerator-Based Atomic Physics Techniques and Applications*, ed. by Shafroth, S. and Austin, J., American Institute of Physics, Woodbury, New York, 1997, ch. 5, pp 147-182.
4. Dittner. P. F., et al., Phys. Rev. A **35**, 3668 (1987).
5. Andersen, L. H., Bolko, J., Kvistgaard, P., Phys. Rev. A **41**, 1293 (1990).
6. Danared, H., Stockholm, private communication 1998
7. Gwinner, G., et al., to be published.
8. Linkemann, J., et al., Phys. Rev. Lett. **74**, 4173 (1995).
9. Chen, M. H., Reed, K. J., Moores, D. L., Phys. Rev. Lett. **64**, 1350 (1990).
10. Lotz, W., Z. Phys. **216**, 241 (1968).
11. Kenntner, J., et al., in preparation for publication in Phys. Rev. A.
12. Linkemann, J., et al., in preparation.
13. Savin, D. W., et al., ApJ. Letters **489**, L115 (1997).
14. Gorczyca, T. W., et al., Phys. Rev. A **56**, 4742 (1997).
15. Schippers, S., et al., Hyp. Int. (1998), in print.
16. Badnell, N. R. and Pindzola, M. S., Phys. Rev. A **45**, 2820 (1992).
17. Karasiov, V. V., et al., Phys. Lett. A **161**, 453 (1992).
18. Knapp, D. A., et al., Phys. Rev. Lett. **74**, 54 (1995).
19. Müller, A., et al., Phys. Rev. Lett. **56**, 127 (1986).
20. Bartsch, T., et al., Phys. Rev. Lett. **79**, 2233 (1997).
21. Robicheaux, F. and Pindzola, M. S., Phys. Rev. Lett. **79**, 2237 (1997).
22. Robicheaux, F., Pindzola, M. S., and Griffin, D. C., Phys. Rev. Lett. **80**, 1402 (1998).
23. Schmidt, H. T., et al., Phys. Rev. Lett. **72**, 1616 (1994).
24. Saghiri, A., et al., *Proceedings of the XXth International Conference on the Physics of Electronic and Atomic Collisions*, Vienna, Austria, July 1997, Abstracts of Contributed Papers, TU 143
25. Müller, A., et al., Phys. Scripta **T37**, 62 (1991).
26. Uwira, O., et al., Hyp. Int. **99**, 295 (1996).
27. Gao, H., et al., Phys. Rev. Lett. **75**, 4381 (1995).
28. Baird, S., et al., Phys. Lett. B **361**, 184 (1995).
29. Uwira, O., et al., Hyp. Int. **108**, 149 (1997).
30. Hoffknecht, A., et al., submitted to J. Phys. B
31. Gao, H., et al., Hyp. Int. **99**, 301 (1996).
32. Zwicknagel, G., Spreiter, Q., and Toepffer, C., Hyp. Int. **108**, 131 (1997).
33. Spies, W., et al., Nucl. Instrum. Meth. in Phys. Res. B **98**, 158 (1995).
34. Brandau, C., et al., Hyp. Int. (1998), in print

Dielectronic Recombination and Excitation-Autoionization in Highly Ionized Heavy Elements

P. Mandelbaum, E. Behar, R. Doron, M Cohen, A. Peleg
and J.L. Schwob

Racah Institute of Physics, The Hebrew University of Jerusalem
Jerusalem 91904, Israel

Abstract: New results of computations of dielectronic recombination (DR) and excitation-autoionization (EA) are presented. The detailed calculations for these processes are performed using the HULLAC relativistic atomic code. Calculations of the DR rate coefficients have been performed for Ar-like Tungsten (W^{+56}, ground $3p^6$). Extrapolation methods for including captures onto very high autoionizing configurations are exposed. The EA calculations have been performed for ions along the GeI isoelectronic sequence (ground $3d^{10}4s^24p^2$) . It is found that in this sequence the EA processes might enhance the direct ionization rates by a factor of up to 7.

INTRODUCTION

The importance of electron impact indirect processes, dielectronic recombination (DR) and excitation-autoionization (EA) in highly ionized heavy ions is now recognized. Taking into account these processes in the theoretical computations is necessary to reproduce the observed fractional abundances of the ionic species in hot plasma. In spite of the importance of these atomic processes, few systematic calculations had been performed until the development of new atomic codes, because of the difficulty of the calculations involved, especially for heavy ions with complex atomic structures. The apparition of new efficient codes, in particular HULLAC [1,2], in conjunction with the development of computer power have changed the picture, allowing now the systematic investigation of these processes in complex ions.

First we present the results of level-by-level calculations for the DR rate coefficients for ArI-like tungsten (W^{+56}, ground $3p^6$) . This ion has been chosen because of the importance of tungsten in planed fusion devices.

In the second part of this work, we show results of EA calculations for ions belonging to the GeI-like sequence (ground $3d^{10}4s^24p^2$). These calculations follow the previous systematic investigation of the EA process in the Cu-like ($3d^{10}4s$) [3], Zn-like ($3d^{10}4s^2$) [4] and Ga-like ($3d^{10}4s^24p$) [5] sequences.

CP443, *Atomic Processes in Plasmas:* Eleventh APS Topical Conference
edited by E. Oks and M.S. Pindzola
© 1998 The American Institute of Physics 1-56396-802-9/98/$15.00

DR FOR Ar-like W^{+56}

Level-by-level calculations

The DR of an Ar-like tungsten ion in its ground state $3p^6$ can occur through any K-like $[1s2s2p3s3p]^{17}nln'l'$ inner-shell excited autoionizing level, ending at any final level below the ionization limit. This mechanism can be represented by the main processes:

$$1s^2 2s^2 2p^6 3s^2 3p^6 + e^- \leftrightarrow [1s2s2p3s3p]^{17} nln'l' \qquad (1)$$

$$[1s2s2p3s3p]^{17} nln'l' \rightarrow 1s^2 2s^2 2p^6 3s^2 3p^6 nl \qquad (2a)$$

$$[1s2s2p3s3p]^{17} nln'l' \rightarrow [1s2s2p3s3p]^{17} nln''l'' \qquad (2b)$$

Process (1) is the electron capture, which is reversible by autoionization, whereas (2a) and (2b) are the *resonant* and *nonresonant* radiative stabilizing transitions, respectively. The total rate coefficient for the DR of the ion in its ground state is given by:

$$\alpha_k^D = \sum_d \alpha_{kd}^D = \sum_d \beta_{kd} B^D(d) \qquad (3)$$

where the summation is on every autoionizing excited state d of the K-like ion. The capture rate coefficient β_{kd} (in $cm^3 s^{-1}$) and the branching ratio for effective dielectronic recombination $B^D(d)$ for such a level d are given by:

$$\beta_{kd} = 1.656 \times 10^{-22} (kT_e)^{-3/2} \frac{g_d}{g_k} A_{dk}^a \exp(\frac{-E_{kd}}{kT_e}) \qquad (4)$$

and:

$$B^D(d) = \frac{\sum_i A_{di} + \sum_{d'>k} A_{dd'} B^D(d')}{\sum_{k'} A_{dk'}^a + \sum_i A_{di} + \sum_{d'} A_{dd'}} \qquad (5)$$

In these formulae, E_{dk} is the energy difference between the level d and the first ionization limit k, and kT_e is the electron temperature, both given in eV units. A^a_{dk} is the coefficient for autoionization from level d to level k expressed in s^{-1}. A_{xy} is the Einstein coefficient for spontaneous emission from level x to level y expressed also in s^{-1}. d' is a lower autoionizing level of the K-like ion, whereas i is a non-autoionizing one. $B^D(d)$ is defined by the recursive equation (5). Only for the lower autoionizing configuration complexes in each complex series were the radiative decays to autoionizing levels taken into account. For the higher complexes, these decays were neglected, leading to the approximation:

$$B^D(d) \cong \frac{\sum_i A_{di}}{\sum_{k'} A_{dk'}^a + \sum_i A_{di}} \qquad (6)$$

All the atomic quantities appearing in these formulae were computed with the HULLAC code. These level-by-level computations were performed for the autoionizing inner-shell excited configuration complexes: $3s^2 3p^5 4ln'l'$ ($4 \leq n' \leq 17$), $3s3p^6 4ln'l'$ ($4 \leq n' \leq 12$), $3s^2 3p^5 3dn'l'$ ($8 \leq n' \leq 18$), $3s3p^6 3dn'l'$ ($7 \leq n' \leq 18$), $2p^5 3s^2 3p^6 3dln'l'$ ($3 \leq n' \leq 8$), $2s2p^6 3s^2 3p^6 3dln'l'$ ($3 \leq n' \leq 5$) and $3p^5 5l5l'$. Details of these calculations will be given elsewhere [6]. It appears that the main DR channels are through the $3s^2 3p^5 4ln'l$ and $3s^2 3p^5 3dn'l$ complexes.

Extrapolation methods

A. $3s^2 3p^5 4ln'l'$ configurations

For this series we used an extrapolation method to compute the non negligible contribution of high n' complexes (n'>17) to the total DR rate coefficient. According to a method first proposed by Hahn [7], the total DR rate coefficient for a complex series is :

$$\sum_{n'=n_0}^{\infty} \alpha^D(n') = \sum_{n'=n_0}^{n_s} \alpha^D(n') + \sum_{n'=n_s+1}^{\infty} \left(\frac{n'}{n_s}\right)^{-3} \alpha^D(n_s) \qquad (7)$$

$\alpha^D(n')$ is the total rate coefficient of a configuration complex defined by n'. n_0 represents the lowest value of n' for which the corresponding complex contains levels above the first ionization limit. n_s is the lowest n' value for which $\alpha^D(n')$ starts to scale accurately enough as n'^{-3}. In figure 1, we give the DR coefficients through the complexes $3s^2 3p^5 4ln'l'$ as a function of the principal quantum number n' for $4 \leq n' \leq 17$ at three different electron temperatures. The dotted curves indicate the n'^{-3} grid as a reference for the scaling of the rate coefficients. For $n' \leq 14$ the rate coefficients decrease more steeply than n'^{-3}, whereas for $n' \geq 14$ the n'^{-3} behavior takes place as expected. There are two reasons for the fast decrease of the rate coefficient for intermediate n' values. The main reason is the progressive closing of the radiative stabilizing channels to $3s^2 3p^5 3dn'l'$, due to the rising of these levels above the ionization limit. For $n'=14$ all these stabilizing channels lie above the ionization limit, resulting in a steep drop in the DR coefficient through $3s^2 3p^5 4ln'l'$ for $n'=14$. The second reason is the opening of strong autoionization channels to the excited $3p^5 4l$ levels of the Ar-like ion. From this, one can infer that the usual extrapolation method as define by equation (7) is valid for $3s^2 3p^5 4ln'l'$ with $n' \geq 14$.

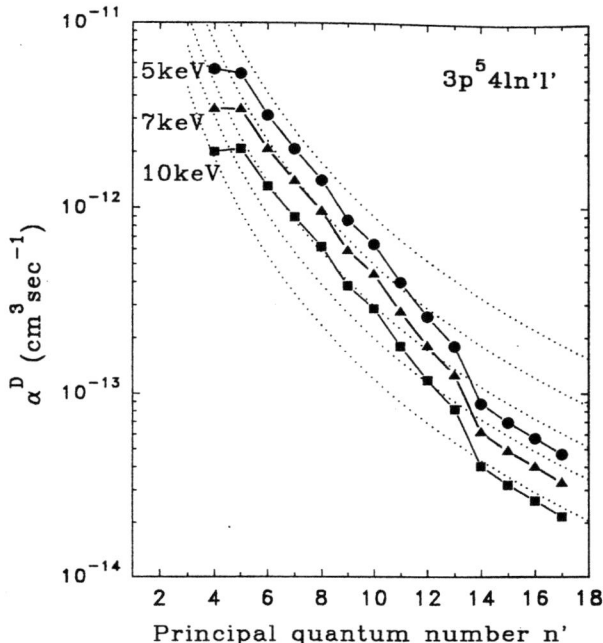

FIGURE 1. DR rate coefficients for $3s^2 3p^5 4ln'l'$ as a function of n'

B. $3s^2 3p^5 3dnl$ configurations

In Figure 2 the DR coefficients of the complexes $3s^2 3p^5 3dnl$ with $9 \leq n \leq 18$ at five different electron temperatures are plotted against n. One can see that in the range $9 \leq n \leq 4$ there are irregularities in the behavior of the rate coefficients, while for $n \geq 15$ the decrease is smooth but slower than n^{-3}. This behavior arises from the fact that the autoionization coefficient ΣA^a_{dk} is high compared to ΣA_{di} for the dominant levels of each complex. Now, the sum of the Einstein coefficients for spontaneous radiative decays that involve only the $3d$ excited electron, is independent of n. Thus, in this case the DR rate coefficient for the level d will be approximately independent of n until n is large enough so that $\Sigma A^a_{dk} \leq \Sigma A_{di}$. Only then, for very high n values, the DR rate coefficient of the level d will decrease steeply as n^{-3}. Thus, in this case one cannot apply the same extrapolation method as presented before. Instead, a more elaborate method based on level-by-level extrapolation is needed in order to avoid calculations for too high n. The method, also proposed by Hahn, relies on the fact that, for intermediate n, even if the DR rate coefficients do not yet scale as n^{-3}, the radiative and autoionization coefficients for individual levels already have a smooth and known dependence on n. It is necessary to distinguish here between two different

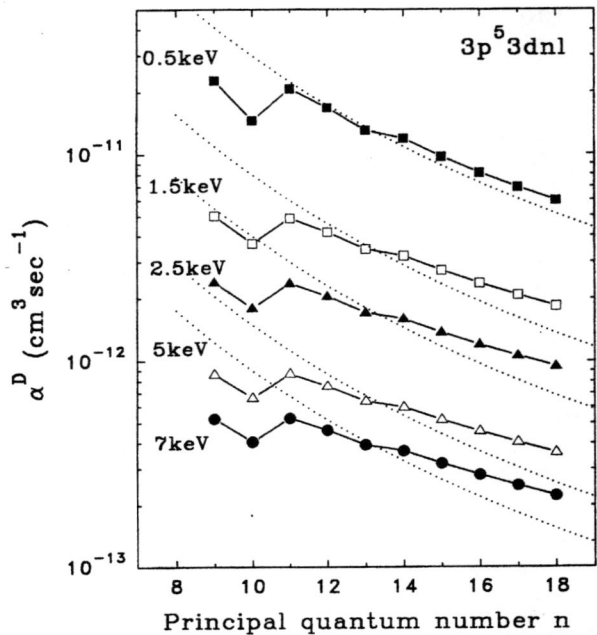

FIGURE 2. DR rate coefficients for $3s^2 3p^5 3dnl$ as a function of n

kinds of rate coefficients. On one hand the rate coefficients β^C_{kd}, A^a_{dk}, and A^{hi}_{di} for transitions that do involve the nl excited electron : they scale as n^{-3}. On the other hand, The radiative rate coefficients A^{lo}_{di} for transitions which involve the 3d excited electron are constant with n. Consequently, the total rate coefficient for DR through the whole complex series can be written as follows:

$$\sum_{n'=n_0}^{\infty} \alpha^D(n') = \sum_{n'=n_0}^{n_s} \alpha^D(n') = + \sum_{d \in 3p^5 3dn_s l'} \beta^C_{kd} \times \sum_{n'=n_s+1}^{\infty} \left(\frac{n'}{n_s}\right)^{-3} \frac{\sum_i A^{lo}_{di} + \left(\frac{n'}{n_s}\right)^{-3} \sum_{i'} A^{hi}_{di'}}{\left(\frac{n'}{n_s}\right)^{-3} \sum_{k'} A^a_{dk'} + \sum_i A^{lo}_{di} + \left(\frac{n'}{n_s}\right)^{-3} \sum_{i'} A^{hi}_{di'}} \quad (8)$$

Computed in this manner, it was found that the contribution of the $n' \geq 18$ complexes is more than 40% of the total DR rate coefficient of the whole $3s^2 3p^5 3dn'l'$ complexes at electron temperature higher than 500eV.

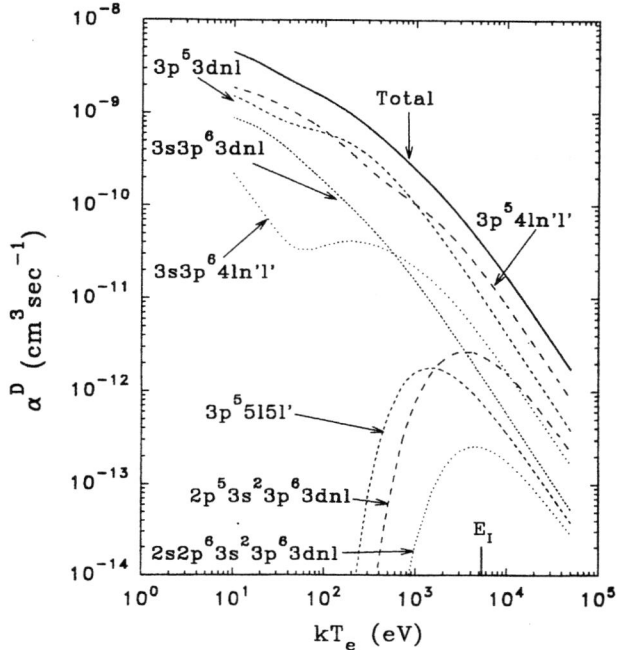

FIGURE 3. Total DR rate coefficients Ar-like tungsten

Total DR rate coefficient

The total DR rate coefficient, obtained by adding the DR contributions of all the significant complex series is given as a function of the electron temperature in Figure 3. As can be seen the $3s^23p^53dnl$ and $3s^23p^54ln'l'$ give the main contribution to the total DR rate coefficient.

EA IN THE Ge-like SEQUENCE

The present work is a continuation of a systematic investigation of EA performed for ions in isoelectronic sequences neighboring the NiI sequence. Indeed, the EA processes are particularly important for these sequences, due to the presence of the ten $3d$ inner-shell electrons [8]. Indeed, it as been shown that for the CuI (ground $3d^{10}4s$, ref. 3) and ZnI ($3d^{10}4s^2$, ref. 4) EA can enhance direct ionization rate coefficients by a factor of up to 17 and 15, respectively, for ions with $Z \approx 55$ at electron temperatures $kTe=0.3E_1$ (where E_1 is the first ionization energy) .This prompted us to extend this investigation to ions of the Ge-like sequence (ground $3d^{10}4s^24p^2$), which have a more complex structure .

Theoretical method

The rate coefficient S^{EA}_C for excitation-autoionization from a level g of the ground configuration $3d^{10}4s^24p^2$ of a Ge-like ion to a level k of the Ga-like ion through inner-shell excitation to an intermediate autoionizing level j of the Ge-like configuration C is given by:

$$S^{EA}_C(T_e) = \sum_{j \in C} B^a_j Q_{gj}(T_e) \quad (9)$$

where Q_{gj} is the electron-impact excitation rate coefficient from the ground level g to level j of the intermediate configuration. B^a_j is the effective branching ratio for autoionization from the level j defined by the recursive expression:

$$B^a_j = \left[\frac{\sum_k A^a_{ik} + \sum_{n<i} A_{in} B^a_n}{\sum_k A^a_{ik} + \sum_{n<i} A_{in}} \right] \quad (10)$$

This model allows one to take into account all the possible secondary autoionizations following cascading. The model developed includes the ground configuration $3d^{10}4s^24p^2$ of the Ge-like ion from which an electron could be excited to the $3d^94s^24p^2nl$ ($4 \le n \le 8$, $0 \le l \le 4$) inner-shell autoionizing configurations. Radiative decay is allowed between the levels of these configurations and also to $3d^{10}4s^24pnl$ This model allows one to take into account all the possible secondary autoionizations($4 \le n \le 8$, $0 \le l \le 4$) and to $3d^{10}4s4p^3$. The Ga-like configuration included are : $3d^{10}4s^2nl$ ($n=4,5$; $0 \le l \le 4$), $3d^{10}4s4p4l$ ($1 \le l \le 3$), $3d^{10}4s4d^2$ and $3d^{10}4p^3$. Excitation from the $3p$ shell is neglected. The model includes 5082 levels. Allowing configuration mixing provides the correction for the overestimation of the $3d$-nl excitation (5) as computed in the isolated configuration framework..

Results

Table 1 give the EA rate coefficients S_{EA} for ten ions along the isoelectronic sequence at three electron temperatures corresponding to $0.3E_1$, $0.5E_1$ and E_1. The temperature of most abundance of the ion at coronal equilibrium varies from about $0.3E_1$ to $1.5E_1$ as Z increases from 40 to 90. The direct ionization rate coefficients S_{direct} are also displayed for comparison. These rate coefficients have been calculated using the Lotz formula. In this Table $X[-Y]$ means $X \times 10^{-Y}$.

TABLE 1: EA and direct ionization rate coefficients (in $cm^3 s^{-1}$) for selected Ge-like ions

Ion	$kTe=0.3E_1$		$kTe=0.5E_1$		$kTe=E_1$	
	S_{EA}	S_{Dir}	S_{EA}	S_{dir}	S_{EA}	S_{Dir}
Kr^{+4}	0.47 [-10]	0.42 [-09]	0.36 [-09]	0.13 [-08]	0.16 [-08]	0.51 [-08]
Zr^{+8}	0.76 [-10]	0.12 [-09]	0.44 [-09]	0.39 [-09]	0.16 [-08]	0.15 [-08]
Mo^{+10}	0.81 [-10]	0.75 [-10]	0.43 [-09]	0.25 [-09]	0.14 [-08]	0.99 [-09]
Pd^{+14}	0.71 [-10]	0.37 [-10]	0.34 [-09]	0.13 [-09]	0.10 [-08]	0.50 [-09]
Xe^{+22}	0.54 [-10]	0.13 [-10]	0.21 [-09]	0.47 [-10]	0.54 [-09]	0.18 [-09]
Ce^{+26}	0.43[-10]	0.91 [-11]	0.16 [-09]	0.31 [-10]	0.39 [-09]	0.12 [-09]
Pr^{+27}	0.40 [-10]	0.83 [-11]	0.15 [-09]	0.28 [-10]	0.36 [-09]	0.11 [-09]
Eu^{+31}	0.21[-10]	0.59 [-11]	0.79 [-10]	0.20 [-10]	0.20 [-09]	0.80 [-10]
Ho^{+35}	0.13[-10]	0.44 [-11]	0.48 [-10]	0.15 [-10]	0.12 [-09]	0.59 [-10]
W^{+42}	0.26 [-11]	0.27[-11]	0.12 [-10]	0.94 [-11]	0.33 [-10]	0.37 [-10]

In Figure 4, the ionization enhancement ratio R of the total ionization rate coefficient ($S_{direct} + S_{EA}$) to the direct ionization S_{direct} is given for the three electron temperatures considered in Table 1 as a function of Z. The curves show a maximum of 6.5 at Z≈60. For low Z the 3d→4l excited configurations are too high to be efficiently excited at the electron temperatures involved. For intermediate Z, these configurations are becoming closer to the ionization limit contributing efficiently to the EA processes. For higher Z, these configurations become progressively below the ionization limit, and closing of the autoionization channels occurs. Also, the branching ratio for autoionization decreases because of the increase of the radiative decay rate coefficients. In conclusion, for most ions the EA rate coefficients are very significant. On the other hand, because of the peculiar behavior of the 3d→4l excitation channels, these rates can be computed only by detailed level-by-level calculations.

REFERENCES

1. Klapisch, M., Schwob, J.L., Fraenkel, B.S. and Oreg, J. *J. Opt. Soc. Am.* **67**, 148 (1977)
2. Oreg, J., Goldstein, W.H., Klapisch, M. and Bar-Shalom, A., *Phys. Rev.* **A44**,1750 (1991).
3. Mitnik, D., et al., *Phys. Rev.* **A53**, 3178 (1996)
4. Mitnik, D., et al., *Phys. Rev.* **A55,** 307 (1997)
5. Oreg, J, et al., *Phys. Rev.* **A44**, 1741 (1991)
6. Peleg, A., Behar,E., Mandelbaum,P. and Schwob J.L., Accepted for publication *Phys. Rev* **A** (1998)
7. Hahn,Y., *Adv. At. Mol. Phys.* **21**, 123 (1985)
8 Mittnik, D. et al., *Phys. Rev* **A50**,4911 (1994)

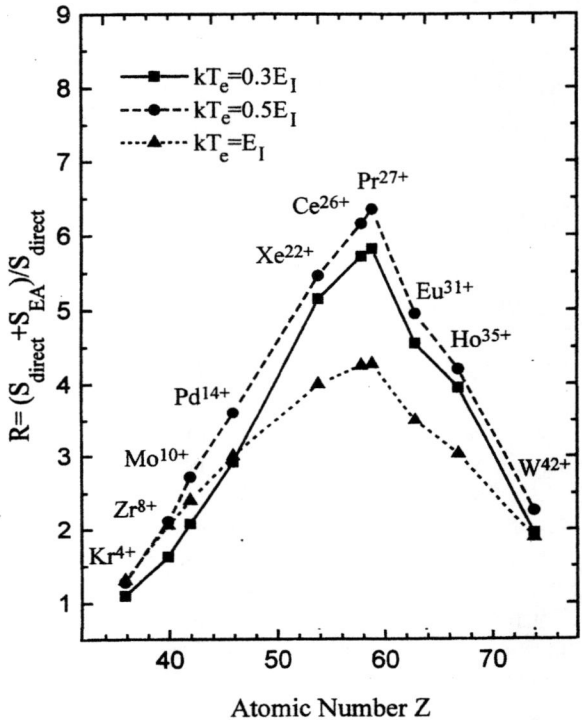

FIGURE 4. Ratio R of the total ionization rate coefficient ($S_{direct} + S_{EA}$) to the direct ionization S_{direct} as a function of Z.

Non-Perturbative vs. Perturbative Methods for Photorecombination

T.W. Gorczyca,[1] N.R. Badnell,[2] F. Robicheaux,[3] and M.S. Pindzola[3]

[1] Department of Physics, Western Michigan University, Kalamazoo, Michigan 49008-5151
[2] Department of Physics and Applied Physics, University of Strathclyde, Glasgow G4 0NG, UK
[3] Department of Physics, Auburn University, Auburn, Alabama 36849-5311

Abstract. The vast majority of photorecombination cross sections available in atomic data bases have been computed using second-order perturbation theory. Numerous benchmarking studies comparing experimental measurements to these theoretical results have established their general validity, for *total* cross sections. Partial recombination cross sections, on the other hand, are more likely to exhibit strong interference effects, necessitating a higher-order theoretical treatment. While it is possible to include third-order terms and higher within perturbative methods, an appealing alternative is to use non-perturbative close-coupling techniques. The use of a radiative optical potential within the close-coupling method was first introduced by Robicheaux *et al.*, and the general method was subsequently incorporated into the Belfast R-matrix codes to treat radiation-damped electron-impact excitation and photorecombination. This talk will focus on the implementation of the radiative optical potential R-matrix method for the computation of accurate photorecombination cross sections, and also on the comparison of these results to perturbative and experimental results. In particular, the general lack of interference effects in *total* recombination cross sections and the widespread importance of radiation damping on the computed cross sections will be demonstrated. In an interesting example, however, strong interference effects in the *total* cross section are found.

INTRODUCTION

Photrecombination consists of the capture of an incident electron by an atomic ion followed by photon emission. It is an important process in astrophysical and laboratory plasmas because of its role in influencing the ionization balance and the emitted spectra, so modeling of these plasmas relies heavily on accurate photorecombination data – these are only available from theoretical calculations for the most part. The direct process involving just the incident electron and the final photon is referred to as radiative recombination (RR), whereas when the ionic target electrons take part, it is referred to as dielectronic recombination (DR). But of course, these two paths are quantum mechanically indistinguishable, so that they contribute coherently to the computed photorecombination cross section. As an example of these alternate pathways, consider the photorecombination of He-like Fe^{24+}:

$$e^- + Fe^{24+}(1s^2) \xrightarrow{A_a} Fe^{23+}(1s2pn\ell)$$
$$\searrow_{RR} \quad \downarrow A_r$$
$$Fe^{23+}(1s^2 n l) + h\nu \ . \qquad (1)$$

The DR process here consists of dielectronic capture (the inverse of autoionization), with rate $A_a = \Gamma_a/\hbar$, followed by radiative stabilization, with rate $A_r = \Gamma_r/\hbar$.

PERTURBATIVE (DISTORTED-WAVE) METHOD

In the independent-processes isolated-resonance distorted-wave (IPIRDW) approximation, the total photorecombination cross section is written as an incoherent sum

$$\sigma_{PR} = \sigma_{RR} + \sigma_{DR} , \qquad (2)$$

where the cross term between the two transition matrix elements has been neglected. If this cross term is large, there arises *interference* effects, which we return to later. The DR cross section in Eq. (2), for a single, isolated resonance, is given by

$$\sigma_{DR} = \frac{2\pi^2}{k^2} \frac{g_f}{g_i} \Gamma_a \left[\frac{\Gamma_r/2\pi}{(E - E_R)^2 + \left(\frac{\Gamma_a + \Gamma_r}{2}\right)^2} \right] , \qquad (3)$$

where k^2 is the incident electron energy (in Rydbergs), g_i and g_f are the statistical weights of the initial and final states, and Γ_a and Γ_r are the autoionization and radiative widths, computed using distorted waves. The energy-averaged expression contributing to convoluted cross sections and rate coefficients is

$$\langle \sigma_{DR} \rangle = \frac{2\pi^2}{k^2} \frac{g_f}{g_i} \Gamma_a \left[\frac{\Gamma_r}{\Gamma_a + \Gamma_r} \right] . \qquad (4)$$

This basic perturbative approach is used by several groups [1–5] for the calculation of DR cross sections.

The expressions used in Eqs. (3) and (4) define what is meant by "radiation damping" [6,7]. That is, the *total* width $\Gamma_{tot} = \Gamma_a + \Gamma_r$ is used in the denominator, or equivalently, the radiative contribution to the resonance profile is included. Theoretical predictions made using resonance-continuum states that do not include the radiative width, such as standard close-coupling (R-matrix) methods, give an (unphysical) *undamped* cross section $\langle \sigma \rangle^{undamped}$. By considering Eq. (4) with and without the radiative width Γ_r in the denominator, this is related to the physically meaningful damped cross section $\langle \sigma \rangle^{damped}$ by the expression

$$\frac{\langle\sigma\rangle^{undamped}}{\langle\sigma\rangle^{damped}} = 1 + \frac{\Gamma_r}{\Gamma_a} . \tag{5}$$

Typically, at low ionic charge Z and low principle quantum number n (of the resonance), the autoionization rates of the prominent resonances dominate the radiative rates by orders of magnitude, so these two expressions are essentially equivalent. But radiation damping needs to be considered whenever the condition $\Gamma_a \gg \Gamma_r$ does not hold. The scaling of these widths with n and Z,

$$\Gamma_a \sim Z^0 n^{-3} \quad \text{and} \quad \Gamma_r \sim \begin{cases} Z n^0 & (\Delta n_c = 0) \\ Z^4 n^0 & (\Delta n_c > 0) \end{cases}, \tag{6}$$

where Δn_c is the change in principal quantum number in the radiative decay of the core, shows that for high enough Z and/or n, the condition $\Gamma_a < \Gamma_r$ is satisfied, so that radiation damping becomes important. Furthermore, for any extremely narrow resonances, e.g. LS-forbidden or high-ℓ ones where $\Gamma_a \ll \Gamma_r$, damping is also a prominent effect ($\frac{\langle\sigma\rangle^{undamped}}{\langle\sigma\rangle^{damped}} \gg 1$) regardless of Z and n.

We now examine the IPIRDW approximation. First, interference between the RR and DR pathways in Eq. (1) has been neglected. Second, it is possible for two neighboring resonances to interfere - this usually occurs only for nearly degenerate resonances. Of course, both effects can be included by considering higher-order perturbative terms, but this is achieved at a significant computational expense. A third shortcoming of the IPIRDW method is that the distorted waves used for the bound ($n\ell$) and continuum ($\epsilon\ell$) orbitals are not as flexible as, say, a close-coupling basis, so that autoionization widths and radiative widths may not be described as accurately. It is thus desirable to investigate the importance of effects that are not included in the IPIRDW approximation by comparing to non-perturbative methods.

CLOSE-COUPLING (R-MATRIX) METHOD

While it is possible to include interference effects in higher-order perturbative treatments, and distorted-wave basis sets can be suitably enlarged, an appealing alternative to perturbative approaches is to use instead a non-perturbative, or close-coupling, formulation. The R-matrix method is rightly considered the most accurate and efficient method in this respect, but like all other close-coupling methods, radiation damping effects are not usually included. The dipole matrix elements that enter photoionization expressions (and therefore photorecombination expressions, by detailed balance)

$$\sigma \sim |\langle \psi_i | \vec{D} | \psi_f \rangle|^2 , \tag{7}$$

use a final state wavefunction ψ_f that is usually computed in the absence of any radiation potential, so that the resulting resonance widths are equal to the autoionization widths alone. In other words, the radiative width is not included in the resonance profile, so that, by considering Eq. (5), these *undamped* cross sections may be too large.

We have overcome this limitation by adding a *radiative optical potential* to the close-coupling Hamiltonian [8], given by

$$V_{rad} = -i \sum_b \vec{D}|b\rangle \cdot \langle b|\vec{D} ,\qquad(8)$$

where $|b\rangle$ are the bound, recombined wavefunctions, $\vec{D}$ is the dipole operator (equal to $\sqrt{\frac{2\omega^3}{3c^3}}\vec{r}$ and $\sqrt{\frac{2\omega}{3c^3}}\vec{\nabla}$ in the length and velocity gauges, respectively), and the emitted photon energy $\omega = E - E_b$ equals the difference between the scattering and bound-state energies. This optical potential was derived from first principles, and was shown to be equivalent to 1) the time-dependent derivation of Davies and Seaton [6] and 2) the Feshbach projection operator approach [9,10]. We incorporated this potential into the Belfast R-matrix codes [11,12] that include Breit-Pauli potentials and compute dipole matrices necessary for photoionization calculations [13]. At this point, we mention that any close-coupling method requires evaluation of the detailed cross section at a sufficiently large number of scattering energies in order that all the resonance features are resolved. Using a constant energy interval, this number can be quite large when narrow resonances need to be resolved, as will be discussed later.

APPLICATION OF THE RADIATIVE OPTICAL POTENTIAL R-MATRIX METHOD

The first studies using this new method focused on electron-impact excitation. In the case of Ti^{20+} [14], it was found that radiation damping is important since undamped results (those from a standard R-matrix calculation without the radiative optical potential) showed much stronger resonance features than were observed experimentally [15]. By also comparing our damped R-matrix and perturbative results, it was discovered that the disagreement between the earlier distorted-wave results and experiment [15] was due to the restricted distorted-wave basis used. This indicated that the flexibility of the basis can be important for describing low-lying resonances, and that disagreement between perturbative results and experiment was due to deficiencies in the perturbative basis set rather than any interference effects. A second electron-impact ionization study showed that for Fe^{25+} and Mo^{41+}, radiation damping was significant in even the lowest resonances [16], approaching an order of magnitude. This suggested that earlier Dirac R-matrix calculations [17], which neglected radiation damping effects, probably severely overestimated the resonance contribution to the collisional rate coefficients of various highly-charged hydrogenic ions.

The method was next used to calculate DR cross sections of Ar^{15+}. A comparison between two independent R-matrix methods and two independent perturbative methods showed excellent agreement in all cases, as is seen in Fig. 1 (the perturbative DRFEUD results were found to be in good agreement with experiment in an earlier study [19]). The agreement between R-matrix and perturbative results indicated that resonance interference effects on the total cross section were negligible. More importantly, in that study, two interesting findings emerged in bringing those four calculations into agreement. First, it was necessary to use *identical* atomic structure so that the computed widths Γ_a and Γ_r nearly the same. Second, it was extremely important that the numerous narrow resonances were fully resolved in the R-matrix calculations to give reliable convolved cross sections. Using a constant-energy mesh, the cross section had to be computed at millions of energies. Fortunately, this was executed more efficiently using multi-channel quantum defect theory (MQDT), where the slowly-varying unphysical scattering quantities were interpolated from a much smaller number of computed energy points. But it

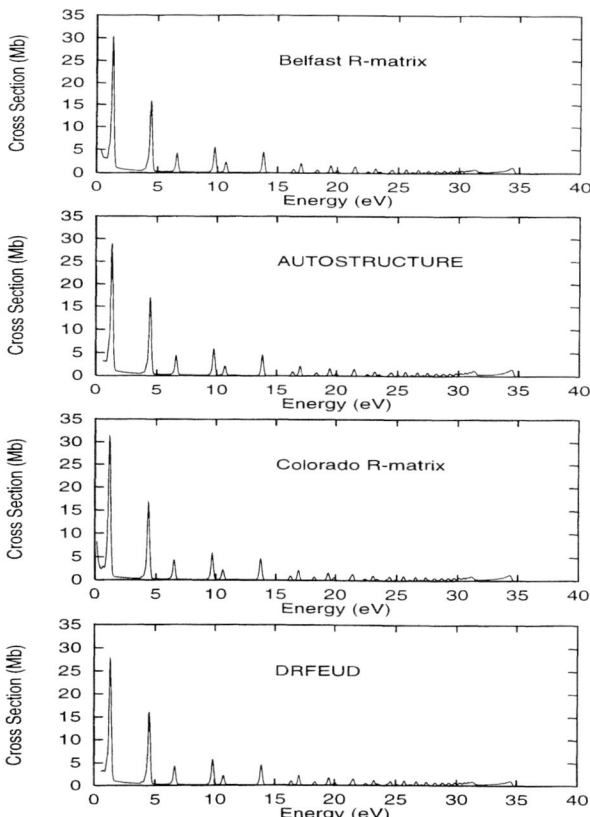

FIGURE 1. Comparison of R-matrix (Belfast and Colorado versions) and perturbative (AUTOSTRUCTURE and DRFEUD) photorecombination results for Ar^{15+}.

was interesting to see that the radiative optical potential R-matrix method and perturbation theory could be brought into close agreement in the first place, because it indicated that interference effects were negligible.

In this $e^- + Ar^{15+}$ system, radiation damping was not particularly important because the core radiative decay was of a $\Delta n_c = 0$ nature. But the earlier assessment of radiation damping effects in highly-charged systems [16,20] indicated that damping effects were quite strong for $\Delta n_c > 0$ decay. It therefore came as quite a surprise when a recent letter by Zhang and Pradhan [21] compared, for photorecombination of Fe^{24+}, *undamped* R-matrix and experimental results, and claimed "good agreement" between the two. They also compared these *undamped* R-matrix and (damped) perturbative results, and suggested that the substantial differences between these results were due to "resonances whose profiles overlap and interact strongly", i.e. interference effects. These puzzling assertions warranted a closer look at the two effects of interest in the photorecombination of Fe^{24+}, namely, radiation damping and interference [22]. We first compared undamped and damped R-matrix cross sections for the KLL resonances (see Fig. 2), and found huge damping effects. In order to determine a quantitative value for the effect of radiation damping, we performed R-matrix calculations *without* the radiative optical potential and used detailed balance; one important discovery emerged. It was necessary to use mil-

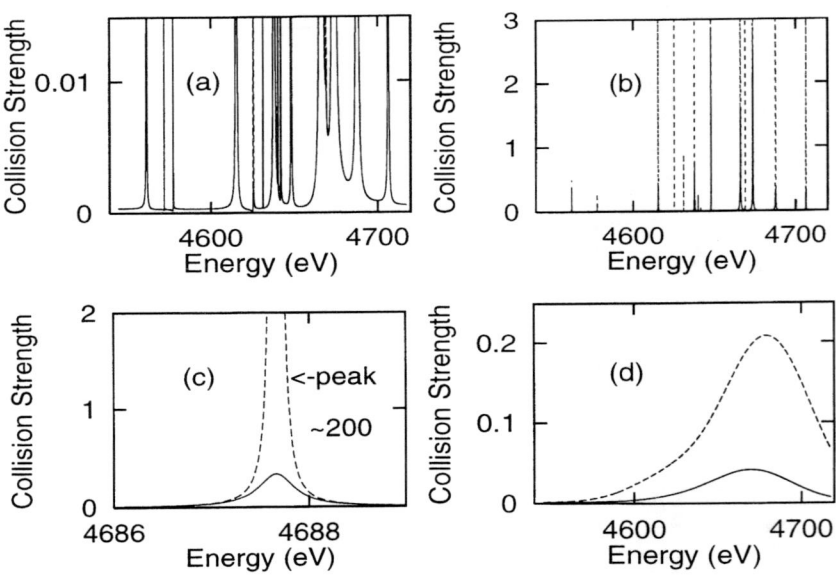

FIGURE 2. Photorecombination collision strengths for the KLL resonances of Fe^{24+}: Dashed line, inverse photoionization method (without damping); solid line, radiative optical potential method (equivalent to inverse-photoionization method with damping); (a), (b), and (c) are unconvoluted, (d) is convoluted with a 50 eV FWHM Gaussian.

TABLE 1. Photorecombination rate coefficients for Fe^{24+} at $T = 2$ keV (10^{-13}cm^{3}s^{-1}).

KLn	Exp[a]	Undamped BPDW[b]	Undamped BPRM[b]	Damped BPDW[b]	Damped BPRM[b]	Damped RDW[c]	Undamped BPRM[c]
KLL	2.48	18.6	17.5	2.417	2.455	2.464	1.784
KLM		30.2	30.2	1.033	1.117	1.469	1.147
KLN		40.9	40.6	0.380	0.377	0.362	0.529
KLO		57.5	57.1	0.180	0.179	0.174	0.307
KLP		78.5	78.2	0.100	0.096	0.097	0.223

[a] Beiersdorfer et al., [25]
[b] This work
[c] Zhang and Pradhan, [21]

lions of energy points, just for the KLL resonances, to fully resolve the *undamped* R-matrix results. A coarser mesh completely missed a myriad of extremely narrow, but extremely high, resonances that contribute significantly to the *undamped* cross section (see Fig. 2). To alleviate this difficulty of resolving extremely narrow resonances, we also incorporated a method of analytically preconvoluting the cross section within an MQDT formalism [23,24]. This allowed a more efficient method for fully-resolving the higher (and narrower) KLn resonances. Resolved results are presented in Table I, showing that, for this system (as well as almost every other we have looked at), R-matrix and perturbative results, whether both damped or both undamped, are in excellent agreement. This indicates that interference effects are negligible for Fe^{24+}. Also listed in Table I are the results of Zhang and Pradhan [21], who, from a later description of that work [26], apparently used only "several thousand" energy points. Thus it is unlikely that they resolved all the numerous narrow resonances, which we find to be much smaller than the spacing $\Delta E \sim (6000 \text{ eV} - 4500 \text{ eV})/10^4 = 0.15$ eV. Their *undamped* R-matrix results, while not in line with our *damped* ones, are lower than our undamped ones, orders of magnitude for the higher-n resonances, and the only explanation we can see for this is that the lack of resolution of these resonances severely underestimates the importance of radiation damping. We now discuss how this underestimation can occur within a model problem.

To illustrate the role of narrow resonances in damped vs. undamped close-coupling photorecombination studies, consider the case of a very strong resonance with autoionization width Γ_a^s, and a very weak resonance with autoionization width Γ_a^w. If they have comparable radiative widths that are much smaller than the strong autoionization width but much greater than the weak autoionization width ($\Gamma_a^s \gg \Gamma_r^s \sim \Gamma_r^w \gg \Gamma_a^w$), then undamped and damped cross sections differ significantly. By considering Eq. (4) and the discussion that follows, the two are related by $\langle\sigma\rangle^{undamped} \sim 2\langle\sigma\rangle^{damped}$. But for low resolution, where the sharp, narrow features of the weak resonance cannot be resolved, only the strong resonance will contribute to the *unresolved* cross section, so that $\langle\sigma\rangle^{undamped}_{unresolved} \sim \langle\sigma\rangle^{damped}$. Neglecting, by simply not resolving, the unphysical contribution from extremely narrow resonances most likely explains why previous close-coupling photoionization/recombination studies [21,26] underestimated the importance of radiation damping.

Preconvolution of Photorecombination Cross Sections

A key development that has enabled us to unambiguously resolve the (undamped) narrowest resonances has been the preconvolution method [23,24] mentioned earlier. This method essentially applies a Lorentzian convolution to the analytic MQDT expression for undamped photorecombination, arbitrarily broadening the resonances while preserving the strength (area). Resolution is then a much simpler task. We find that only thousands of energies are required for a suitably-chosen finite broadening width that is neither too narrow to make resolution a problem nor too broad that the MQDT scattering parameters vary appreciably over this width. The preconvolution method was applied to the *resolved* undamped R-matrix results shown in Table 1, where these results were found to be in excellent agreement with undamped perturbative results. But we give a further comparison between the two for the actual (convoluted) resonance profiles in Fig. 3 below, showing the excellent agreement on this scale. Again, we emphasize that our statements regarding 1) the importance of radiation damping effects and 2) the earlier underestimation of these effects [21,26], initially rested on our converged calculations that used millions of energies, but more importantly now rests on our ability to unambiguously resolve the resonances with this preconvolution technique. The agreement between

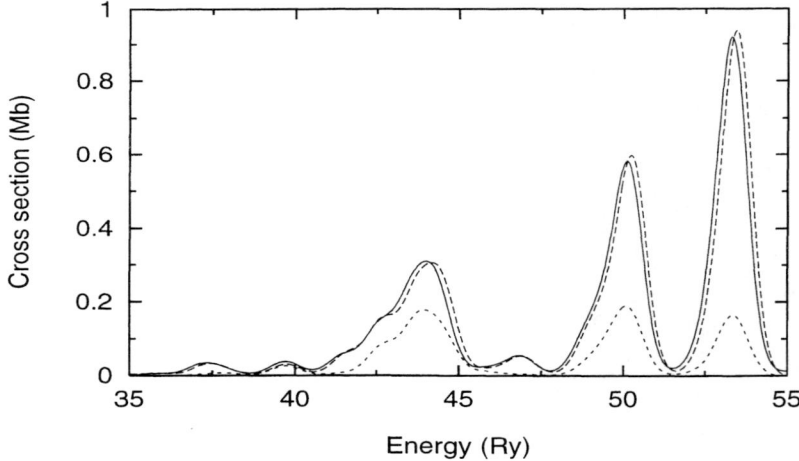

FIGURE 3. Photorecombination of Fe^{16+} for the LMn ($n = 4 - 6$) resonances. Short-dashed curve, radiation-damped perturbation theory; full curve, undamped perturbation theory; long-dashed curve, undamped preconvoluted R-matrix theory. All cross sections have been convoluted with a 1 Ryd FWHM Gaussian function.

undamped R-matrix and perturbative results in Fig. 3 means that perturbation theory, a much more efficient method, can be used to assess damping effects in any ion.

Survey of Computed Radiation Damping Effects

We find, using damped and undamped perturbative methods, that damping in He-like ions occurs even for ionic charges as low as $Z = 4$ in C^{4+} [24]. We show in Fig. 4 undamped and damped perturbative cross sections. They are compared to experiment [27] and also to the results from an undamped R-matrix calculation [26]. The latter undamped results, while differing somewhat from experiment, do not give nearly the large overestimate that we find, probably because only "several thousand" energies were used [26].

Given that radiation damping effects are particularly severe for He-like ions, even at low stages of ionization, we next demonstrate the widespread effects of radiation damping for multi-electron isoelectronic sequences (Fig. 5 also shows this for a Ne-like system). By performing damped vs. undamped perturbative calculations, we determined damping factors $\frac{\langle\sigma\rangle^{undamped}}{\langle\sigma\rangle^{damped}}$ for various systems (see Table 2). Although the photon energies are reduced in going to higher-electron systems, thereby reducing the radiative rates, we still find that damping effects are non-negligible in every sequence. These appreciable damping factors highlight the importance of considering radiation damping effects in a variety of ionic systems,

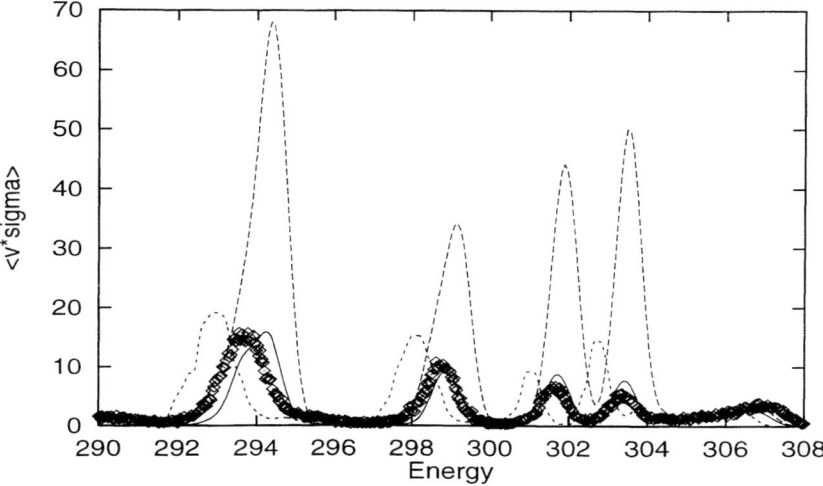

FIGURE 4. Photorecombination of C^{4+} for the KLn resonances, dominated by $n = 4 - 7$. Full curve, damped perturbation theory; long-dashed, undamped perturbation theory; short-dashed, undamped R-matrix theory of Pradhan and Zhang[26]; experiment from Mannervik, et al.[27]

TABLE 2. Importance of radiation damping for low-lying photorecombination resonances.

Type	Ion	$r = \frac{\langle\sigma\rangle^{undamped}}{\langle\sigma\rangle^{damped}}$			
		LMn (1s2$\ell n\ell'$)			
		$n=3$	$n=4$	$n=5$	$n=6$
Li-like	Fe^{23+}	1.6	1.7	3.1	6.3
	Mg^{9+}		1.1	1.3	2.0
Be-like	Fe^{22+}	1.2	1.5	2.7	4.7
	Mg^{8+}			1.3	1.6
B-like	Fe^{21+}	1.4	2.4	3.6	5.8
	Mg^{7+}	1.1	1.2	1.4	2.0
C-like	Fe$^+$	1.2			
N-like	Fe^{19+}	1.3			
	Ne^{3+}				1.5
O-like	Fe^{18+}	1.2			
F-like	Fe^{17+}	1.3	1.9	3.2	
Ne-like	Fe^{16+}	1.9	2.4	4.1	
Na-like	Fe^{13+}		1.1	1.2	1.3
		MNn (2$\ell 2\ell'n\ell''$)			
		$n=4$	$n=5$	$n=6$	
Ar-like	Zr^{22+}	2.7			
	Fe^{8+}	1.2			
K-like	Zr^{24+}	1.2	1.8	2.0	
Ni-like	Zr^{12+}	1.5			
	Se^{6+}	1.2			

including all ionic stages of Fe.

INTERFERENCE BETWEEN DR AND RR PATHWAYS

We return now to interference effects. Although they are commonplace in photoionization processes, they are, for the most part, insignificant in photorecombination calculations [28], as we show for a representative case. An excellent early discovery of strong interference effects was in the outer-shell photoionization of the argon ground state:

$$h\nu + \text{Ar}(3s^23p^6) \longrightarrow \text{Ar}(3s3p^6np) \searrow \downarrow \text{Ar}^+(3s^23p^5) + e^-(\epsilon s, \epsilon d) . \quad (9)$$

Experimental observation [29] and many-body perturbation theory predictions [30] of strong interference effects - window resonances - were made over 25 years ago. Therefore, by detailed balance, similar strong interference effects are expected in the *partial* recombination of Ar$^+$ into the ground state of Ar. However, when viewing this process in reverse, the intermediate resonance states $3s3p^6np$ can radiatively decay via the core to excited Ar states as well. Furthermore, since there are no restrictions on the symmetry of the (incoming) scattering wavefunction, triplet continua, including triplet resonances, can be populated, as can many others. But consideration of the decay pathways for ^1P and ^3P scattering symmetries alone,

$$3s^23p^5\epsilon s, \epsilon d(^1P,^3P) \to 3s3p^6np(^1P,^3P) \to 3s^23p^6(^1S)$$
$$\to 3s^23p^5np(^1S)$$
$$\to 3s^23p^5np(^3S) \,, \qquad (10)$$

shows that radiative decay to final states other than the ground state needs to be considered. By focusing on the contributions of each alternate pathway, it is easy to see why the strong interference effects in partial photorecombination to the ground state are masked by the core decays, where the outer electron simply acts as a spectator. We present R-matrix results for photorecombination of Ar^+ in Fig. 5, showing that when all three decay pathways in Eq. (10) are included, the resonance profile is essentially Lorentzian for all n. To summarize, even though strong interference effects may exist in photoionization, and therefore partial recombination, observance of such effects in *total* recombination are less likely. Indeed, the only experimental photorecombination study claiming to show interference effects was for *partial* recombination in highly-charged uranium ions [31], where the emitted photons were energy-resolved.

Strong Interference Effects in Photorecombination of Sc^{3+}

Recently, however, we have identified a system for which strong interference effects might be detected in *total* recombination [32]. Consider the following process:

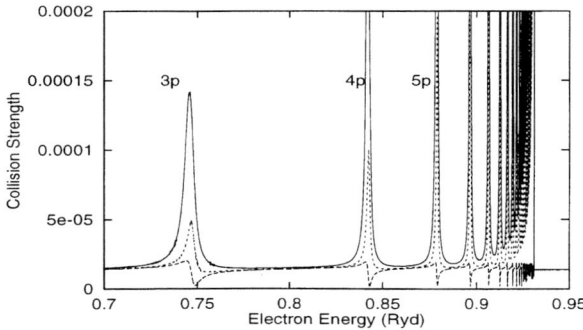

FIGURE 5. Partial photorecombination of Ar^+. Long-dashed curve includes radiative decay to only the $3s^23p^6(^1S)$ final state; short-dashed curve also includes radiative decay to the $3s^23p^5np(^1S)$ final state; solid line further includes radiative decay to the $3s^23p^5np(^3S)$ final state.

$$e^- + Sc^{3+}(3p^6) \rightarrow Sc^{2+}(3p^5 3d^2 \, ^2P^o, ^2F^o)$$
$$\searrow \quad \downarrow$$
$$Sc^{2+}(3p^6 3d \, ^2D) + h\nu \, . \qquad (11)$$

This is special in two respects. First, these low-lying resonances can only decay to the ground state, making interference more likely, as in the partial recombination of Ar$^+$ above. Second, the upper-lying $3p^5 3d^2(^2F^o)$ resonance has a very large autoionization width (similar broad features were seen in photoionization of neutral scandium [33,34]). We present R-matrix cross sections in the near threshold region in Fig. 6, showing that this broad resonance interferes considerably with the RR background, leading to a noticeably asymmetric profile. Given this theoretical prediction, an experimental study at the Test Storage Ring in Heidelberg then looked for this asymmetric profile [35]. The experimental resolution was low due to an unusually high background signal, however, so it was difficult to unambiguously detect asymmetric features (see Fig. 6). Nevertheless, there appears to be some evidence that the rather broad theoretically-predicted $3p^5 3d^2(^2F)$ resonance is perturbed by another resonance, leading to a dip in this broad feature. Later measurements on the isoelectronic Ti^{4+} system [36], where much better resolution was achieved, suggested that this broad feature straddled the zero-energy threshold, making identification of the asymmetric feature difficult. Our preliminary R-matrix results, which were fairly complex to begin with, first predicted that this broad resonance occurred at 4 eV above the Ti$^{4+}(3s^2 3p^6)$ threshold. An even larger configuration-interaction treatment gave instead a position of 1 eV above thresh-

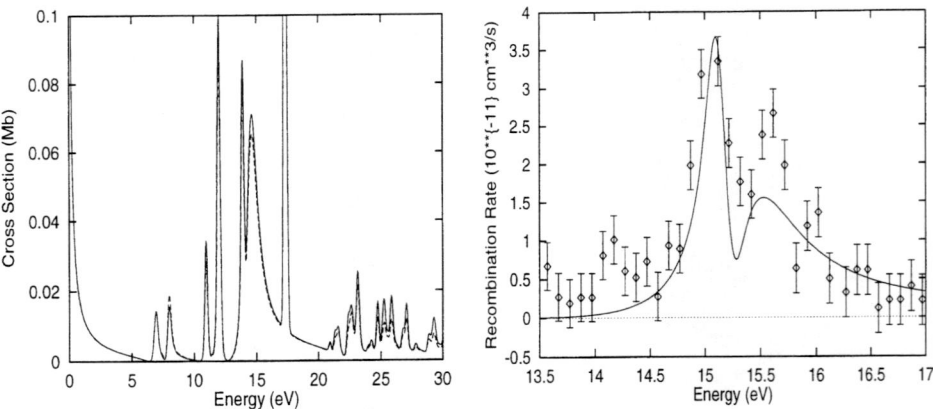

FIGURE 6. Photorecombination of Sc^{3+}. On left are R-matrix results[32] in the length (solid) and velocity (dashed) gauges, showing a noticeable interference between the $3s^2 3p^5 3d^2(^2F)$ resonance (at 15 eV) with the background. On right are later R-matrix rate coefficients that now show additional interference with the $3s^2 3p^5 3d 4s(^2F)$ resonance; experimental results are from Schippers, et al.[36]

old. The difference in these R-matrix results therefore emphasizes three important aspects of photorecombination studies. First, the sensitivity to the atomic basis of low-lying resonance positions near threshold means that it is important to converge atomic structure calculations. Second, experimental guidance as to the positions is quite helpful. Third, the unambiguous identification of interference effects in photorecombination is non-trivial even when likely systems have been identified.

CONCLUSION

We have compared independent perturbative and R-matrix approaches for the calculation of photorecombination cross sections, allowing the assessment of the primary difference between the two, namely, interference effects. In almost every comparison, results from both methods were in excellent agreement *provided that* 1) radiation damping effects were included in the R-matrix method, 2) the resonances were fully resolved in the R-matrix method, and 3) both methods used the same atomic structure. We find that the first of these, radiation damping effects, is particularly important for a wide variety of atomic systems, emphasizing that close-coupling (R-matrix) calculations performed in the absence of radiation damping will usually predict unphysically large *fully-resolved* cross sections. Computed differences between close-coupling and perturbative results should first rule out such likely causes before attributing these differences to interference effects. Partial cross sections, on the other hand, and in special cases even total cross sections, might show actual interference effects, however, so a non-perturbative method is still desirable for these cases, as well as being a beneficial independent check in its own right. Just as importantly, we find extreme sensitivity to the atomic basis in the theoretically predicted positions of low-lying photorecombination resonances. This emphasizes that sophisticated atomic structure techniques must be utilized and that experimental studies are invaluable in guiding theoretical efforts, especially for the determination of low-lying resonance positions.

REFERENCES

1. R. D. Cowan, *The Theory of Atomic Structure and Spectra* (University of California, Berkeley, 1981).
2. D. C. Griffin, M. S. Pindzola, and C. Bottcher, *Phys. Rev. A* **31**, 568 (1985).
3. M. H. Chen, *Phys. Rev. A* **31**, 1449 (1985).
4. N. R. Badnell, *J. Phys. B* **19**, 3827 (1986).
5. J. Oreg, W. H. Goldstein, M. Klapisch, and A. Bar-Shalom, *Phys. Rev. A* **44**, 1750 (1991).
6. P. C. W. Davies and M. J. Seaton, *J. Phys. B* **2**, 757 (1969).
7. R. H. Bell and M. J. Seaton, *J. Phys. B* **18**, 1589 (1985).
8. F. Robicheaux, T. W. Gorczyca, M. S. Pindzola, and N. R. Badnell, *Phys. Rev. A* **52**, 1319 (1995).

9. H. Feshbach, *Ann. Phys.* (N.Y.) **5**, 357 (1958); **19**, 287 (1962).
10. K. J. LaGattuta, *Phys. Rev. A* **36**, 4662 (1987)
11. K. A. Berrington, W. B. Eissner, and P. H. Norrington, *Comput. Phys. Commun.* **92**, 290 (1995).
12. P. G. Burke and K. A. Berrington, *Atomic and Molecular Processes: An R-matrix Approach* (IOP Publishing, Bristol, 1993).
13. N. S. Scott and K. T. Taylor, *Comput. Phys. Commun.* **25**, 347 (1982).
14. T. W. Gorczyca, F. Robicheaux, M. S. Pindzola, and N. R. Badnell, *Phys. Rev. A* **52**, 3852 (1995).
15. S. Chantrenne, P. Beiersdorfer, R. Cauble, and M. B. Schneider, *Phys. Rev. Lett.* **69**, 265 (1992).
16. T. W. Gorczyca and N. R. Badnell, *J. Phys. B*, L283 (1996).
17. R. Kisielius, K. A. Berrington, and P. H. Norrington, *J. Phys. B* **28**, 2459 (1995).
18. T. W. Gorczyca, F. Robicheaux, M. S. Pindzola, and N. R. Badnell, *Phys. Rev. A* **54**, 2107 (1996).
19. S. Schennach, A. Müller, O. Uwira, J. Haselbauer, W. Spies, A. Frank, M. Wagner, R. Becker, M. Kleinod, E. Jennewein, N. Angert, P. H. Mokler, N. R. Badnell, and M. S. Pindzola, Z. Phys. D **30**, 291 (1994).
20. K. Sakimoto, M. Terao, and K. A. Berrington, *Phys. Rev. A* **42**, 291 (1990).
21. H. L. Zhang and A. K. Pradhan, *Phys. Rev. Lett.* **78**, 195 (1997).
22. T. W. Gorczyca and N. R. Badnell, *Phys. Rev. Lett.* **79**, 2783 (1997).
23. F. Robicheaux, *Phys. Rev. A* **48**, 4162 (1993).
24. N. R. Badnell, T. W. Gorczyca, and A. D. Price, *J. Phys. B* **31**, L239 (1998).
25. P. Beiersdorfer, T. W. Phillips, K. L. Wong, R. E. Marrs, and D. A. Vogel, *Phys. Rev. A* **46**, 3812 (1992).
26. A. K. Pradhan and H. L. Zhang, *J. Phys. B* **30**, L571 (1997).
27. S. Mannervik, S. Asp, L. Broström, D. R. DeWitt, J. Lidberg, R. Schuch, and K. T. Chung, *Phys. Rev. A* **55**, 1810 (1997)
28. M. S. Pindzola, N. R. Badnell, and D. C. Griffin, *Phys. Rev. A* **46**, 5725 (1992).
29. R. P. Madden, D. L. Ederer, and K. Codling, *Phys. Rev.* **177**, 136 (1969).
30. H. P. Kelly and R. L. Simons, *Phys. Rev. Lett.* **30**, 529 (1973).
31. D. A. Knapp, P. Beiersdorfer, M. H. Chen, J. H. Scofield, and D. Schneider, *Phys. Rev. Lett.* **74**, 54 (1995).
32. T. W. Gorczyca, M. S. Pindzola, F. Robicheaux, and N. R. Badnell, *Phys. Rev. A* **56**, 4742 (1997).
33. W. R. S. Garton, E. M. Reeves, F. S. Tomkins, and B. Ercoli, *Proc. R. Soc. London Ser. A* **333**, 1 (1973).
34. F. Robicheaux and C. H. Greene, *Phys. Rev. A* **48**, 4429 (1993).
35. S. Schippers, T. Bartsch, C. Brandau, J. Linkemann, A. Müller, A. A. Saghiri, and A. Wolf, *Hyperfine Interactions*, in press (1998).
36. S. Schippers and A. Müller, private communications.

ATOMIC PROCESSES IN INERTIAL FUSION PLASMAS

ATOMIC PHYSICS IN INERTIAL CONFINEMENT FUSION (ICF)

C.J. Keane
Office of Inertial Fusion and the National Ignition Facility
U.S. Dept. of Energy

Atomic physics and the related field of plasma spectroscopy have long played important roles in the US Inertial Confinement Fusion (ICF) Program. The US ICF Program is currently constructing the National Ignition Facility (NIF), which is expected to start operations at the end of 2001. Advanced pulsed power experiments relying on the recent advances on the Z machine at Sandia National Laboratories are also planned. This talk will review the played by atomic physics to date in these ICF and comment on future needs of ICF in atomic physics. Advances in experimental , theoretical, and computational techniques should lead to an increasingly important role for atomic physics in ICF.

Emission and absorption based plasma spectroscopic diagnostics have a long history of use in ICF for diagnostics of plasma properties such as T_e and N_e. Initially such diagnostics relied on relatively straightforward methods such as simple K-shell line ratios and gross linewidths. Advanced experimental and modeling capabilities have led to new diagnostics relying of detailed shapes of line profiles and ratios of entire clusters of lines. Future experiments will require development beyond these capabilities, largely at higher photon energies. For example, on NIF it would be desirable to determine the temperature and density profiles through the DT "spark-plug" and fuel regions to assess whether a true "spark-plug" has been formed. Advanced diagnostics using atomic-physics based systems such as x-ray lasers will also become more important.

More recently, advanced ICF target designs that rely critically on detailed atomic physics phenomena have been developed. This is true for both direct and indirect drive ICF. As an example, for indirect drive mixtures of hohlraum wall materials ideal for generating maximum x-ray drive have been developed using state of the art LTE opacity codes. For direct drive, that use of soft x-ray emission from a high-Z material embedded in the capsule to mitigate the acceleration phase Rayleigh-Taylor instability appears promising. The latter case is particularly stressing as it is NLTE and relies on producing x-ray photons of the right energy at the correct place and time in the implosion.

In summary ,with continuing advances in experimental, theoretical, and computational capabilities, it is expected that atomic physics will play an increasingly larger role in ICF. In particular, while the diagnostic role of atomic physics will remain important it is expected that atomic physics will also have an increasingly important impact in ICF target design. This talk will discuss these issues and also discuss connections between atomic physics research and the recently started DOE Inertial Fusion Science in Support of Stockpile Stewardship Grants Program.

CP443, *Atomic Processes in Plasmas:* Eleventh APS Topical Conference
edited by E. Oks and M.S. Pindzola
© 1998 The American Institute of Physics 1-56396-802-9/98/$15.00

Invited talk abstract prepared for submittal to the
Atomic Processes in Plasmas
Auburn, Alabama
March 23-26, 1998

ATOMIC PROCESSES IN INERTIAL FUSION PLASMAS*

C. A. Back, N. C. Woolsey, O. L. Landen,
S. B. Libby, and R. W. Lee
*Lawrence Livermore National Laboratory, P.O. Box 808, L-473,
Livermore CA 94551*

Implosion experiments using the Nova laser have progressed from single experiments with measurements of peak densities to a series of reproducible experiments with measurements of peak densities and peak temperatures. We will review the experiments to date which use 570 µm diameter capsules filled with D_2 doped with 0.1 % Ar. These implode at about 1.7 ns when heated by a scale-1 hohlraum irradiated with ~ 30 kJ of laser light. We are continuing to develop these experiments with the goal of producing a stable hydrodynamic implosion with measured temperature and density gradients. With these measurements, we are able to provide a testbed for detailed spectroscopic studies. We will also discuss the radiation flux inside the hohlraum where these implosions take place. Experimental data recently obtained in the XUV x-ray regime offer a look at exciting new studies of atomic processes in inertial fusion plasmas.

*Work performed under the auspices of the U.S. D.O.E. by LLNL under contract number W-7405-ENG-48.

ATOMIC AND NUCLEAR PROCESSES PRODUCED IN ULTRA-HIGH INTENSITY LASER IRRADIATION OF SOLID TARGETS

M H Key , M D Cable , T E Cowan , K G Estabrook , B A Hammel , S P Hatchett , E A Henry , D E Hinkel , J D Kilkenny , J A Koch , W L Kruer A B Langdon , B F Lasinski R W Lee , B J MacGowan. , A MacKinnon[1] , J D Moody , M J Moran , A.A Offenberger[2] D M Pennington , M D Perry, T J Phillips , T C Sangster , M S Singh , M A Stoyer , M Tabak G L Tietbohl , M Tsukamoto[3] , K. Wharton , S C Wilks.

Lawrence Livermore National Laboratory, P.O. Box 808, L-473, Livermore CA 94550
e-mail key1@llnl.gov

[1] Visiting from Blackett Laboratory , Imperial College of Science ,Technology and Medicine London SW7 2AZ, United Kingdom
[2] Visiting from Department of Electrical Engineering , University of Alberta , Edmonton, Alberta, T6G 2G7, Canada
[3] Visiting from the Joining & Welding Research Institute, Osaka University, Ibaraki, Osaka 567, Japan

Irradiation of solid targets at ultra-high intensity up to 10^{20} Wcm^{-2} produces plasmas of extremely high energy density which are of interest as the ignition spark in fast ignition [1] or as sources of intense MeV X-rays but which also are rich in interesting physics at the atomic and nuclear level . Recent results of experiments with the Nova Petawatt laser facility will be reviewed .

[1] M Tabak J Hammer M E Glinsky W L Kruer S C Wilks J Woodworth E M Campbell M D Perry . Ignition and high gain with Ultra powerful lasers . Phys Plasmas, 1, 1626, (1994)
[2] M. D. Perry, "Crossing the Petawatt Threshold" Science & Technology Review page 4, December 1996.

Work performed under the auspices of the US Department of Energy by the Lawrence Livermore National Laboratory under Contract No. W-7405-ENG-48

SPECTROSCOPY AND DIAGNOSTICS II

Spectroscopic Studies on Well-Diagnosed Pinch Columns

Th. Wrubel*, S. Büscher*, I. Ahmad[†], and H.-J. Kunze*

*Institute for Experimental Physics, Ruhr-University-Bochum
44780 Bochum, Germany
[†]Department of Physics, Quaid-i-Azam University, Islamabad, Pakistan

Abstract. We report on laser light scattering diagnostics and spectroscopic studies of plasmas of a gas-liner z-pinch where Thomson scattering is performed radially resolved while the resolution of spectroscopy is along the z-axis. The gas-liner pinch is used for accurate measurements of line shapes of multiply ionized atoms and can be operated in a stable and an unstable condition. Implications for the line profiles measured are discussed.

INTRODUCTION

Well-diagnosed plasmas are needed to check theoretical calculations of intensity and profile of lines emitted by atomic and ionic species [1]. The independence of the diagnostic is thereby of utmost importance to ensure reliability. In order to investigate Stark-broadened profiles the intermediate density regime is preferable because on the one hand Stark-broadening dominates Doppler-broadening and on the other hand the density is not too high to apply an independent diagnostic, e.g., Thomson scattering. The understanding of Stark-broadening is especially important for diagnostics of high density plasmas [2], for opacity studies [3], and for x-ray laser investigation [4].

Plasmas in the intermediate density regime are available in the gas-liner pinch facility at Bochum which is a large aspect ratio gas-puff z-pinch. With this device we are capable to investigate the interaction of atoms and multiply ionized ions with plasma particles for stable and unstable conditions. We used stable conditions in a lot of investigations on line profiles and stimulated theory leading to improvements in theoretical calculations (see, e.g., Refs. [5,6] presented at this conference).

All observations confirm the stability and homogeneity of the plasma after maximum pinch compression if the density of the test gas ions is kept at a low level (up to 5 %) depending on the mass of these ions. In order to study pinch dynamics and properties of instabilities radially resolved Thomson scattering and axially resolved spectroscopic measurements of the Stark-width of Kr III were performed.

For the stable cases examples of line profile measurements and also of line intensity ratios will be discussed. Axially resolved Kr III spectra will be shown. The test gas ions are well localized in the homogeneous central part of the plasma. At times slightly before maximum pinch compression the impurities are extremely concentrated so that the temperature decreases locally due to the increased radiation loss while the electron density increases. This can lead to a radiative instability which has certainly implications on the interpretation of spectroscopic measurements which always integrate along the line-of-sight. When the test gas concentration is increased above a specific limit depending on the ion species a severe macroscopic hydrodynamic instability develops resulting in non-equilibrium plasmas along the z-axis. The question will be briefly addressed how the Rayleigh-Taylor instability might be related with the radiative instability.

EXPERIMENTAL AND DIAGNOSTIC SETUP

The device. The gas-liner pinch is a large aspect ratio z-pinch (5 cm electrode spacing, 18 cm diameter) producing a plasma column of 5 cm in length and 1 to 2 cm in diameter (see Figure 1). A capacitor bank of 11.1 μF is charged to (25-42) kV resulting in a maximum discharge current of (0.3-0.5) MA with a time period of about 6 μs. Depending on discharge conditions and the types of gases inserted,

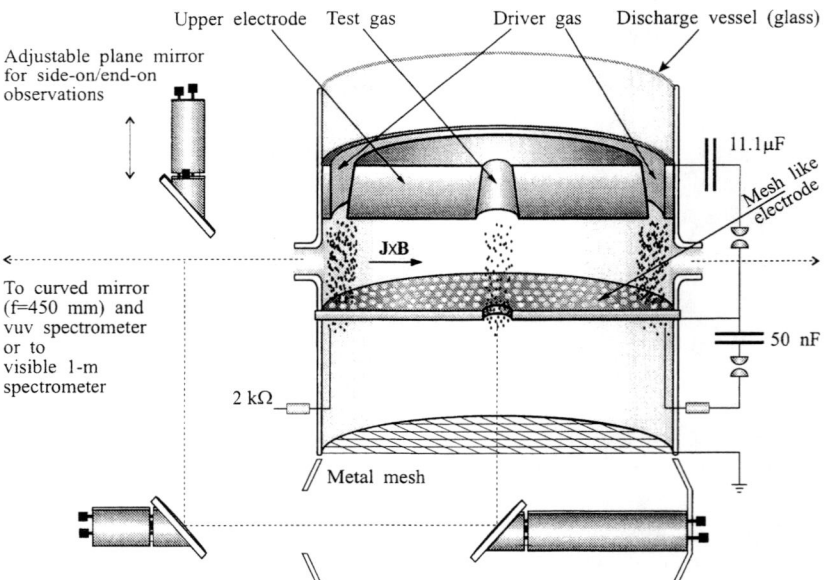

FIGURE 1. Schematic of the discharge vessel. The mirrors shown give the opportunity to investigate the plasma also end-on.

the typical electron densities range from 0.5×10^{18} to 5×10^{18} cm^{-3} and electron temperatures from 3 to 50 eV. Maximum compression is reached near the time of maximal current and the life-time of the high density plasma is 300-400 ns.

The plasma is observable through four ports in the glass chamber, one serving for spectroscopic investigation, three for Thomson scattering diagnostics (see Figure 2). For spectroscopy two different spectrographs are available: one vuv-spectrograph (McPherson, model 225) equipped with a 1200 lines/mm curved grating (blazed at 120 nm) and a MCP (microchannel plate) in the exit plane sensitive in the wavelength region of (30-200) nm. The plasma is imaged onto the entrance slit of the spectrograph via a curved mirror (f=450 mm). For the visible wavelength region the radiation is imaged (1:1 magnification optics) onto the entrance slit of a 1-m Czerny-Turner spectrograph (Spex M1000) which can be equipped with different gratings. The spectral range of this spectrograph was extended down to at least 150 nm using a helium or argon atmosphere in the spectrograph and employing MgF$_2$ optics. For both spectrographs side-on observation is possible and also end-on via

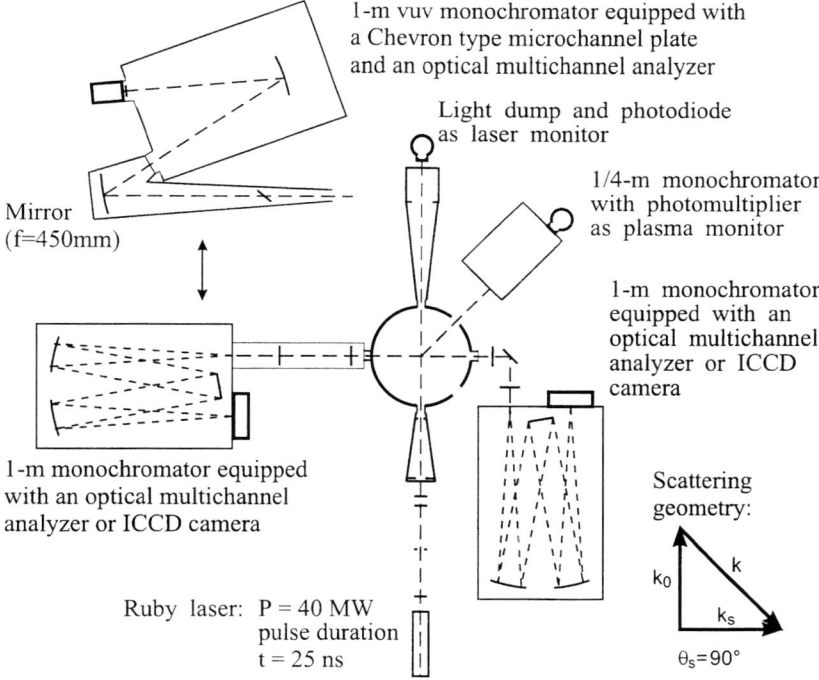

FIGURE 2. Schematic of the experimental setup. The vuv-spectrograph can be installed instead of the second vis-spectrograph. With helium or argon flow through one of the visible spectrographs its wavelength range of can be extended down to at least 150 nm. The scattering vector **k** has the direction as indicated.

three mirrors (see Figure 1).

Through the opposite port the plasma is diagnosed with a 1-m spectrograph (Spex 1704) via 90°-Thomson scattering. The laser beam (Korad K1-Q, 1J, 25 ns (FWHM), 40 MW, 694.3 nm) is focused into the center of the plasma and its time evolution is monitored by a fast photodiode (FND 100). The laser beam passes an assembly of diaphragms and is finally absorbed by a light dump both in order to avoid stray light. The reproducibility of the discharge and its time evolution is checked with a plasma monitor which consists of a 1/4-m monochromator and a photomultiplier observing continuum radiation at 520 nm. This monitor signal is also used as time reference.

Working scheme. The special feature of the device are two independently acting gas inlet systems, both equipped with a fast electromagnetic valve. The main gas (usually hydrogen or helium) initially forms a hollow cylinder near the wall. After preionization the hollow plasma shell is driven onto the axis by discharging the capacitor bank and thus forming the plasma column. The second valve injects the gas of interest (called test gas) along the axis of the discharge tube. If properly done the test gas ions are confined along the axis of the plasma column, where — in the case of low test gas concentration — the plasma is homogeneous. Cold boundary layers are thus missing which otherwise could cause self-absorption of lines or would complicate evaluation of Stark-width because of the necessity of an Abel inversion. In order to achieve homogeneous and stable plasma conditions the amount of test gas must not exceed (1-5) % of the driver gas depending on the charge of the ions. On the other hand increasing the percentage of the test gas, unstable plasma conditions can be cultivated in order to study instabilities as will be discussed later. Up to now the device is restricted to investigations on atoms/ions which can be introduced as gas or as gaseous compound. First attempts to inject atoms of solids into the plasma by laser ablation have been done.

THOMSON SCATTERING DIAGNOSTIC

In order to study plasma parameter and pinch dynamics we employed radially resolved Thomson scattering [7]. The scattered light of the ruby laser system is focused onto the entrance slit (50 μm width and 14 mm height) of the spectrograph. Every position along the height of the slit corresponds to a radial position in the plasma whereby the rotation of the image is achieved by two mirrors. The plasma parameters like electron density and temperature, ion temperature and density and temperature of the test gas ions are, therefore, radially resolved. In addition, the projection of bulk velocity, i.e. macroscopic plasma motion, along the scattering vector **k** is measured via Doppler-shift of the Thomson spectra. With the 1:1 magnification optics the size of the rotated slit and the diameter of the laser focus determine the investigated plasma volume, i.e. 50 μm height, 14 mm length and 1 mm depth. An ICCD-camera (Princeton Instruments) with 384×578 pixels is mounted at the exit plane of the spectrograph. A grating with 1200 lines/mm blazed at 1000 nm is fitted to the spectrograph giving a linear reciprocal dispersion of 6.3 pm/pixel in second order. The apparatus profile of the system has a width (FWHM) of 2.5 pixel and was determined from a Rayleigh scattering spectrum. The size of each ICCD-pixel is 23 μm resulting in a maximal spatial resolution of 60 μm when taking into account the apparatus profile.

An example of a measured ICCD spectrum is shown in Figure 3 (a). The radial resolution is determined by three backlighted slits positioned at the geometrical center of the discharge vessel. The slanting of this Thomson scattering spectrum reflects the decompression velocity of the plasma outside the center. By averaging 20 neighbored radial channels the signal-to-noise ratio is sufficiently improved and the theoretical form factor of Evans is fitted [8,9]. A detailed description of the fitting procedure and the calibration via Rayleigh scattering on propane is given in Ref. [10,7]. Examples of radially resolved plasma parameters obtained from this procedure at different times of the discharge are plotted in Figures 4, 5.

Time evolution and pinch dynamics. Two of the plasma parameters are especially suited to investigate the dynamic of the pinch, i.e. the time evolution of the electron density (Figure 4 (a)) and of the bulk velocity (Figure 4 (b)). During the implosion phase the electron density forms a hollow profile. At times near maximum pinch compression the electron density rises and is rather homogeneous (to within 8 %) over a central diameter of 8 mm. In the decompression phase the electron density relaxes in a steady profile with smaller maximum density.

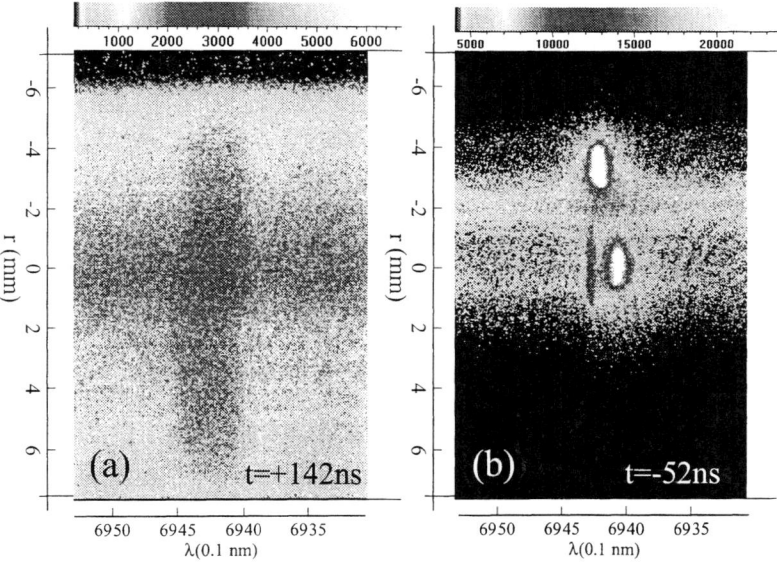

FIGURE 3. Two radially resolved ICCD Thomson spectra. (a) corresponds to a low test gas injection, (b) to a high test gas injection where instabilities occur.

The projection of the bulk velocity onto the scattering vector **k** can also be seen in Figure 4 (b). The center of the pinch according to the velocity is shifted slightly off the geometrical center. During the whole time of the discharge the center is at rest and, therefore, a macroscopic Doppler-shift does not affect spectroscopic investigations. The compression velocity (5×10^4 m/s) according to these measurements agrees with former measurements of the velocity of the magnetic piston [10,11]. The plasma is at rest at maximum pinch compression within the error bars and the decompression velocity is a factor of 2 smaller than the compression velocity. Summing up we can state that the time evolution of both electron density and bulk velocity is as expected from the working scheme of a gas-puff z-pinch. This data can serve to check MHD-codes radiation included which model z-pinch plasma formation and the yield of K-shell emission [12], and which do not give yet satisfactory explanation.

Location of impurity ions. The radial resolved distribution of the impurity concentration and of the electron temperature at times shortly before maximum pinch compression is shown in Figure 5 where we used argon as test gas. The test gas ions are concentrated in a plasma volume of about 2 mm in diameter and they are shifted slightly out of the geometrical center of the discharge, but are still within the region of a homogeneous electron density. Note, that the location of the impurities matches with the resting part of the plasma. As pointed out above the absence of macroscopic Doppler-motion is important for spectroscopic investigations. The impurity concentration shown in that figure is 5 % of the local electron density, but when averaging over the whole scattering volume it reduces to less than 1 % of n_e. At later times the impurity peak smears out but is still within the homogeneous part of the plasma.

The electron temperature normally increases with increasing density. Its tempo-

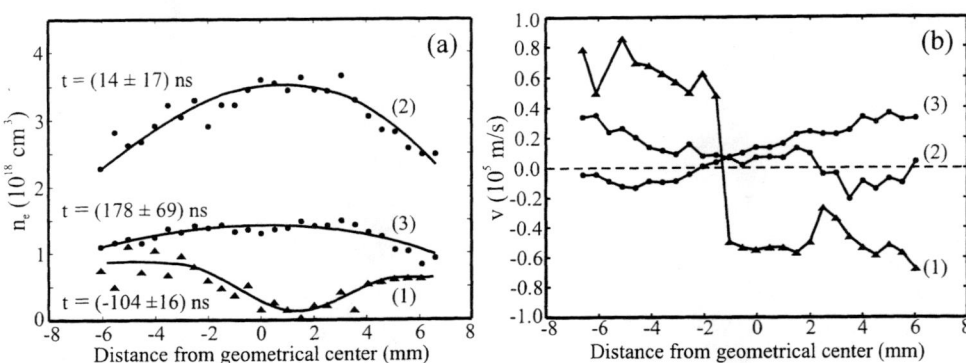

FIGURE 4. Time evolution of the radially resolved electron density (a) and the bulk velocity (b) during different time phases of plasma formation. Curves 1-3 represent the mean of up to 5 single shots. We averaged 20 neighbored channels resulting in a radial resolution of 0.46 mm. The time $t = 0$ ns corresponds to maximum pinch compression.

ral behaviour is similar to that of the density and it is constant in the central part of the plasma. Nevertheless, at times shortly before maximum pinch compression and in cases where the impurity ions are located in a very narrow central plasma region we find that the electron temperature is significantly decreased exactly at the position of the impurity ions (see Figure 5). We suppose that this is caused by enhanced radiative losses which is a function of Z: Due to the high spectroscopic charge number of the impurity ions ($Z \gtrsim 4$) and their high local concentration the plasma cooles locally and the electron density is increased. This becomes obvious with a refined averaging procedure described in Ref. [13] where the spatial resolution is better then 0.14 mm. This behaviour can be explained by the following scenario: High-Z impurities locally cool the hydrogen plasma, the temperature decreases and thus the plasma pressure, and as consequence the surrounding plasma particles stream into this volume, increasing the density. The refined analysis also shows a filamentation of the plasma parameters on a small scale (≈ 0.5 mm) which will be further discussed in Ref. [13].

This extreme localization of the impurity implies that indeed spectroscopic investigations before maximum compression have to be considered with special care. Thomson scattering not resolving the narrow impurity peak would give an averaged temperature, which is too high, and an averaged density, which is too low. Therefore, spectroscopic investigations are usually performed after maximum pinch compression.

Large impurity concentration. When increasing the amount of injected test gas the condition changes from a stable, homogeneous and reproducible plasma confinement to an unstable formation which has no high reproducibility anymore. In contrast to the low injection case the high-density plasma center is varying about 2 mm from shot to shot. The test gas dominates the plasma formation: with and without injection of test gas the resulting plasma is drastically changed. This behaviour is also mirrored in the Thomson spectra. An example is shown in Figure 3 (b). The scattered radiation can no longer be described by thermalized theoretical

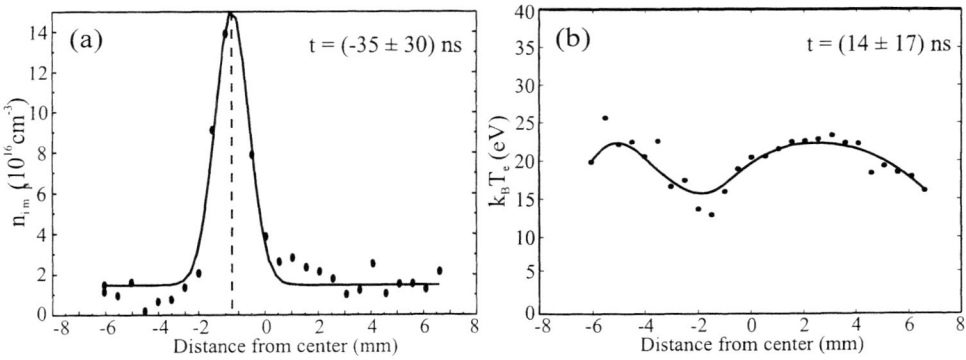

FIGURE 5. Radially resolved impurity concentration (a) and electron temperature (b).

Thomson scattering spectra; it seems to be light scattering at an over-thermally exited ion-acoustic wave in a shock front running through the plasma.

The dependence of the onset of this regime on the mass of the test gas was systematically investigated. The higher the mass of the impurity ions and the higher the amount injected, the earlier unstable plasma conditions occur as will be discussed later. For that reason, when investigating spectral line shapes of high-Z ions special effort has been made to keep the test gas concentration as low as possible. On the other hand, instabilities can be cultivated to study their influence on line shapes, plasma formation or the instability itself.

Instabilities play a dominant in role z-pinch plasmas [12,14,15]. Attempts had been made to correct models for z-pinches, however, this was only successful in a phenomenological treatment [12]. Viscosity, heat conductivity and electrical resistivity had to be increased by large factors to describe the plasma formation in such MHD-models. These factors are attributed to the existence of instabilities. For a further experimental investigation of the instabilities axial resolved diagnostics of the plasma parameter is necessary. Since Thomson scattering along the z-axis is up to now not available, spectroscopic diagnostics via the Stark-width must be employed. Of course the Stark-width must be calibrated before it can be applied.

SPECTROSCOPIC MEASUREMENTS

For the investigation of the axially resolved density we chose krypton as test gas. Three components of the $5s^5S^o - 5p^5P$ Kr III quintet ($J = 2 - 1, \lambda = 335.2$nm; $J = 2 - 2, \lambda = 332.6$nm; $J = 2 - 3, \lambda = 324.6$nm) were studied with respect to their Stark-width, width ratio and intensity ratio [16]. Special problems with krypton come up because energy levels are poorly known especially for high principal quantum numbers. Therefore, theoretical calculations of Stark-width are unreliable at present.

We employed a 1200 lines/mm grating blazed at 750 nm in a 1-m spectrograph (Spex M1000) in first, second and third order and an ICCD camera giving a linear reciprocal dispersion of 18.2, 8.7 and 5.2 pm/channel, respectively. Two examples of spectra in second order are shown in Figure 6. In first order all three components are observable in one spectrum. Therefore, first order measurements were used to determine the line intensity ratios. The experimental ratios are $(2.64 \pm 8\%) : 1.7 : (0.94 \pm 13\%)$ and agree within the error bars with the ratios 2.5 : 1.7 : 1 predicted by LS-coupling. The results of the second order measurements also prove these ratios. For such transitions between levels of high principal quantum number predictions of LS coupling should be valid because the valence electron is far away from the parent electrons and the core. In addition, the ratios confirm that the lines are emitted optically thin.

The Stark-width of the three multiplet components were determined from third order measurements where the relative contribution of apparatus and Doppler-profile is minimal. The widths agree within their error bars, again indicating —

even more sensitive than the intensity ratio — the low optical thickness of the multiplet [17]. The lines were measured at different densities and the linear density dependence [18] was checked. The resulting scaling law will be shown and discussed in detail elsewhere [16]. For low test gas injection neither the intensity ratio nor the absolute intensity or width change along the z-axis and the emission is thus homogeneous. We conclude that the plasma parameter are constant in axial direction indicating the absence of macroscopic instabilities.

Large amount of injection In contrast, when large amounts of test gas are injected the emission of the plasma becomes inhomogeneous (Figure 6 (b)). The intensity is inhomogeneously distributed along the z-axis. The analysis of the Stark-width shows that high electron densities occur at such positions where the intensity of the lines is increased. At axial positions where the emission is decreased, two explanations are conceivable: Either the temperature is too low (or too high) to have krypton ions in the Kr III ionization stage in a sufficient fraction or the concentration of all krypton impurity ions is varying along the z-axis. Whether a temperature effect or a distribution effect is responsible cannot be decided by these measurements. For that reason axially resolved Thomson scattering will be carried out to measure the variation of the temperature directly.

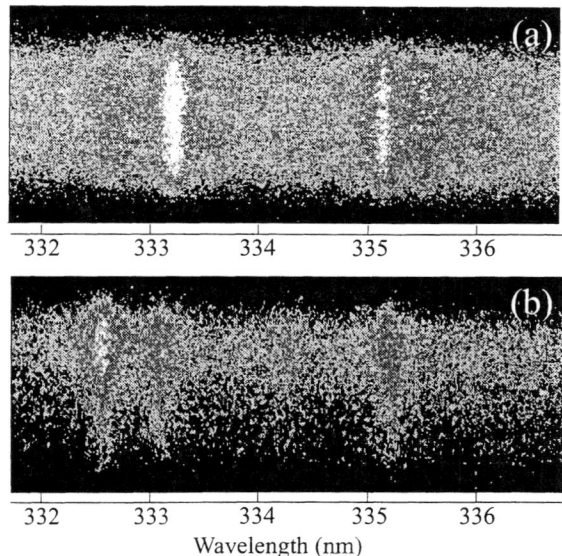

FIGURE 6. Axially resolved spectra of Kr III $5s^5 S^0 - 5p^5 P$ quintet components for small test gas concentration (a) and for a high one (b). The Stark-width along the axial position indicate a strong variation of the electron density which can be explained by instabilities.

DISCUSSION AND CONCLUSION

The spectroscopically observed variation of the electron density is most probably caused by Rayleigh-Taylor instabilities growing in the compression phase of the plasma. These instabilities have also been observed with a pin hole camera in Ref. [19] where they were correlated with inhomogeneous line radiation. It is well known that Rayleigh-Taylor instabilities are intrinsic to common z-pinch devices [12,14,15]. One necessity for the development of this kind of instability is a sufficient high acceleration. In order to suppress instabilities it is favorable to have small amplitudes of the initial perturbations. This is ensured by the preionization and the initial location of the hydrogen gas near the glass wall of the chamber. If the amount of injected test gas is small or even missing, the instability does not grow seriously until maximum pinch compression. For a large amount of test gas our present picture is that the high deceleration of the imploding plasma shell by the heavy test gas yields enhanced growth rates of the instability. Furthermore, as it is stated in Ref. [14], accumulation of mass increases the growth rate.

As discussed above the local reduction of the temperature observed with radially resolved Thomson scattering can be attributed to radiative instabilities. In addition, a variation of the radial density has been found at such conditions, which might be the result of filamentation in the plasma. It is obvious that care has to be exercised when interpretating spectroscopic data even if Thomson scattering is carried out which is not locally resolved. A whole series of line profile measurements at stable plasma conditions (low test gas concentrations) were performed and the results gave new impetus to new theoretical calculations which will be discussed in other talks at this meeting [5,6].

ACKNOWLEDGEMENT

This studies were supported by the Sonderforschungsbereich 191 (A6) of the DFG. One of us (I.A.) thanks the DAAD for support.

REFERENCES

1. H.R. Griem, *Principles of Plasma Spectroscopy*, Cambridge Monographs on Plasma Physics Bd. 2., Cambridge University Press, Cambridge (1997)
2. B. A. Hammel, C. J. Keane, T. R. Dittrich, D. R. Kania, J. D. Kilkenny, R. W. Lee, and W. K. Levedahl, J. Quant. Spectrosc. Radiat. Transfer **51**, 113 (1994)
3. F. J. Rogers and C. A. Iglesias, Astrophys. J. Suppl. Ser. **79**, 507 (1992)
4. J. C. Moreno, H. R. Griem, R. W. Lee, and J. F. Seely, Phys. Rev. A **47**, 374 (1993)
5. R. Stamm et. al, in *Proceedings of the 11th APS Topical Conference on Atomic Processes in Plasmas*, eds. E. Oks and M. S. Pindzola, March 22-26, 1998, Auburn, (AIP Press 1998)

6. W. Lee, S. Glenzer, S. Alexiou, J. Nash, A. Osterheld, and Y. Ralchenko, in *Proceedings of the 11th APS Topical Conference on Atomic Processes in Plasmas*, eds. E. Oks and M. S. Pindzola, March 22-26, 1998, Auburn, (AIP Press 1998)
7. Th. Wrubel, I. Ahmad, S. Büscher, and H.-J. Kunze, in *Proceedings of the 4th International Conference on Dense Z-Pinches, May 28-30, 1997, Vancouver*, CP409, eds. N.R. Pereira, J.Davis, and P.E. Pulsifer (AIP Press 1997), p. 455
8. D. E. Evans, Plasma Phys. **12**, 573 (1970)
9. H.-J. Kunze, *The Laser as a Tool for Plasma Diagnostics*, in *Plasma Diagnostics*, ed. W. Lochte-Holtgreven, North Holland Publ., Amsterdam, p. 550 (1968)
10. Th. Wrubel, S. Glenzer, S. Büscher, and H.-J. Kunze, J. Atmos. Terr. Phys. **58**, 1077 (1996)
11. A. W. DeSilva, T. J. Baig, I. Olivares, and H.-J. Kunze, Phys. Fluids B **4**, 458 (1992)
12. J. W. Thornhill and K. G. Whitney, Phys. Plasmas **1**, 321 (1994)
13. Th. Wrubel, I. Ahmad, S. Büscher, and H.-J. Kunze, to be published
14. J. S. De Groot, A. Toor, S. M. Goldberg, M. A. Liberman, Phys. Plasmas **4**, 737 (1996)
15. L. I. Rudakov and R. N. Sudan, Physics Reports **283**, 253 (1997)
16. I. Ahmad, Th. Wrubel, S. Büscher, and H.-J. Kunze, to be published
17. Th. Wrubel, S. Glenzer, S. Büscher, and H.-J. Kunze, AIP Conference Proceedings # **386**, Spectral Line Shapes, eds. M. Zoppi and L. Ulivi (AIP Press, New York 1997) Florenz, Vol. 9, p. 71
18. M. Baranger, Phys. Rev. **112**, 855 (1958)
19. S. Glenzer and H.-J. Kunze, Phys. Rev. E **49**, 1586 (1994)

Models for Stark Broadening Applied to Plasma Diagnostics

R. Stamm, A. Calisti, S. Ferri, M. Koubiti, T. Meftah, L. Mouret,
C. Mossé, F. Reva, and B. Talin

Laboratoire de Physique des Interactions Ioniques et Moléculaires, UMR 6633, Université de Provence, centre Saint Jérôme, 13397 Marseille, France

Abstract. For the case of line shapes dominated by Stark broadening several models have been recently developed with the aim of being used in line shape codes valid for a wide range of plasma conditions. Several of these models are described, and their results are compared to experimental results. Plasma conditions are identified for which the standard Stark broadening approximations are not valid.

INTRODUCTION

Plasma diagnostic achieved by a study of the emitted line shape is a commonly used technique in plasma physics and the physics of stellar atmospheres (1). The focus of this paper will be on the modeling of Stark broadening processes, using a semi-classical picture for the emitter-perturber interaction. The modeling of the radiative properties of an emitter in the plasma requires a knowledge of atomic physics and a description of the interaction of the emitter with charged particles. The availability of powerful workstations has favored the emergence of models which are both accurate and numerically very efficient, in order to be used in line shape codes. After a brief review of today's most commonly employed models, comparisons between different models and experimental results will be discussed. Although one of the strength of these models is their ability to treat complex atomic transitions, we shall illustrate the different regimes of the emitter-perturber interactions by comparing to hydrogen or hydrogenic ions line shapes obtained in well diagnosed plasmas.

The spectral line shape is usually given as the Laplace-Fourier transform of the dipole autocorrelation function:

$$I(\omega) = \frac{1}{\pi} \operatorname{Re} \int_0^\infty C_{dd}(t) e^{i\omega t} dt, \qquad (1)$$

where, in Liouville space formalism (2), the dipole correlation function is given by:

$$C_{dd}(t) = \langle\langle \mathbf{d}^* | \{U(t)\} | \mathbf{d}\rho_0 \rangle\rangle. \tag{2}$$

In this expression, $\{U(t)\}$ denotes an average of the emitters evolution operator over the charged perturbers, $\mathbf{d}$ and ρ_0 are respectively the dipole, and the equilibrium density matrix for the quantum system considered. The evolution of $U(t)$ is given by a Liouville equation with a time dependent interaction potential $l_F = -\mathbf{d}.\mathbf{F}(t)$, where $\mathbf{F}(t)$ is the sum of the electric fields due to all the perturbing particles. At this point, two different point of view which are equivalent when formulated precisely, can be used to evaluate the average of the emitters evolution operator. The first point of view is to consider that the broadening results from a collisional exchange of internal energy of the emitter with the individual particles of the plasma. The total field $\mathbf{F}(t)$ is then viewed as a function of the coordinates of the m charged particles which produce the field,

$$\mathbf{F}(t) = \mathbf{F}[\mathbf{r}_1(t), \mathbf{r}_2(t), \ldots \mathbf{r}_m(t)]. \tag{3}$$

This is the particle point of view which is the starting point for the collisional impact theory, or other kinetic theory approaches. The other point of view is the field point of view, which consists in averaging over all possible realizations of the microfield. We need here only the mean value of the evolution operator $U(t)$, but even this simple statistical property requires the knowledge of high orders of the microfield moments, as can be readily verified by an expansion of $U(t)$ as a Magnus series (3). The choice of one or the other of these two point of view is dictated by the unavoidable further approximations which have to be done if a quantitative solution is requested. Two approximations are usually made for what we may call the standard Stark broadening model. The first approximation consists in neglecting entirely the motion of the charged perturbers, which amounts to use the field :

$$\mathbf{F}(t) = \mathbf{F}[\mathbf{r}_1(0), \mathbf{r}_2(0), \ldots \mathbf{r}_m(0)]. \tag{4}$$

In the standard model, this static approximation is applied to the ionic component of the plasma. This static ion model is formulated in the field point of view, and requires the static microfield distribution function $P(F)$, a well known statistical quantity for plasmas (4). The other approximation made in the standard model is the impact theory generally applied for the electronic component of the plasma. The impact theory starts from a particle point of view, and relies on the exclusion of strong simultaneous interactions between the emitter and the charged perturbers. In that sense, the impact theory is a binary collision approach requiring the validity condition $\rho_w \ll d$, where ρ_w is the strong collision radius (Weisskopf radius), and $d = N_e^{-1/3}$ is the average distance between the electrons. If this validity condition is satisfied, a perturbation theory to the second order in the emitter-perturber interaction potential is the correct starting point for the impact theory. Roughly speaking, the reversed inequality $\rho_w \gg d$ would allow the use of the static approximation mentioned before. Although this standard model is very successful for many types of plasmas, observational conditions exist today for which either one or the other of these two approximations is not valid. A plasma diagram for Stark broadening can be used to visualize the plasma conditions for which the impact, static or none of these two approximations is valid. This is done for the Balmer α (H_α) line on fig.1 on a decimal logT-logN_e plot on which two lines are drawn for each type of particles (ion solid lines and electron dotted lines), corresponding to a ratio λ of the Weisskopf radius (for hydrogen $\rho_w = \dfrac{\hbar n^2}{m_e v}$, where n is the principal quantum number of the upper level, and v is the perturber average velocity) to the average distance between particles equal to 0.1 (upper line) or 10 (lower line).

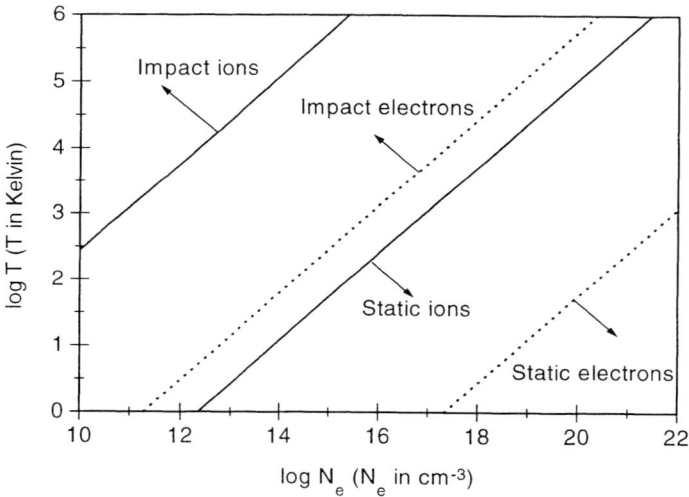

FIGURE 1. Plasma diagram for Stark broadening in the case of H_α.

Because λ is a measure of how many simultaneous strong collisions occur, the upper left part of the $\lambda=0.1$ line roughly corresponds to the validity conditions of the binary approximation, and the lower right part of the $\lambda=10$ line to that of the static approximation. For each type of particle it can be seen on fig.1 that there is a wide strip of plasma conditions which cannot be treated by one of the limiting approximations. For such cases, the main difficulty is due to the dynamic many body problem of the emitter interacting with a large number of ions or electrons. In the two last decades, several models have been developed which provide an approximate solution for this problem. In the following, several of these approaches will be presented and discussed. The results of a numerical model developed in our laboratory will be discussed and compared to laboratory plasma experiments and line shape simulation calculations.

MODELS FOR STARK BROADENING

Model Microfield Method

Initially proposed by Brissaud and Frisch (5), the Model Microfield Method (MMM) has been the first approach taking the field fluctuations into account by using the microfield point of view. The basic idea of the MMM is to replace the complex true microfield, by a simpler stochastic process which allows an analytical calculation of the line profile. Most of the MMM calculations already performed use the "Kangaroo process", which assumes that the microfield is stepwise constant, and jumps to a new value with a field dependent rate ν(F). This simple model leads to the following expression for the line profile :

$$I(\omega) = \left\{\tilde{U}_S(\Omega)\right\}_F + \left\{\nu\tilde{U}_S(\Omega)\right\}_F \left\{\nu - \nu^2\tilde{U}_S(\Omega)\right\}_F \left\{\nu\tilde{U}_S(\Omega)\right\}_F , \qquad (5)$$

where $\tilde{U}_S(\Omega)$ is the Laplace transform at $\Omega=\omega+i\nu$ of the evolution operator $U_S(t)$ in a static field F. The symbol $\{...\}_F$ denotes an average over the static microfield distribution P(F). The jumping rate is obtained by fitting the MMM field correlation function $C_{FF}(t)$ to an expression obtained with the kinetic theory (5). The static microfield distribution P(F) and the field correlation function $C_{FF}(t)$ are the only statistical quantities used by the MMM, which appears as a model able to describe both the static and impact limit, and the intermediate cases. It is interesting to note that the MMM can be applied simultaneously to the ionic and electronic component, allowing a unified treatment of Stark broadening (6).

Line Shape Simulation Techniques

Line shape simulation techniques consist in solving numerically the Schrödinger equation for the emitters evolution operator submitted to the microfield of typically one hundred to one thousand particles (7). This microfield is obtained either by a sampled set of particles moving on straight lines (8,9), or by a molecular dynamics technique applied to the charged particules (10). Such calculations are usually performed to provide reference line shapes (9), because no approximation is made for the evolution of the quantum emitter submitted to the microfield of a large number of particles. There are in general restricted to simple atomic system, due to the intensive computer work required by the solution of the Schrödinger equation for a complex atomic transition. A recent exception concerns the hydrogen line without fine structure, for which extensive calculations of the hydrogen Lyman , Balmer, (9) and Paschen (11) series have been performed by using an efficient algorithm for solving the emitters evolution equation. This equation makes use of the decoupling of the Schrödinger equation obtained by the use of the Runge-Lenz vector $\vec{M}_n$, which is related to the restriction of the position operator $\vec{r}_n$ in the subspace of principal quantum number n by:

$$\vec{r}_n = -\frac{3}{2} n a_0 \vec{M}_n , \qquad (6)$$

where a_0 is the Bohr radius. Using the Runge-Lenz operator together with the angular momentum operator $\vec{L}_n$, two commuting operators $\vec{J}_n(1)$ and $\vec{J}_n(2)$ may be defined in the form:

$$\begin{aligned}\vec{J}_n(1) &= \left(\vec{L}_n + \vec{M}_n\right)/2\hbar \\ \vec{J}_n(2) &= \left(\vec{L}_n - \vec{M}_n\right)/2\hbar .\end{aligned} \qquad (7)$$

The use of these two vectors allows the factorization of the emitters evolution operator in a product of upper and lower state evolution operators. A very efficient solution for these evolution operators can be found, after their expansion in powers of the product $\vec{\alpha}.\vec{J}$, where the elements of the operator $\vec{\alpha}$ are the Euler-Rodrigues parameters (8). This solution has a number of advantages : It is simple to numerically evaluate, and gives a very accurate result. It also has a property of numerical stability which is useful for a numerical integration over long time configurations.

The recent simulations by V. Cardenoso and M. Gigosos use this technique (9,11) for a two component simulation (ions and electrons), and thus treat in a unified manner the effect of the ions and the electrons.

Frequency Fluctuation Model

Although being related to the MMM, and some aspect of the simulation calculation, the Frequency Fluctuation Model (FFM) is a different and new model which has been developed in our laboratory as a line shape code, with the aim of treating complex atomic transitions encountered in various type of plasmas (2). Like in the other models already presented, the average operator $\{U(t)\}$ is obtained from an average over all the time dependent microfield configurations. The starting point of our model is a stochastic Liouville equation for the emitter evolution operator submitted to the interaction potential l_F=-**d**.**F**(t) of the atomic dipole with the electric field F :

$$\frac{dU_F(t)}{dt} = i(L_0 + l_F)U_F(t), \qquad (8)$$

where L_0 is the atomic Liouville operator which contains the transition frequencies of the unperturbed atom, and the spontaneous emission rates for each radiative transition.

If the electron broadening can be treated in the impact approximation, the corresponding relaxation rate may be added in L_0 also. However, like the MMM or the simulation, the FFM can potentially be applied to both ionic and electronic component.

The first step of the FFM consists in averaging over a set of constant microfields weighted with the static microfield distribution P(F). For each microfield, the evolution operator in each field F is numerically diagonalized, leading to a large number of Stark components. This large number corresponds to the number of eigenstates of the atom in a constant field, times the number of fields chosen for representing the static microfield distribution P(F). The line shape is then obtained as a sum over generalized lorentzians :

$$I(\omega) = \sum_k \frac{c_k(\omega - f_k) + a_k \gamma_k}{(\omega - f_k)^2 + \gamma_k^2}, \qquad (9)$$

where the Stark component k is defined by two complex numbers, the generalized intensity $a_k + i c_k$ and the generalized frequency $f_k + i \gamma_k$.

In the second step of the FFM, we replace the complete set of Stark components by a smaller ensemble which preserves the smallest details of the spectral profile. This can be done for the cases where several spectral components appear within the narrowest observable spectral width, which is usually determined by the relaxation processes. For such cases one radiative channel is defined for replacing a whole subsets of Stark components. A set of radiative channels is obtained by a numerical sampling, and may be viewed as a coarse-grained distribution of the Stark components.

In the last step of the FFM, the radiative channels are mixed by a stationary Markov process, assuming that the frequency fluctuation of the radiative channels is the observable effect of the microfield fluctuations. The chosen Markov process is based on a unique fluctuation rate ν, obtained for instance from a molecular dynamic simulation for the microfield fluctuation rate. If we consider r radiative channels, the Markov process is entirely defined by two quantities, the instantaneous probability of states p_1, p_2,...,p_r, and the transition rate between these states W_{ij}. These transition rates are chosen to be proportional to the fluctuation rate ν and are defined by :

$$W_{i,j} = \nu p_i \ (i \neq j), \quad W_{i,i} = -\nu(1 - p_i). \qquad (10)$$

In the FFM, the probabilities p_i are defined to be the normalized real part of the corresponding radiative channel intensity. The quantities W and p follow the sum rules required by normalization and detailed balance :

$$\sum_i W_{i,j} = 0, \quad \sum_j W_{i,j} p_j = 0. \qquad (11)$$

The FFM spectrum associated with the stochastic mixing of the r radiative channels appears as a sum over final states f, and an average over initial states i :

$$I(\omega) = \frac{1}{\pi} \text{Im} \sum_{i,f} p_i \left(\langle\langle D | [\omega - L'_0 + iW]^{-1} | D \rangle\rangle \right)_{i,f} \quad , \quad (12)$$

where L'_0 is a diagonal operator whose eigenvalues are the complex frequencies $f_k + i\gamma_k$, and the operator D is an effective dipole defined by its component D_i for the channel i:

$$D_i = \sqrt{\left(1 + i\frac{c_i}{a_i}\right) \sum_{k=1}^{r} a_k} \quad . \quad (13)$$

The equations (10-13) constitute the starting point of our line shape code, which has been written and is maintained by always employing fast numerical techniques. The ability of our model to generate from an atomic database all the dipole elements required by a line shape calculation is an important feature of the FFM code which allows for a given case to retain only the atomic states really relevant for Stark broadening.

RESULTS AND DISCUSSION

Fine structure

In weakly to moderately dense laboratory plasmas, fine structure splitting may affect the line width and shape of Stark profiles even for low Z ions. An example is the H_α line of HeII, for which experimental profiles have been obtained in discharge plasma conditions at an electron density 5.5×10^{16} cm^{-3}, and a temperature of 44000 K (12). On fig. 2 we have used our code to investigate on this line shape the effect of fine structure and ion dynamics. The two profiles on the right of the figure are the static (dashed line) and dynamic ion (short dashed line) cases without fine structure. For these two profiles a large effect of ion dynamics is observed, since it leads to an increase of the line width by a factor of two. The inclusion of fine structure in the static ion case (dotted line), results also in a doubling of the width of the line. The effect of ion dynamics on the profile with fine structure (dash-dotted line) is a further increase of the width by 45%. A final convolution with Doppler broadening leads to a smooth profile (solid line) with a FWHM of 0.32 Angström. This value is similar to what has been found by a recent calculation using the MMM for the ions, and a Green's function technique for the electrons (13). Both theoretical values are however 30% narrower than the experimental profile. For these plasma conditions a binary approach for the electrons is well justified, and the discrepancy with the experiments may probably be explained by a lack of broadening of the FFM and MMM applied to the ionic component.

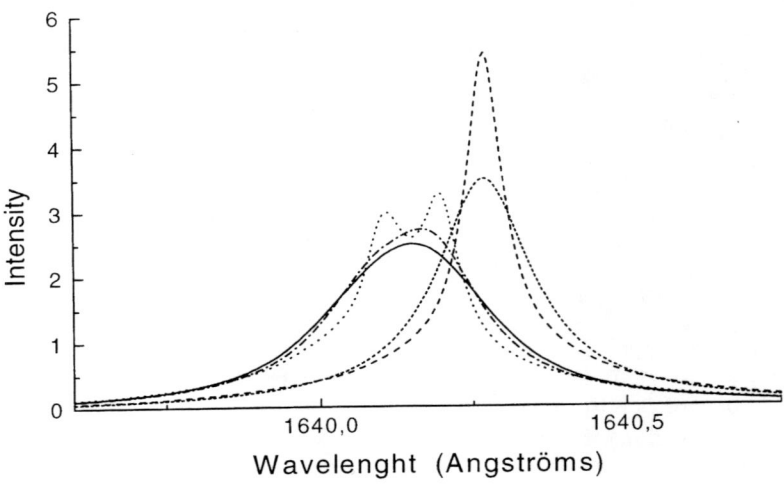

FIGURE 2. Influence of different effects on the line shape of H_α of HeII for the conditions of the experiment of Piel and Slupek (12). The two profiles on the right are the static (dashed line) and dynamic ion (short dash) cases without fine structure. The three profiles on the left include fine structure in the static ion (dotted line), dynamic ion (dash-dotted line), and dynamic ion plus Doppler broadening case (solid line).

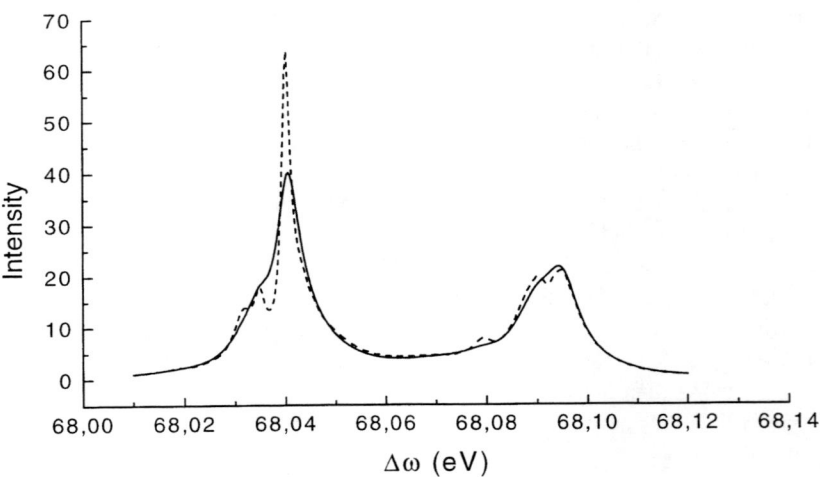

FIGURE 3. Static ion profile (dashed line), and dynamic ion profile (solid line) of CVI H_α for $N_e = 5 \times 10^{18}$ cm^{-3}, and a temperature of 20 eV.

The influence of fine structure is more relevant for emitters with a higher charge. A well known example is the H_α line of hydrogenic carbon, which has been discussed in the frame of X-ray laser research. We have plotted this line on fig.3 for a density $N_e=5\times10^{18}$ cm^{-3}, and a temperature of 20 eV, in a pure carbon plasma. Fine structure is clearly visible on fig.3, the static ion case allowing even to see some details in the level structure. The two main lines which remain in the dynamic ion case on fig.3 correspond mainly to the large energy separation of the $2p_{3/2}$ and the $2s_{1/2}$, $2p_{1/2}$ levels.

Comparison of the FFM with the simulation

Comparisons of the FFM with the recent ion plus electron simulation (9) of hydrogen profiles for a density range between 10^{14} and 10^{18} cm^{-3} reveal systematic tendencies. On fig.4, we have plotted the FWHM of Hydrogen Paschen α (P_α) at T=10000 K in a proton plasma, as a function of electronic density. In the plotted density range, we expect the ionic component to act from near impact type broadening at low density, to almost static broadening at the highest density. It can be seen from fig.4 that there is a good overall agreement between the FFM and the simulation. Two different tendencies appear at low and high density. The lack of broadening of the FFM at low density may be explained by an imprecise treatment of ionic binary collisions. Work is in progress in our laboratory for improving the FFM in this near impact regime.

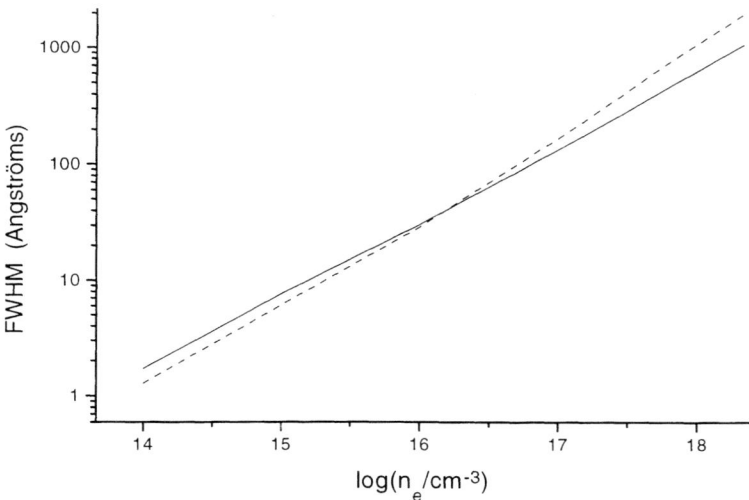

FIGURE 4. FWHM of Hydrogen P_α at T=10000 K in a proton plasma, as a function of electronic density. The results of the simulation (9) (full line) are compared to the FFM (dashed line).

For the highest densities calculated, the FFM widths are broader than the simulation as can be seen on fig.4. For such conditions, the static ion approximation is well justified, and the ionic contribution is certainly accurately treated, but it is now the validity of the electronic impact model which is questionable. In the case of P_α, the parameter $\lambda \approx 0.4$ for the highest density calculated, which means that simultaneous strong collisions effects are expected, invalidating the use of a binary impact operator. This regime will be further discussed when comparing to low temperature experiments.

Comparison with experiments

Gas liner pinch experiments

The high temperatures obtained in pinch devices are usually such that the binary approximation for electrons is well justified for low lying lines ($\lambda < 0.1$). For not too high densities, an effect of ion dynamics on the line profile is generally observed.

The shift and width of the H_α line of hydrogen in a dense helium plasma has been measured recently in a gas-liner pinch (14). We have compared on fig. 5 the width of H_α

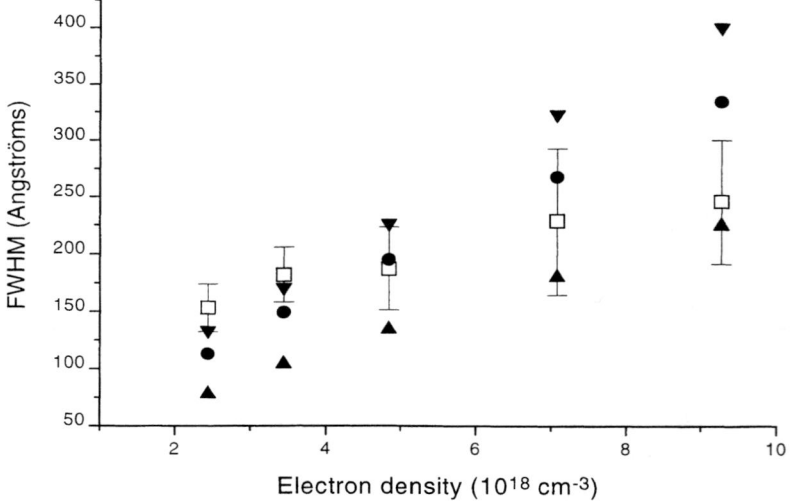

FIGURE 5. Full width of H_α as a function of electron density in the conditions of the experiment of Bödekker et al. The experiment (squares) is compared to the calculations of Kepple and Griem neglecting (down triangles), and taking into account the interference term (up triangles), and our code (circles).

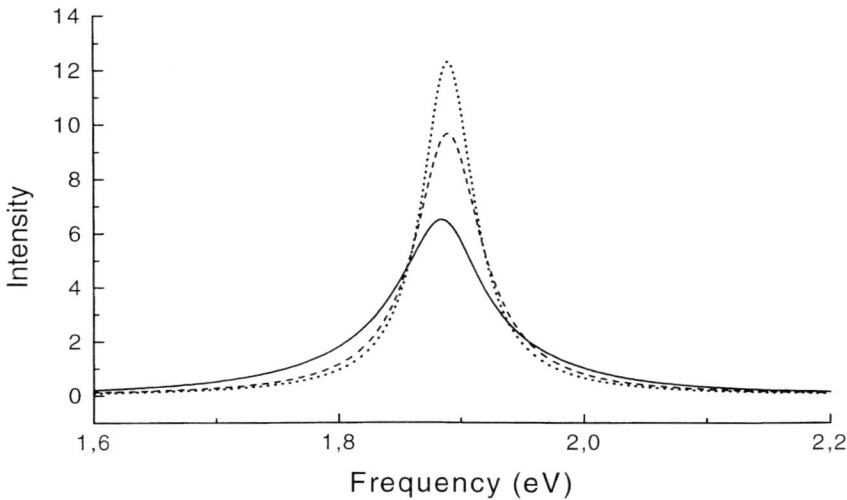

FIGURE 6. Profile of H_α for $N_e=9.27\ 10^{18}$ cm^{-3}, and a temperature of 10 eV. The profile including the effect of ions and electrons (full line) is compared to the profiles broadened by electrons only with (dashed line) and without (dotted line) the effects of inelastic collisions.

as a function of electronic density between 2.44×10^{18} cm^{-3} and 9.27×10^{18} cm^{-3}, with corresponding temperatures varying between 7 and 10 eV. The widths obtained by our code are in between the calculations neglecting the upper-level lower-level interference term, and calculations including those, but neglecting inelastic collisions, quadratic Stark effect, and ion dynamics. These effects explain most of the differences between our results and the calculations of Kepple (15) including interference effects. For the highest density observed, our calculations predict a width which is above the experimental measure. For this high density, the effect of ion dynamics results only in a 3% increase of the width. The important broadening mechanisms for these conditions are the inelastic collision contribution to the electron impact operator, and the quadratic ionic Stark effect due to n=4 and n=5 levels. An illustration of the amplitude of these effects is given on fig.6 for a density of $N_e=9.27\ 10^{18}$ cm^{-3}, and a temperature of 10 eV. Quadratic Stark effect and inelastic electronic effect represent a significant amount of the broadening due to ion and electrons (full line profile). If ion broadening is completely ignored we obtain the dashed line profile by retaining inelastic electron broadening, and the 20% narrower dotted line profile for elastic broadening only. It is interesting to note that compared to the electron broadened profiles, the quadratic Stark effect shifts the line to the red by 5% of the linewidth, and provokes a small asymmetry of the line. The ionic quadratic Stark effect due to the n=4 and n=5 depends on the energy difference with each radiative level. This results in an asymmetric red shift of the line. We did not compare to the experimentally observed shift (14), because our calculation does not take account of the electronic shift operator. Also a consistent study of the asymmetry would imply the inclusion of ionic quadrupole interactions, which have been omitted here.

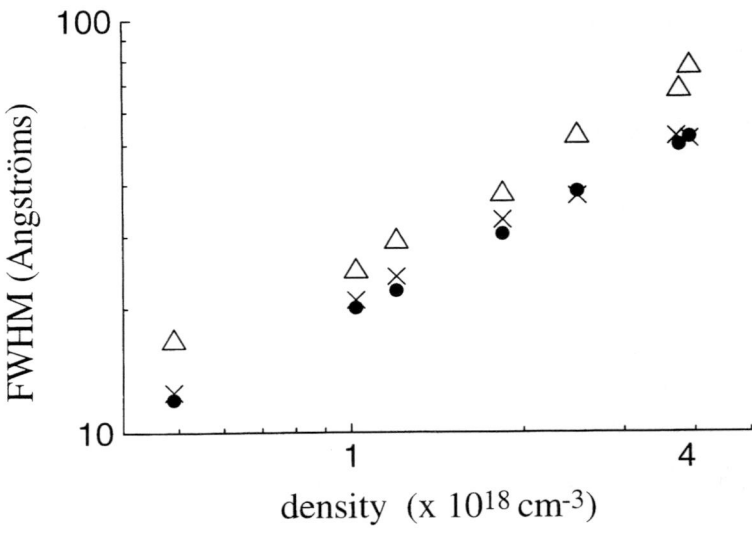

FIGURE 7. FWHM of HeII P_α at T=3-5.4 eV in a proton plasma, as a function of electronic density. The experimental results (triangles) of Büscher et al (17) are compared to the simulation (16) (crosses) and the FFM (circles).

Other interesting measures performed in the gas-liner pinch concern the Paschen α and β line of HeII (17) in a plasma dominated by protons at N_e=0.4-4x10^{18} cm^{-3} and T=3-5.4 eV. Although there is a good overall agreement between the experiment and the FFM results, it can be seen on fig.7 that recent calculations of the P_α with the FFM code (circles) predict widths which are about 20 to 30% narrower than the experiment (triangles). Our calculations have recently been confirmed by simulations(crosses) using the model previously described (9). This remaining discrepancy, which affects mainly the P_α line, is not well understood, and may be due to an unidentified broadening mechanism.

Low temperature experiments

By using devices such as capillary discharges, or wall stabilised discharge plasmas, rather high densities can be reached, while keeping the temperature close to one eV. For such condition the λ electronic ratio is no longer negligible compared to one, and non binary electronic effect are expected. A useful reference is then again provided by the full ion plus electron simulation (9), for which none of the approximations of the standard model is made. We have first plotted on fig.8 a comparison of the experimental

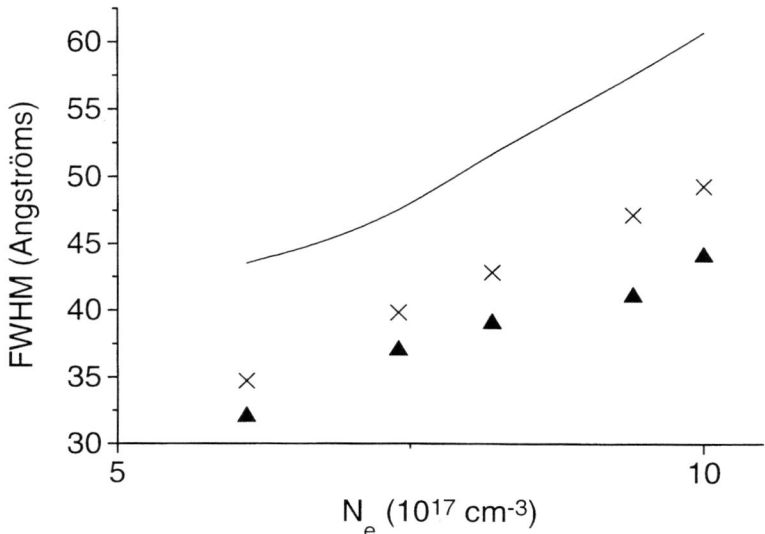

FIGURE 8. Full width of H_α at $T \approx 1.5$ eV in an argon plasma, as a function of electron density in the conditions of the experiment of Vitel (18). The experiment (triangles) is compared to the simulation of Gigosos et al (9) (crosses), and to the calculations of the FFM with a binary electronic collision operator (solid line).

and calculated Full Widths at Half Maximum of the hydrogen H_α in an argon plasma, as a function of density for the conditions of the experiment (triangles) of Vitel (18), which have been done for temperatures between 16000 K and 18700 K. The simulation calculation (crosses) is an a rather good agreement (within 10%) with the experiment, while the standard model (solid line), which is obtained here by the FFM in the static ion limit, and using an electronic binary impact approximation, predicts widths which are 40% larger.

Similar conditions are found in the recent measure of the hydrogen P_α width by Döhrn et al (19) done for temperatures between 11720 K and 13900 K. A difference increasing with density, and reaching 30% at the highest density measured, is seen on fig.9 between the standard model (solid line) and both the experiment (triangles) and the simulation (crosses). For the conditions of figs.8-9, the electronic λ value reaches values up to 0.2, indicating that the use of a binary collision operator can severely overestimate the electronic broadening. A possible improvement for treating the electronic component for such case, is to generalize the FFM to the two components of the plasma. For these conditions, we are actually in the same situation as for the ion dynamic problem, namely with simultaneous effects of strong collisions on the emitter, and any solution of the many body ion dynamic problem should also be applicable to the electrons.

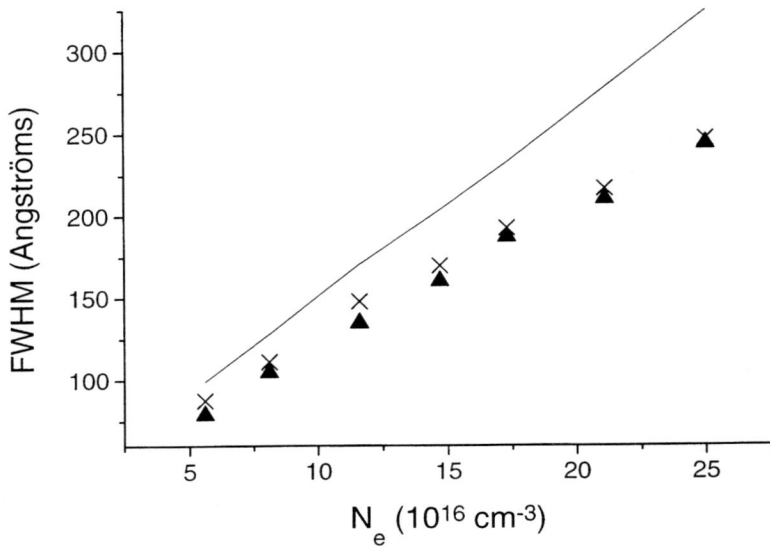

FIGURE 9. Full width of P_α at T≈1 eV in an argon plasma, as a function of electron density in the conditions of the experiment of Döhrn et al (19). The experiment (triangles) is compared to the simulation of Gigosos et al (9) (crosses), and to the calculations of the FFM with a binary electronic collision operator (solid line).

CONCLUSION

We have briefly described several Stark broadening models used in some of the line shape codes which are available today. The main emphasis has been centered on the Frequency Fluctuation Model developed at the Université de Provence for the diagnostic of various plasma conditions. The FFM code accepts atomic data from standard atomic structure code, enabling an analysis of the effects of fine structure or inelastic collisions. Comparisons of the FFM with the profiles observed in well diagnosed hydrogen and hydrogenic ion plasmas reveal a good overall agreement of our model with the experiments. For the ionic component, discrepancies with the experiments remain in the near impact regime, calling for a further improvement of the model. Comparisons to simulations and experiments in high density and low temperature cases indicate that a binary collision operator for the electrons may seriously overestimate the broadening for such conditions. An alternative approach to a binary collision operator may then be a model unifying the FFM model to both the ion and electron component. This would allow an application of our code to explore the conditions found in recent observations of the edge of tokamak plasmas, or of various discharge plasmas at low temperature.

The Laboratoire de Physique des Interactions Ioniques et Moléculaires is "Unité Mixte de Recherche associée au CNRS et à l'Université de Provence".

REFERENCES

1. Griem, H. R., *Plasma Spectroscopy*, MacGraw-Hill, New York(1964).
2. Talin, B., Calisti, A., Godbert, L., Stamm, R., Lee, R. W., and Kein, L., *Phys. Rev. A* **51**, 1918(1995).
3. Magnus, W., *Comm. Pure Appl. Math.* **7**, 649 (1954).
4. Iglesias, C., Lebowitz, J., and MacGowan, D., *Phys. Rev. A* **28**, 1667(1983).
5. Brissaud, A., and Frisch, U., *J. Quant. Spectrosc. Radiat. Transfer* **11**, 1767(1971).
6. Stehlé, C., *Physica Scripta* **65**, 183 (1996).
7. Stamm, R., and Voslamber, D., *J. Quant. Spectrosc. Radiat. Transfer*, **22**, 499(1979).
8. Gigosos, M., and Cardenoso, V., *J. Phys. B.* **22**, 1743(1989).
9. Gigosos, M., and Cardenoso, V., *J. Phys. B.* **29**, 4795(1996).
10. Stamm, R., Talin, B., Pollock, E. L., and Iglesias, C. A., *Phys. Rev. A* **34**, 4144(1986).
11. Cardenoso, V., and Gigosos, M., *J. Phys. B.* **30**, 3361(1997).
12. Piel, A., and Slupek, J., *Zs. Naturforsch.*, **39a**, 1041 (1984).
13. Günter S., and Könies A., *Spectral Line Shapes* 9, 99 (1996), AIP conference proceeding 386.
14. Böddeker, S., Günter, S., Könies, A., Hitzschke, L. and Kunze, H. J., *Phys. Rev. E* **47**, 2785 (1993).
15. Kepple, P. C., and Griem, H. R., *Phys. Rev.* **173**, 317(1968).
16. Gigosos, M., private communication.
17. Büscher, S., Glenzer, S., Wrubel, Th., and Kunze, H.-J., *J. Phys. B.: At. Mol. Opt. Phys.*, **29**, 4107(1996).
18. Vitel, Y., *J. Phys. B.: At. Mol. Phys.*, **20**, 2327(1987).
19. Döhrn, A., Nowack, P., Könies, A., Günter, S., Helbig, V., *Phys. Rev. E*, **53**, 6389(1996).

IONIZATION AND RECOMBINATION II

Recombination at Ultra-low Energies

M. R. Flannery and D. Vrinceanu

School of Physics, Georgia Institute of Technology, Atlanta, GA 30332-0430[1]

Abstract. Three-body electron-ion recombination is described at ultralow electron temperatures T_e. At $4K$, the initial stage involves extremely rapid collisional capture into high Rydberg states $n > 200$ with high angular momentum $l \approx n - 1$ at a rate $\sim T_e^{-4.5}$. This is followed by extremely slow collisional-radiative decay. The key collisional mechanism appears to be collisional l-mixing of the Rydberg atoms A(n) by ions and electrons until sufficiently low l's are attained so as to permit relatively rapid radiative decay to the lowest electronic levels. This sequence is in direct contrast to the sequence of much slower collisional capture at higher T_e followed by the much faster decay of A(n) by electron collisions to lower levels where radiative decay completes the recombination. At ultra-low temperatures, the rate limiting sequence is therefore collisional l-mixing followed by radiative decay in contrast to recombination at much higher energies and electron densities $N_e \sim 10^8$ cm^{-3}, where the rate limiting step is the initial collisional or radiative capture at intermediate $T_e (\sim 1$ eV$)$ and higher T_e (~ 10 eV) respectively.

An exact classical solution for l-mixing is obtained.

INTRODUCTION

Three-body collisional capture

$$e^- + A^{Z+} + e^- \longrightarrow A^{(Z-1)+}(n) + e^- \qquad (1)$$

and radiative capture

$$e^- + A^{Z+} \longrightarrow A^{(Z-1)+}(n) + h\nu \qquad (2)$$

in an electron gas at cryogenic temperatures $T \sim 4K$ and lower are considered. Simple classical expressions for the cross section σ and rate $\alpha \sim Z^3 T_e^{-4.5}$ of (1) are derived. We shall also provide an exact classical solution for the angular momentum mixing $n, l \to n, l'$ of Rydberg levels by the time varying electric field $\mathcal{E} = -[Z_1 e/R^3(t)]\mathbf{R(t)}$ generated by distant collisions with an ion of charge $Z_1 e$. The work has particular significance to the production of anti-hydrogen at 4 K and to proposed experiments on electron recombination with Rb$^+$ and Xe$^+$ at 5 mK.

---

[1] This Research is supported by the U.S. Air Force Grant No. F49620-96-1-0142

RADIATIVE RECOMBINATION

The quantal cross section for photoionization of $A(n,l)$, in conventional notation [1], is

$$\sigma_I(h\nu) = \left\{\left[\frac{64\alpha}{3\sqrt{3}}(\pi a_0^2)\frac{n}{Z}\right]\left(\frac{I_n}{h\nu}\right)^3\right\}\frac{2l+1}{n^2}G_{nl}(h\nu) \equiv {}_K\sigma_n(h\nu)\frac{2l+1}{n^2}G_{nl}(h\nu)$$

where ${}_K\sigma_n(h\nu)$ is the semiclassical (Kramers) photoionization cross section [2] and the bound-free Gaunt factor $G_{nl} \sim 1$ and $\omega^{-(l+1/2)}$ as $\omega = h\nu/I_n$ tends to 1 and ∞ respectively. The former limit is appropriate to ultracold ejected electrons while the latter limit indicates that low l' are favored. The detailed balance relation is

$$g_e g_A^+ k_e^2 \sigma_R(E) = g_\nu g_A k_\nu^2 \sigma_I(h\nu)$$

where the statistical weights g_i are $\{2, 1, 2, 2(2l+1)\}$ for e, A^+, $h\nu$ and A, respectively and where $k_\nu^2/k_e^2 = \alpha^2(h\nu)^2/(4EI_H) = (v_0/c)^2(I_H/4E)$ with $I_H = (e^2/2a_0)$. The cross section for radiative recombination of electrons with energy E is then

$$\sigma_R(n) = \left\{\left(\frac{8\pi a_0^2 \alpha^3}{3\sqrt{3}}\right)\frac{2Z^2 I_H}{E}\right\}\left[\frac{2}{n}\frac{I_n}{(I_n+E)}\right]G_n(E+I_n)$$

which is much smaller by a factor $\mathcal{O}(\alpha^2)$ than σ_I. G_n is the l-averaged Gaunt factor $n^{-2}\sum_{l=0}^{n-1}(2l+1)G_{nl}$. The corresponding rate is

$$\alpha_{RR}(n, T_e) = \langle v_e \rangle\left(\frac{8\pi a_0^2 \alpha^3}{3\sqrt{3}}\right)\left[\frac{2Z^2 I_H}{kT_e}\right]\frac{2}{n}F_n(b_n)$$

with

$$F_n(b_n) = b_n e^{b_n}\int_1^\infty \frac{G_n(\omega)}{\omega}e^{-b_n\omega}\,d\omega$$

where $b_n = I_n/kT_e$ and $\omega = h\nu/I_n$. For ultracold recombination, $G_n \sim 1$ so that

$$\alpha_{RR}(n, T_e \to 0) = \left\{\langle v_e\rangle\left(\frac{8\pi a_0^2 \alpha^3}{3\sqrt{3}}\right)\left(\frac{2Z^2 I_H}{kT_e}\right)\right\}\frac{2}{n}b_n e^{b_n}E_1(b_n)$$

$$\stackrel{\text{def}}{\equiv} \alpha_0(T_e)\frac{2}{n}b_n e^{b_n}E_1(b_n) \qquad (3)$$

the semiclassical (Kramers) rate [1] where

$$\alpha_0(T_e) = 5.966\ 10^{-13}\left(\frac{300}{T_e}\right)^{1/2}Z^2\ \text{cm}^3\text{s}^{-1}$$

and where E_1 is the first exponential integral $\int_b^\infty e^{-x}\,dx/x$. Since $be^b E_1$ increases from 0 to 1 as b increases from 0 to ∞, (3) shows that Radiative Recombination favors transitions to low n.

Ultracold Radiative Recombination into the high level n_0 determined by $kT_e = I_n$ then proceeds at the rate

$$\alpha_{RR}(n) = 7.12 \; 10^{-13} \left(\frac{300}{T_e}\right)^{1/2} \frac{Z^2}{n} \; \text{cm}^3\text{s}^{-1}$$

The rate into all levels $n' \geq n$ is

$$\alpha_{RR}(n' \geq n) = \alpha_0(T_e) \int_0^{b_n} e^b E_1(b) \, db$$

$$= \alpha_0(T_e)[\gamma + \ln b_n + e^{b_n} E_1(b_n)]$$

where $\gamma = 0.5772$ is the Euler's constant. For $b_n = 1$, then

$$\alpha_{RR}(n' \geq n_0) = 7 \; 10^{-13} \left(\frac{300}{T_e}\right)^{1/2} Z^2 \; \text{cm}^3\text{s}^{-1}$$

The rate into all levels $n' \leq n$ including the effect of cascades [3] is

$$\alpha_{RR}(n' \leq n) = 3.80 \; 10^{-12} \left(\frac{300}{T_e}\right)^{0.63} Z^2 \; \text{cm}^3\text{s}^{-1}$$

COLLISIONAL CAPTURE RATE

The forward recombination and reverse ionization rates α_R and α_I of process (1) are interconnected by detailed balance

$$\tilde{N}_e \tilde{N}(n) \alpha_I(n) = \tilde{N}_e \tilde{N}_+ \alpha_R(n) \tag{4}$$

where the Saha-Boltzmann distribution $\tilde{N}$ of bound levels n in collisional equilibrium with the continuum state distributions $\tilde{N}_e$ and $\tilde{N}_+$ is

$$\frac{\tilde{N}(n)}{\tilde{N}_+ \tilde{N}_0} = \frac{h^3}{(2\pi m_e kT_e)^{3/2}} \, n^2 \exp(I_n/kT_e) \quad ; \quad I_n = Z^2 \frac{I_H}{n^2} \tag{5}$$

where I_n is the ionization energy level n. It is useful to rewrite (5) in the equivalent forms:

$$\frac{\tilde{N}(n)}{\tilde{N}_+ N_e} = Z \, (\pi R_e^2)^{3/2} \, (n^2/n_0^2) \exp(n_0^2/n^2) \quad ; \quad I_n/kT_e = n_0^2/n^2 \tag{6}$$

in terms of the characteristic Coulomb Radius $ZR_e = Z(e^2/kT_e) = 2n_0^2(a_0/Z)$, and

$$\frac{\tilde{N}(n)}{\tilde{N}_+ N_e} = \frac{Z}{2} \, (\pi R_e^2)^{3/2} \, b_n^{-5/2} \frac{db_n}{dn} \, \exp(b_n) \quad ; \quad b_n = I_n/kT_e \tag{7}$$

where $db_n/dn = (2b_n)/n$. The distribution (6) exhibits a minimum at $n = n_0$ which corresponds to $b_n = 1$.

The rate for ionization of $A(n)$ by electrons of energy E is

$$\alpha_I(n) = \langle v_e \sigma_I(E) \rangle = \langle v_e \rangle \int_0^\infty \epsilon\, e^{-\epsilon}\, \sigma_I(\epsilon)\, d\epsilon \quad ; \quad \epsilon = E/kT$$

A. Detailed Balance Rate: The Thomson collisional cross section [4]

$$\sigma_I(E) = \pi e^4 \frac{Z^2}{E}\left(\frac{1}{E} - \frac{1}{I_n}\right) = \pi R_e^2 \frac{Z^2}{\epsilon}\left(\frac{1}{b_n} - \frac{1}{\epsilon}\right) \tag{8}$$

for electron ejection from the Rydberg $A(n)$, yields the rate of ionization

$$\alpha_I(n) = Z^2 \pi R_e^2 \langle v_e \rangle \frac{e^{-b_n}}{b_n}\left(1 - b_n e^{b_n} E_1(b_n)\right)$$

From detailed balance (4) and (7), the rate of collisional capture into level n is then:

$$\alpha_{CC}(n) = Z^3 (\pi R_e^2)^{5/2} \langle v_e \rangle\, N_e\, (n b_n^{5/2})^{-1}[1 - b_n e^{b_n} E_1(b_n)] \tag{9}$$

The factor multiplying [ ] in (9) varies as $Z^3 n^4 T_e^{-2}$. Since $b e^b E_1(b) \to 1 - b^{-1}$ for very large b then $\alpha_{CC} \sim Z^3 n^6 T_e^{-1}$ at ultracold energies when $kT_e \ll I_n$.

Recombination into all levels $\Sigma \leq n$ proceeds at a rate

$$\alpha_{CC}(\Sigma) = \sum_{n'=1}^{n} \alpha_{CC}(n') \approx \int_1^n \alpha_{CC}(n')\, dn' = \int_{b_n}^{b} \alpha_{CC}(n) \frac{n}{2b_n}\, db_n \tag{10}$$

where $b = I_1/kT_e$ can be replaced by infinity since $b_1 \gg b_n$. Then

$$\alpha_{CC}(\Sigma) = \frac{Z^3}{2}(\pi R_e^2)^{5/2}\, \langle v_e \rangle\, N_e I(b_n) \tag{11}$$

where

$$I(b_n) = \int_{b_n}^{\infty} \frac{1 - b_n e^{b_n} E_1(b_n)}{b^{7/2}}\, db$$

Since

$$R_e^{5/2}\langle v_e \rangle \equiv \frac{8.094\ 10^{-9}}{T(K)^{4.5}} = 5.769\ 10^{-20}\left(\frac{300}{T(K)}\right)^{4.5}$$

$$= \frac{5.144\ 10^{-27}}{E_{eV}^{4.5}} = \frac{4.070\ 10^{-32}}{E_{Ry}^{4.5}}$$

for T_e expressed in different units, then

$$\alpha_{CC}(\Sigma) = 6.71 \; 10^{-20} N_e \left(\frac{300}{T_e}\right)^{4.5} R(b_n) \quad (12)$$

which exhibits the characteristic $Z^3 T_e^{-4.5}$ dependence [5]. The function is $R(b_n) = I(b_n)/I(1)$. Since the equilibrium distribution (7) exhibits a maximum at $kT_e = I_n$, where $n = n_0$ or $b_n = 1$, there is a bottleneck at $n = n_0$ and it may be argued that all levels $n \geq n_0$ are collisionally connected with the continuum and may be regarded as subject to re-ionization alone, while levels $n \leq n_0$ are connected mostly with lower bound levels and are therefore subject only to further de-excitations. Then R in (12) is unity.

In terms of this bottleneck level n_0,

$$ZR_e = 2n_0^2(a_0/Z) \quad ; \quad \langle v_e \rangle = \left(\frac{2}{\pi}\right)^{1/2} (Zv_0/n_0)$$

Then (9) provides the CC-rate into level n_0 as

$$\alpha_{CC}(n_0) = 64 \; \pi^2 \; Z^{-6} \; n_0^8 (a_0^2 v_0)(N_e a_0^3) \; \{[1 - eE_1(1)] = 0.4\}$$

and (11) the CC-rate into all levels $n \leq n_0$ as

$$\alpha_{CC}(n \leq n_0) = 32 \; \pi^2 n_0^9 (a_0^2 v_0)(N_e a_0^3) \; \{I(1) = 0.1329\} \quad (13)$$

The strong n_0-dependencies are therefore highlighted. Since $n_0^2 \sim Z^2/kT_e$, the $Z^3 T_e^{-4.5}$ dependence of (11) is recovered from (13). Although the Thomson cross section (8) can be replaced by a variety of cross sections derived from more elaborate binary-encounter classical theory, the key dependence $Z^3 T_e^{-4.5}$ is still preserved. Allowance for the discrete n' summation rather than the continuous integration in (10) shows that $\alpha_R(n)/2$ is simply added to (11).

B. Direct CC Rate: In the three-body CC (1), some of the energy gained by the incident (test) electron 1 moving with total energy $E_1 = (3/2)kT_e$ and kinetic energy $T_1(R) = 3/2kT_e + |V_{13}(R)|$ in the field of the ion 3 with charge Ze is lost upon repulsive Coulomb collision with an ambient (field) electron 2 of average energy $E_2 = (3/2)kT_e$ in an electron gas bath at temperature T_e. For (1-2) scattering by angle Ψ in the (1-2) CM, the energy in the laboratory (fixed ion 3) frame lost by 1 to 2 is

$$\Delta E = [T_1(R) - E_2] \sin^2 \frac{\Psi}{2} = |V_{13}(R)| \sin^2 \frac{\Psi}{2}$$

In order that the total energy of 1 is changed from E_1 to all bound energies $E_1' \leq -\epsilon_0 kT_e$, the (1-2) scattering must satisfy

$$|V_{13}(R)| \sin^2 \frac{\Psi}{2} \geq (\frac{3}{2} + \epsilon_0) kT_e \quad (14)$$

In a repulsive Coulomb (1-2) encounter

$$\sin^2 \frac{\Psi}{2} = \frac{b_c^2}{b^2 + b_c^2} \quad , \quad b_c = \frac{e^2}{2E_{rel}}$$

where E_{rel} is the energy of (1-2) relative motion which, on average, is $3/2kT_e$. Condition (14) then provides the averaged cross section

$$\sigma_{12}(R) = \frac{1}{9}\pi R_e^2 \left(\frac{R_0}{R} - 1\right) \quad , \quad R_0 = \frac{ZR_e}{\epsilon_0 + 3/2}, \quad R_E = \frac{e^2}{kT_e} \tag{15}$$

for the formation of bound states with energy $E_1' \leq -\epsilon_0 kT_3$ by (e - e) collisions at test-electron - ion separation $R \leq R_0$. For $R \geq R_0$, σ_{12} vanishes. When $\epsilon_0 = 0$, R_0 is the Thomson trapping radius [1]. The collisional capture cross section at total energy

$$E = \frac{1}{2}m_e v^2(R) + V_{13}(R) = \frac{1}{2}m_e \dot{R}^2 + V_{13}(R) + 2m_e E \frac{\rho^2}{R^2}$$

is

$$\sigma_{CC}(E) = 2\pi \int_0^\infty \rho d\rho \int_{-\infty}^\infty W(R, E) \, dt \tag{16}$$

where $W(R, E) \, dt$ is the probability of the energy-reducing (1-2) collision occuring within the time interval $dt = dR/v_R(R)$ about $R(t)$ of (1-3) trajectory, so that

$$\sigma_{CC}(E) = 4\pi \int_0^\infty \rho d\rho \int_{R_i(\rho)}^\infty W(R, E) \, \frac{dR}{\dot{R}}$$

where R_i is the orbit's pericenter at which $\dot{R} = 0$. On reversing the order of integration,

$$\sigma_{CC}(E) = 2\pi \int_{R_i(0)}^\infty W(R, E) \, dR \int_0^{\rho_1} \frac{d\rho^2}{\dot{R}(E, E)}$$

where $\rho_1^2(R, E) = R^2[1 - V(R)/E]$ gives the maximum impact parameter ρ_1 that can just acces the point R. For purely atractive interactions $R_i(0) = 0$ and $V = E$ at $R_i(0)$ when repulsion occurs at zero impact parameter. Since

$$\int_0^{\rho_1} \frac{d\rho^2}{\dot{R}(R, E)} = \frac{2R^2}{v(R)}[1 - \frac{V_{13}(R)}{E}]$$

then

$$\sigma_{CC}(E) = 4\pi \int_{R_i(0)}^\infty \frac{W(R, E)}{v(R)}[1 - \frac{V_{13}(R)}{E}] \, dR$$

The (1-2) collision probability within an element ds of the (1-3) orbit is

$$W(R, E) \, dt = \frac{ds}{\lambda} = N_e \sigma_{12}(R) v(R) \, dt$$

where λ is the free path length towards collision. Hence (16) reduces exactly to

$$\sigma_{CC}(E) = 4\pi N_e \int_0^{R_0} \sigma_{12}(R) \left(1 - \frac{V_{13}(R)}{E}\right) R^2 \, dR$$

which with (15) and $V_{13}(R) = -Ze^2/R$, yields

$$\sigma_{CC}(E = \epsilon kT) = \frac{4}{27}\pi^2 Z^3 R_e^5 N_e \left[\frac{1 + 3(\epsilon_0 + 3/2)/\epsilon}{2(\epsilon_0 + 3/2)^3}\right] \quad (17)$$

The factor $Z^3 R_e^3$ originates from the reaction volume R_0^3 while the additional R_e^2 originates from the (1-2) scattering cross section (15).

The associated collisional capture rate

$$\alpha_{CC} = \langle v_e \rangle \int_0^\infty \sigma_{CC}(\epsilon)\epsilon e^{-\epsilon} \, d\epsilon$$

is then

$$\alpha_{CC}(\epsilon_0) = \frac{4}{27}\pi^2 Z^3 \langle v_e \rangle R_e^5 N_e \left[\frac{3}{2} \frac{\epsilon_0 + 11/6}{(\epsilon_0 + 3/2)^3}\right]$$

The rate for production of all levels $\leq -kT_e$ is therefore

$$\alpha_{CC}(1) = 2.29 \, 10^{-20} \left(\frac{300}{T_e}\right)^{4.5} N_e \quad (18)$$

which is in excellent agreement with the Monte-Carlo result of Mansbach and Keck [5] and a factor of 3 lower than the detailed balance rate (12) derived from the Thompson ionization cross section (8). This simple result $\sim Z^3 T_e^{-4.5}$ is preserved even in elaborate classical treatments of the ionization [6] or of the recombination [7] with correct averaging of the dynamical quantities.

The frequencies $\nu_{CC} = \alpha_{CC} N_e$ and $\nu_{RR} = \alpha_{RR} N_e$ are compared in Table (1) at various T_e for typical $N_e \approx 10^8$ cm^{-3}. Since $\nu_{CC}/\nu_{RR} \sim N_e/T_e^4$ it is seen that Collisional Capture dominates Radiative Recombination at cryogenic and lower T_e by large factors $\sim 10^7$. Also CC $\approx$ RR at $T_e \sim 300$K while CC $\ll$ RR at 1eV.

TABLE 1. Collisional Capture and Radiative Recombination frequencies for different temperatures

T_e	$\nu_{CC}(s^{-1})$	$\nu_{RR}(s^{-1})$
4 K	6.3×10^4	5.8×10^{-3}
300 K	2.3×10^{-4}	3.8×10^{-4}
1 eV	2.0×10^{-11}	3.9×10^{-5}

RADIATIVE DECAY RATES

Level i decays radiatively to level j at a rate

$$A(i \to j) = \frac{1}{6} A_0 (E_{ij}/I_H)^3 (S_{ij}^{au}/g_i)$$

where the line strength S_{ij} in a.u. is $\sum_{\alpha,\alpha'} |\langle nlm|\vec{r}|n'l'm'\rangle|^2$ and $A_0 = \alpha^3/t_0 = 1.604\ 10^{10}\ s^{-1}$ where t_0 is the a.u. of time $(e^2/a_0)/\hbar$. Level nl decays to all lower levels s at frequency

$$\nu_r(nl) = \sum_{s<n} A(nl \to s, l+1)$$

$$= \frac{2}{3} Z^4 A_0 \left[\frac{1}{n^3(l+\frac{1}{2})^2} \to \begin{cases} n^{-5} & , \text{for } l = n-1 \\ 4/9\ n^{-3} & , \text{for } l = 1 \end{cases} \right]$$

The circular ($l = n-1$) states, mainly populated by collisional capture (CC), are therefore very long lived and decay at frequency

$$\nu_r(n, l = n-1) \approx \frac{10^{10}}{n^5} \ll \nu_r(np \to 1s) \approx \frac{4}{9} \frac{10^{10}}{n^3}\ (s^{-1})$$

very much less than that for the ($np \to 1s$) transition. The levels $n_0 \approx 200$ populated by CC at 4K have therefore a long radiative lifetime $\tau_r \sim 32$ s when compared with $2\ 10^{-3}$ for the ($n_0 p \to 1s$) transition. The overall rate of recombination via capture into level n followed by radiative decay is

$$\alpha \sim \frac{\nu_C \nu_r}{\nu_C + \nu_r}$$

where $\nu_C = \nu_{CC} + \nu_{RR}$. Since the radiative decay frequencies for $n \approx n_0$ are much less than the frequency $\alpha_{CC} N_e^2$ for collisional capture at ultralow temperatures, the overall recombination rate is controlled by the limiting rate ν_r. The question now addressed is the further effect of collisions between Rydberg atoms and ambient electrons and ions.

INELASTIC COLLISIONS AND STARK MIXING

Three important frequencies or timescales in the collision, depicted in Figure (1) are:

(A) The projectile Rotation (Collision) Frequency ω_R of the Projectile

$$\omega_R = \dot{\Phi} = \frac{bv}{R^2} \xrightarrow{\text{large b}} \frac{v}{b} \approx \frac{1}{\tau_{coll}}$$

(B) The Transition (Orbital) Frequency ω_n of the Rydberg electron:

$$\omega_{ij} = \frac{E_i - E_j}{\hbar}$$

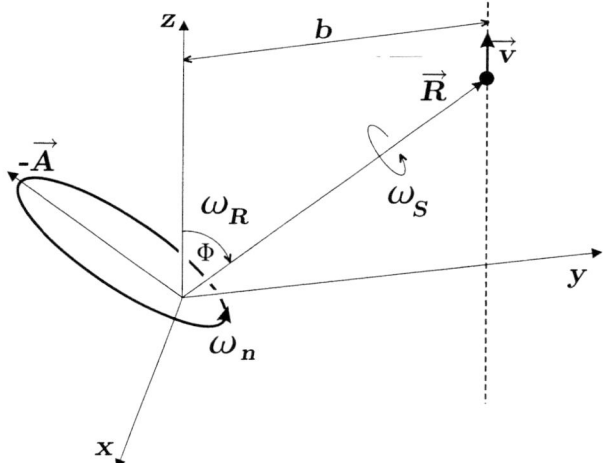

FIGURE 1. The three important frequencies in the Rydberg projectile collision

which for transitions $n \to n-1$ between neightboring levels is simply $\omega_{n,n-1} = \omega_0/n^3 = \omega_n = v_n/a_n$, the orbital frequency, where $a_n = n^2 a_0/Z$ and $v_n = Zv_0/n$.

(C) The Stark Precession Frequency ω_S for precession of **A** about **R**:
Under the ion Electric Field $\mathcal{E} = Z_1 e/R^2$ the Stark frequency

$$\omega_S = \frac{3}{2}\frac{a_n}{n\hbar}(e\mathcal{E}) = \frac{3}{2}\frac{Z_1 a_n v_n}{R^2}$$

provides the precessional frequency of the Runge-Lenz (eccentricity) vector **A** of the Rydberg orbit about the field direction.

By considering the $\exp(i\omega t)$ factor in time dependent perturbation theory, the following types of collisions can be characterized as in [8].

1. *Orbital Adiabatic:* $\omega_R \ll \omega_n$

$$\frac{b}{a_n} \gg \frac{v}{v_n}$$

where the orbital electron adjust itself adiabatically to the ion perturbation and no transitions occur.

2. *Orbital Sudden:* $\omega_R \gg \omega_n$

$$\frac{b}{a_n} \ll \frac{v}{v_n}$$

which causes inelastic $(n \to n')$ transitions

3. *Stark Adiabatic:* $\omega_R \ll \omega_S$

$$\frac{bv}{R^2} \ll \frac{3}{2}\frac{Z_1 a_n v_n}{R^2}$$

$$\frac{b}{a_n} \ll \frac{3}{2}\frac{Z_1 v_n}{v}$$

where elastic collisions occur.

4. *Stark Sudden:* $\omega_R \gg \omega_S$

$$\frac{b}{a_n} \gg \frac{3}{2}\frac{Z_1 v_n}{v}$$

which causes l-mixing $(nl \to nl')$ transitions.

5. *The weak Field Condition* $\omega_S \ll \omega_n$ i.e., the Stark splitting $\hbar\omega_S \ll \Delta E_{n,n-1}$ implies that $e\mathcal{E} \ll 2/3(e^2/a_n^2)$ so that the collision occurs for distant encounters at $R \gg a_n$, the Rydberg radius, for fields $e\mathcal{E} = Z_1 e^2/R^2$. The electron's orbital time is then much shorter than any characteristic time to cause changes to the elliptical orbit. The vectors **A** and angular momentum **L** which are constant for the unperturbed motion become good dynamical variables for the description of the perturbed motion.

With respect to orbital motion, the collision is sudden or adiabatic according to $b < b_C$ and $b > b_C$, respectively, where $b_C = (v/v_n)a_n$. With respect to the Stark frequency, collision is adiabatic or sudden according to $b < b_S$ and $b > b_S$, respectively, where $b_S = (v_n/v)a_n$. The impact parameter b-space can then be partitioned according to:

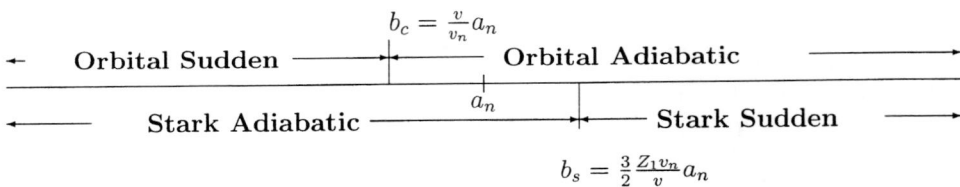

As v decreases, b_S increases outward and b_C increases inward, thereby limiting the extent of the sudden regions where n changes and l-mixing occurs. The (v,b)-phase space can be partitioned into the four characteristic regions illustrated in Figure (2). For $v > v_n$, the (n,l) changing and l changing (Orbital and Stark Sudden) shaded regions overlap and expand, in direct contrast to ultracold speeds $v \leq v_n$ where the Orbital and Stark Adiabatic (clear) regions increase and the shaded regions diminish and do not overlap indicating few collisional changes.

CLASSICAL THEORY OF STARK MIXING COLLISIONS

In addition to the energy E_n and angular momentum **L** of an unperturbed Rydberg electron moving with velocity **v** in an elliptical orbit with eccentricity

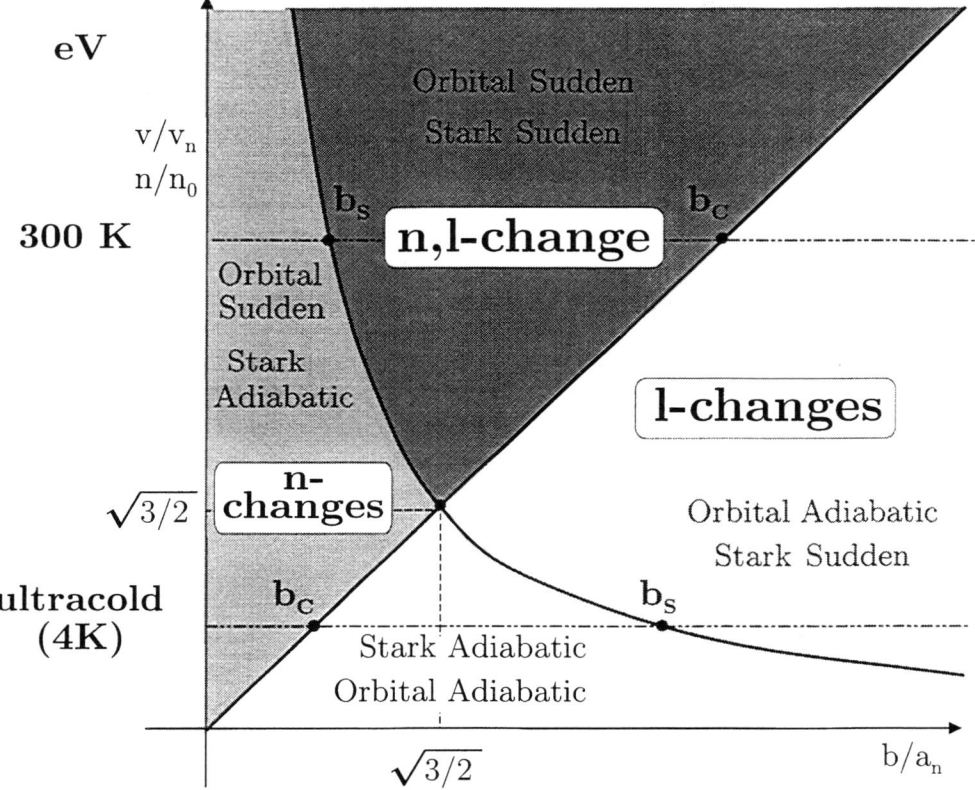

FIGURE 2. Partitioning the v-b phase space map into 4 regions characterized mainly by: (a) energy changes, (b) energy and angular momentum changes, (c) angular momentum changes and (d) no changes under the weak field condition (5)

$\epsilon = (1 - L^2/ma_nk)^{1/2}$ and semi-major axis $a_n = k/2|E_n|$ in the field $-k/r$, the Runge-Lenz (or eccentricity) vector

$$\mathbf{A} = v_n^{-1}[\mathbf{v} \times \mathbf{L} - \frac{k}{r}\mathbf{r}]$$

$$v_n = \langle v^2 \rangle^{1/2} = \frac{Zv_0}{n} \qquad k = Ze^2 \qquad v_0 = \frac{e^2}{\hbar}$$

directed toward the pericenter and normalized to angular momentum units is also conserved in magnitude $A = (ma_nk)^{1/2}\epsilon$ and direction. Moreover $\mathbf{A} \cdot \mathbf{L} = 0$ and $A^2 + L^2 = ma_nk = n^2\hbar^2$. The dipole classically averaged over the orbit time is

$$\mathbf{d} = -e\langle \mathbf{r} \rangle = \frac{3e}{2}(a_n\boldsymbol{\epsilon}) = \frac{3e}{2}\frac{\mathbf{A}}{mv_n}$$

In a.u., $A^2 = n^2 - (l + 1/2)^2$ and the dipole $\mathbf{d} = (3/2)n\mathbf{A}$ is maximum $3/2n^2$ for highly eccentric orbits $l \approx 0$ and is vanishingly small for circular orbits ($l \lesssim n$). In presence of an electric field of intensity $\boldsymbol{\mathcal{E}}$, the angular momentum $\mathbf{L}$ changes at the rate

$$\frac{d\mathbf{L}}{dt} = -e\mathbf{r} \times \boldsymbol{\mathcal{E}}$$

so that the change of $\mathbf{L}$ (classically) averaged over one orbital period T is given by

$$\frac{\Delta \mathbf{L}}{T} = \langle \frac{d\mathbf{L}}{dt} \rangle = -\frac{e}{T}\int_0^T (\mathbf{r} \times \boldsymbol{\mathcal{E}})\,dt$$

Approximation 1: Assume that the perturbation is adiabatic ($\omega_R \ll \omega_n$) with respect to the orbital frequency ω_n. Then $\langle \mathbf{r} \times \boldsymbol{\mathcal{E}} \rangle = \langle \mathbf{r} \rangle \times \langle \boldsymbol{\mathcal{E}} \rangle$ so that

$$\frac{\Delta \mathbf{L}}{T} = -\boldsymbol{\omega}_S \times \mathbf{A}$$

in terms of the Stark frequency

$$\boldsymbol{\omega}_S(t) = \frac{3}{2}\frac{e\langle \boldsymbol{\mathcal{E}} \rangle}{mv_n} = \frac{3}{2}\left(\frac{a_n}{n\hbar}\right)e\langle \boldsymbol{\mathcal{E}} \rangle$$

which is $3n\langle \boldsymbol{\mathcal{E}}_{au} \rangle/2$ in a.u.

Approximation 2: Assume also that $\langle \boldsymbol{\mathcal{E}} \times \mathbf{L} \rangle = \langle \boldsymbol{\mathcal{E}} \rangle \times \langle \mathbf{L} \rangle$ for slowly varying $\boldsymbol{\mathcal{E}}$. It can be shown that the change $\mathbf{A}$ over one period is

$$\frac{\Delta \mathbf{A}}{T} = \langle \frac{d\mathbf{A}}{dt} \rangle = -\boldsymbol{\omega}_S \times \mathbf{L}$$

Approximation 3: Replace the changes $\Delta \mathbf{L}$ and $\Delta \mathbf{A}$ over one period by $\dot{\mathbf{L}}T$ and $\dot{\mathbf{A}}T$ to give the set of coupled equations

$$\frac{d\mathbf{A}}{dt} = -\boldsymbol{\omega}_S(t) \times \mathbf{L} \qquad \frac{d\mathbf{L}}{dt} = -\boldsymbol{\omega}_S(t) \times \mathbf{A}$$

where $\boldsymbol{\omega}_S$ tracks the time variation of the electric field in magnitude and direction. Under the substitution

$$\mathbf{X} = \frac{\mathbf{L}+\mathbf{A}}{2} \qquad \mathbf{Y} = \frac{\mathbf{L}-\mathbf{A}}{2} \qquad (19)$$

the above set of differential equations becomes decoupled to yield

$$\frac{d\mathbf{X}}{dt} = -\boldsymbol{\omega}_S(t)\times\mathbf{X} \qquad \frac{d\mathbf{Y}}{dt} = +\boldsymbol{\omega}_S(t)\times\mathbf{Y} \qquad (20)$$

where $X^2 = Y^2 = (L^2+A^2)/4 = n^2\hbar^2/4$. The classical analysis for constant electric fields is given Born in [9]. For time independent $\boldsymbol{\omega}_S$, both $\mathbf{X}$ and $\mathbf{Y}$ precess with constant frequency ω_S about the (fixed) direction of $\boldsymbol{\omega}_S$. For general time-varying $\boldsymbol{\omega}_S$, the system of differential equations (20) does not have an exact solution. Percival and Richards [8] have used classical perturbation theory to solve (20) and then provided a diffusional theory of angular momentum mixing. Bellomo et al [10] approached the same problem by proceeding via the time evolution propagator $U_{rot}^\pm(t,t_0)$ for $\mathbf{X}$ and $\mathbf{Y}$ in the rotating frame, an approach which results in formulae too complicated for physical changes $\Delta\mathbf{L}$ and $\Delta\mathbf{A}$ to be extracted.

An **Exact Analytical Solution** is however possible when the external time dependence of $\boldsymbol{\omega}_S(t)$ is provided by the electric field $\boldsymbol{\mathcal{E}} = Z_1e(-\hat{R})/R^2$ of a projectile ion or electron of charge Z_1e passing the Rydberg atom at large distances $R\gg a_n$, the condition for the weak field approximation. Then

$$\boldsymbol{\omega}_S(t) = -\frac{3}{2}\left(\frac{Z_1e^2}{mv_n}\right)\frac{\hat{R}(t)}{R^2(t)} = -\frac{3}{2}\left(\frac{Z_1}{Z}\right)\left(\frac{a_nv_n}{R^2(t)}\right)\hat{R}(t)$$

which varies in both magnitude and direction. Since the vector $\mathbf{R}$ rotates at the frequency

$$\boldsymbol{\omega}_R(t) = -\frac{d\Phi}{dt}\hat{\imath} = \frac{vb}{R^2(t)}\hat{\imath}$$

then the vector

$$\boldsymbol{\alpha}(t) = \frac{\boldsymbol{\omega}_S}{\omega_R} = \frac{3}{2}\left(\frac{Z_1}{Z}\right)\frac{a_nv_n}{bv}\hat{R}(t)$$

has the constant magnitude $3/2(Z_1/Z)(\tilde{b}\tilde{v})^{-1}$ for a particle with *reduced* speed $\tilde{v} = v/v_n$ in a given trajectory of reduced impact parameter $\tilde{b} = b/a_n$. Since the value $\alpha = 1$ separates the Stark Sudden ($\alpha\ll 1$) and the Stark Adiabatic ($\alpha\gg 1$) regions at $b = b_S$ (see Figure (2)), we refer to α as the Stark parameter. This suggests that Φ could be a more useful variable instead of time t. Hence the first equation in (20) is

$$\frac{d\mathbf{X}}{d\Phi} = \frac{d\mathbf{X}}{dt}\frac{d\Phi}{dt} = \boldsymbol{\alpha}(0,\sin\Phi,\cos\Phi)\times\mathbf{X} \qquad (21)$$

Since the tip of $\mathbf{X}$ moves on the surface of a sphere, the solution of this equation, the final $\mathbf{X}$ vector (corresponding to $t\to\infty$ or $\Phi = 0$), is a rotation of the initial

X vector (corresponding to $t \to -\infty$ or $\Phi = \pi$). The second equation in (20) has a similar solution. The exact form of the resulting rotation matrices are presented in the Appendix. As a result of the collision the vectors **L** and **A** change to

$$\begin{bmatrix} \tilde{L}_1 \\ \tilde{L}_2 \\ \tilde{L}_3 \end{bmatrix} = \frac{1}{\gamma^2} \begin{bmatrix} 1 + \alpha^2 \cos(\pi\gamma) L_1 & -\alpha^2 \gamma \sin(\pi\gamma) A_2 & \alpha(-1 + \cos(\pi\gamma)) A_3 \\ \gamma^2 \cos(\pi\gamma) L_2 & \gamma \sin(\pi\gamma) L_3 & -\alpha\gamma \sin(\pi\gamma) A_1 \\ \gamma \sin(\pi\gamma) L_2 & -(\alpha^2 + \cos(\pi\gamma)) L_3 & \alpha(1 - \cos(\pi\gamma)) A_1 \end{bmatrix} \quad (22)$$

and

$$\begin{bmatrix} \tilde{A}_1 \\ \tilde{A}_2 \\ \tilde{A}_3 \end{bmatrix} = \frac{1}{\gamma^2} \begin{bmatrix} (1 + \alpha^2 \cos(\pi\gamma)) A_1 & -\alpha^2 \gamma \sin(\pi\gamma) L_2 & \alpha(-1 + \cos(\pi\gamma)) L_3 \\ -\gamma^2 \cos(\pi\gamma) A_2 & \gamma \sin(\pi\gamma) A_3 & \alpha\gamma \sin(\pi\gamma) L_1 \\ \gamma \sin(\pi\gamma) A_2 & -(\alpha^2 + \cos(\pi\gamma)) A_3 & \alpha(1 - \cos(\pi\gamma)) L_1 \end{bmatrix} \quad (23)$$

where $\gamma = \sqrt{1 + \alpha^2}$ and the components of the initial and final vectors (**L**, **A**) and ($\tilde{\mathbf{L}}$, $\tilde{\mathbf{A}}$) are defined in the fixed coordinate frame of Figure (1). It easy to check that the above exact solutions satisfy the invariant relations

$$\tilde{\mathbf{L}} \cdot \tilde{\mathbf{A}} = \mathbf{L} \cdot \mathbf{A} = 0$$

and

$$\tilde{\mathbf{L}}^2 + \tilde{\mathbf{A}}^2 = \mathbf{L}^2 + \mathbf{A}^2 = n^2 \hbar^2$$

The new $(n, \tilde{\mathbf{L}})$ orbit is confined to a plane perpendicular to the new $\tilde{\mathbf{L}}$ and the energy is preserved (n is not changed). The key to the present exact solutions of (20) for a non-uniform rotating field is provided by the recognition that Φ is a more useful variable than time t and by the Rotation Matrix Algebra (in the Appendix).

In the Stark Sudden region, the Stark parameter α is $\ll 1$. In this limit, the solutions have the simpler form

$$\begin{bmatrix} \tilde{L}_1 \\ \tilde{L}_2 \\ \tilde{L}_3 \end{bmatrix} = \begin{bmatrix} (1 - 2\alpha^2) L_1 - 2\alpha A_3 \\ L_2 + \pi\alpha^2/2 \; L_3 \\ (1 - 2\alpha^2) L_3 - \pi\alpha^2/2 \; L_2 + 2\alpha A_1 \end{bmatrix}$$

and

$$\begin{bmatrix} \tilde{A}_1 \\ \tilde{A}_2 \\ \tilde{A}_3 \end{bmatrix} = \begin{bmatrix} (1 - 2\alpha^2) A_1 - 2\alpha L_3 \\ A_2 + \pi\alpha^2/2 \; A_3 \\ (1 - 2\alpha^2) A_3 - \pi\alpha^2/2 \; A_2 + 2\alpha L_1 \end{bmatrix}$$

The magnitude of the final angular momentum (for $\alpha \ll 1$) is then

$$\tilde{\mathbf{L}}^2 = \mathbf{L}^2 + 4\alpha(-A_3 L_1 + A_1 L_3) + 4\alpha^2 (\mathbf{A}^2 - A_2^2 - \mathbf{L}^2 + L_2^2)$$

On retaining only the first order term in α

$$\tilde{\mathbf{L}}^2 \approx \mathbf{L}^2 + 4\alpha (\mathbf{A} \times \mathbf{L}) \cdot \hat{y} = \mathbf{L}^2 + 4\alpha A L \cos \xi$$

where ξ is the angle between $\hat{\mathbf{A}} \times \hat{\mathbf{L}}$ and the Oy axis (Figure (1)).

Since the vectors $\mathbf{A}$ and $\mathbf{L}$ are always orthogonal, the angle ξ is then uniformly distributed over the range $(0, \pi)$ if $\mathbf{L}$ is uniformly distributed on the sphere with radius L, as for an equilibrium distribution of magnetic m-sublevels. The initial L-states then evolves to the final $\tilde{L}$ states with angular momentum distributed according to

$$\mathcal{F}(\tilde{L}) \, d\tilde{L} = \frac{1}{\pi} \, d\xi = \frac{2}{\pi} \frac{\tilde{L}}{\sqrt{(4\alpha AL)^2 - (\tilde{L}^2 - L^2)^2}} \, d\tilde{L}$$

In fact $\mathcal{F}$ can be identified with $W_{L \to \tilde{L}}$, the transition probability density for $L \to \tilde{L}$ transitions.

For arbitrary α, the transition probability W has in general not a simple form. It is however important to recognize that the probability W depends on the projectile velocity v and impact parameter b only via the Stark parameter $\alpha = 3Z_1 n / 2Zvb$.

The cross section for l-mixing in the the Rydberg atom is now

$$\frac{d\sigma}{d\tilde{L}} = 2\pi \int_{b_S}^{\infty} W_{L \to \tilde{L}} \, b \, db = 2\pi \left(\frac{a_n}{\tilde{v}_n}\right)^2 \left(\frac{3Z_1}{2Z}\right)^2 \int_0^1 \frac{1}{\alpha^3} W_{L \to \tilde{L}}(\alpha) \, d\alpha$$

where the α-integral is independent of the speed of the projectile and the Rydberg level n. This general result (where I is the above mentioned integral)

$$\frac{d\sigma}{d\tilde{L}} = 2\pi a_0^2 \left(\frac{3nv_0}{2v}\right)^2 \left(\frac{Z_1}{Z}\right)^2 I$$

is in agreement with the behavior of a normalized Born approximation presented in the recent review [11].

SUMMARY

In addition to providing radiative and collisional rates, the three main points of this paper are:

- Development of a simple classical expression for the cross section (16) and rate (18) of the three-body collisional capture (CC) into level n. The result agrees with the Monte-Carlo trajectory calculations [5] and preserves the $Z^3 T_e^{-4.5}$ dependence of α.

- Presentation of an exact classical analytical solution for the changes $\Delta \mathbf{L}$ and $\Delta \mathbf{A}$ in the angular momentum $\mathbf{L}$ and Runge-Lenz vector $\mathbf{A}$ of a Rydberg atom in the time varying electric field $\mathcal{E} = -Z_1 e \hat{R}(t) / R^2(t)$.

- Pointing out that recombination at ultracold temperatures proceeds by very rapid three-body collisional capture into levels nl, with high l, followed by angular momentum mixing by collisions mainly with ions. Recombination then becomes stabilized by radiative decay of the lower l-levels so produced. This sequence is in direct contrast to recombination at $T_e \gtrsim 1$ eV where collisional and radiative capture are generally the rate limiting steps.

APPENDIX: ROTATION MATRIX ALGEBRA

For a given direction $\hat{n}$, rotate vector $\mathbf{x}$ by angle ϕ about the fixed direction $\hat{n}$ to give
$$\mathbf{x}' = \cos\phi \mathbf{x} + (1 - \cos\phi)(\hat{n} \cdot \mathbf{x})\hat{n} + \sin\phi(\hat{n} \times \mathbf{x}) \equiv \hat{R}(\hat{n}, \phi)\mathbf{x}$$

This rotation corresponds to the matrix operator:

$$\hat{R}(\hat{n}, \phi) = \begin{bmatrix} c + (1-c)n_1^2 & (1-c)n_1 n_2 - sn_3 & sn_2 + (1-c n_1 n_3) \\ (1-c)n_1 n_2 + sn_3 & c + (1-c)n_2^2 & -sn_1 + (1-c)n_2 n_3 \\ -sn_2 + (1-c)n_1 n_3 & sn_1 + (1-c)n_2 n_3 & c + (1-c)n_3^2 \end{bmatrix}$$

where $c = \cos\phi$, $s = \sin\phi$ and $\hat{n} = \hat{n}(n_1, n_2, n_3)$. This matrix can also be written as
$$\hat{R}(\hat{n}, \phi) = \exp\phi(n_1 \hat{I}_x + n_2 \hat{I}_y + n_3 \hat{I}_z)$$

where I_x, I_y, I_z are the infinitesimal generators of the O(3) group, i.e.

$$\hat{I}_x = \begin{bmatrix} 0 & 0 & 0 \\ 0 & 0 & -1 \\ 0 & 1 & 0 \end{bmatrix} \quad \hat{I}_y = \begin{bmatrix} 0 & 0 & 1 \\ 0 & 0 & 0 \\ -1 & 0 & 0 \end{bmatrix} \quad \hat{I}_z = \begin{bmatrix} 0 & -1 & 0 \\ 1 & 0 & 0 \\ 0 & 0 & 0 \end{bmatrix}$$

The equation (21) can now be rewritten as
$$\frac{d\mathbf{X}}{d\Phi} = \alpha(\cos\Phi \hat{I}_z + \sin\Phi \hat{I}_y)\mathbf{X}$$

The transformation ("rotation frame")
$$\mathbf{X}' = \hat{R}(\hat{x}, \Phi)\mathbf{X}$$

eliminates the Φ dependence of the precessional frequency of the vector $\mathbf{X}'$ in the rotating frame since
$$\frac{d\mathbf{X}'}{d\Phi} = (\frac{d\hat{R}}{d\Phi}\hat{R}^{-1} + \alpha\hat{R}(\cos\Phi \hat{I}_z + \sin\Phi \hat{I}_y)\hat{R}^{-1})\mathbf{X}' = (\hat{I}_x + \alpha\hat{I}_z)\mathbf{X}'$$

The general solution for $\mathbf{X}'$ is then
$$\mathbf{X}'(\Phi) = \exp\left((\Phi - \Phi_0)(\hat{I}_x + \alpha\hat{I}_z)\right)\mathbf{X}'(\Phi_0)$$

or in the language of rotation matrices,

$$\mathbf{X}'(\Phi) = \hat{R}\left(\frac{1}{\sqrt{1+\alpha^2}}(\hat{n}_1 + \alpha\hat{n}_3),\ (\Phi - \Phi_0)\sqrt{1+\alpha^2}\right) \mathbf{X}'(\Phi_0)$$

The exact solution for $\mathbf{X}(t)$ in the laboratory frame Figure (1) is

$$\mathbf{X}(\Phi) = \hat{R}(\hat{n}_1, -\Phi)\hat{R}\left(\frac{1}{\sqrt{1+\alpha^2}}(\hat{n}_1 + \alpha\hat{n}_3),\ (\Phi - \Phi_0)\sqrt{1+\alpha^2}\right)\hat{R}(\hat{n}_1, \Phi_0)\mathbf{X}(\Phi_0)$$

Initially $\mathbf{X} = \mathbf{X}_i$ when the projectile approaches from infinity so that $\Phi_i = \pi$. The collision is completed when $\Phi_f = \pi$ so that

$$\mathbf{X}_f = \hat{R}\left(\frac{1}{\sqrt{1+\alpha^2}}(\hat{n}_1 + \alpha\hat{n}_3),\ -\pi\sqrt{1+\alpha^2}\right)\hat{R}(\hat{n}_1, \pi)\mathbf{X}_i$$

Upon replacing α by $-\alpha$ the solution $\mathbf{Y}_f$ of the second equation in (20) is

$$\mathbf{Y}_f = \hat{R}\left(\frac{1}{\sqrt{1+\alpha^2}}(\hat{n}_1 - \alpha\hat{n}_3),\ -\pi\sqrt{1+\alpha^2}\right)\hat{R}(\hat{n}_1, \pi)\mathbf{Y}_i$$

These last two equations and the substitutions (19) provide the relations (22) and (23) between the initial and final angular momentum and Runge-Lenz vectors.

REFERENCES

1. Flannery, M. R., *Electron-Ion and Ion-Ion Recombination in Atomic, Molecular and Optical Physics Handbook*, edited by G. W. F. Drake, New York: AIP Press, 1995, ch. 52, pp. 605-629.
2. Kramer, H. A., *Phil. Mag.* **46**, 836 (1923).
3. Stevefelt, J., Boulmer J., and Delpech, J. F., *Phys. Rev. A* **12**, 1246 (1975).
4. Thomson, J. J., *Phil. Mag.* **32**, 419 (1912).
5. Mansbach, P., and Keck, J., *Phys. Rev.* **181**, 275 (1965).
6. Vriens, L., in *Case Studies in atomic collision physics*, edited by E. W. McDaniel and M. K. C. McDowell, New-York: Elsevier, 1969, p. 337.
7. Flannery, M. R., and Vrinceanu, D., [in preparation].
8. Percival, I.C., and Richards, D., *J. Phys. B: Atom. Molec. Phys* **12**, 2051 (1979).
9. Born, M., *The Mechanics of the Atom*, New York: Ungar, 1960, p. 235.
10. Bellomo, P., Farrelly, D., and Uzer, T., *J. Chem. Phys.* **107**, 2499 (1997);
11. Beigman, I. L., and Lebedev, V. S., *Phys. Rep.* **250**, 95 - 328 (1995).

Recoil Ion Momentum Spectroscopy A 'momentum microscope' to view atomic collision dynamics

R. Dörner[1], V. Mergel[1], H. Bräuning[1,2,3], M. Achler[1],
T. Weber[1], Kh. Khayyat[1], O. Jagutzki[1], L. Spielberger[1],
J. Ullrich[4], R. Moshammer[1,4], Y. Azuma[5], M.H. Prior[3],
C.L. Cocke[2], H. Schmidt-Böcking[1]

[1] *Institut für Kernphysik, Universität Frankfurt, August Euler Str. 6, D60486 Frankfurt, Germany*
[2] *Department of Physics, Kansas State University, Manhattan, KS 66506*
[3] *LBNL, Berkeley, CA 94720*
[4] *Fakultät für Physik, Universität Freiburg, Germany*
[5] *Photon Factory, IMSS, KEK, Tsukuba, Japan.*

Abstract. Recoil ion momentum spectroscopy is a powerful tool for investigating the dynamics of ion, electron or photon impact reactions with atoms or molecules. It allows to measure the three dimensional momentum vector of the ion from those reactions with high resolution and 4π solid angle. It can be easily combined with novel 4π electron momentum analysers and for coincident detection of the projectile. This technique gives a complete image of the square of the correlated many body final state wave function in momentum space (i.e. fully differential cross sections) for the various reactions. The application to photo double ionization of helium by linear and circular polarized light is discussed.

I INTRODUCTION

How do we understand a match on Center Court in Wimbledon? And what's necessary to enjoy a football game? To do so, two ingredients are indispensable. First, we need good detectors: Our eyes, or at least a TV camera, which can see the whole arena and deliver images bright and with high enough resolution to us so that we can follow the action of the players and the motion of the ball. And second, we need at least a rudimentary theory, i.e., the knowledge of some rules of the game, to help us interpret what we see. These two ingredients open the way for a subtle interplay which enhances our understanding. Sometimes we learn the strategy of the game by carefully watching the players, and sometimes more detailed theoretical knowledge helps us to realize things we overlooked before. This article deals with only one side of the coin. It discusses a powerful camera to view the dynamics in the atomic collision arena. For the first time, an imaging technique called Cold Target Recoil Ion Momentum Spectroscopy (COLTRIMS) can give us a broad picture of atomic reactions in momentum space: capturing all aspects of the final state and providing bright, high resolution pictures of the breakup of atoms and molecules (see [1] for a recent review). Like a wide angle camera focused on the Center Court, it does not only show the motion of just one of the players, but captures the whole game with a resolution high enough that we can zoom in on any detail of interest. It delivers multi-dimensional momentum space images of the reaction, yielding the square of the many-body final state wave function.

One subfield of atomic physics has focused not on the dynamics in momentum space, but has achieved tremendous progress in the determination of energy eigenvalues. Laser spectroscopy, high resolution studies at synchrotron radiation sources, measurements in traps, dielectronic recombination studies and x-ray spectroscopy in ion storage rings provide us with unprecedented precise information on important quantities in atomic physics: energy levels and life times. These are time averaged quantities. They teach us the *static* long term properties of the atomic world. In the example of the tennis match, this corresponds to three-hour exposures of a photograph. It shows the probability distribution of the players on the court, like a time independent wave function in coordinate space which yields the eigenvalues. For observing a mountain or appreciating the beauty of a greek temple such long time exposures are clearly adequate, as they allow for extremely sharp pictures. For a tennis match, however, they obviously miss most of the fun. They will never tell us who won Wimbledon. Another very active subfield in atomic physics is working on unraveling the the basic many-body dynamics of atomic processes. Tremendous progress has been made here for example for (e,2e) and $(\gamma, 2e)$ reactions. Using coincident electron detection kinematically complete experiments could bee performed in this field (see eg. [2–9]. for $(\gamma, 2e)$ and [10–14] for (e,2e) and (e,3e)). Each of those experiments concentrates on certain angular or fragment energy ranges. Often a prohibitively large number of experiments would have to be performed to map the complete many particle momentum space. One of the fundamental technical challenges for fully differential experiments is the small solid angle of traditional spectrometers. The momentum space imaging techniques discussed in this paper are a concept to overcome this basic problem, by providing 4π solid angle for the coincident, three dimensional, high resolution momentum measurement of ions and electrons from atomic and molecular reactions.

The final state momenta of a reaction are interrelated by the three momentum and one energy conservation law. Thus for an n-body final state only 3n-4 momentum components have to be measured and the experimentalist is free to choose which to determine. For single ionization by fast charged particle impact for example, the azimuthal and polar scattering angle of the projectile can be measured in coincidence with the three dimensional momentum vector of the recoil ion. The energy loss of the projectile and the momentum of the emitted electron can than be deduced using conservation laws. In many fast ion-atom collisions however the relative momentum change of the projectile (energy loss and scattering angle) is unmeasurable small (typically $\Delta p/p_{projectile} < 10^{-5}$). In this case a coincident momentum measurement of ion and electron(s) provides the only access to fully differential cross sections. The momenta are measured via combined position and time of flight (TOF) measurements. For the TOF either a pulsed beam or a coincidence with a projectile can be used. For photo double ionization, to give a second example, either the two electrons or one electron plus the ion can be measured. In this case again the easiest and most precise measurement can be performed if a pulsed beam enables an absolute TOF measurement of all particles. In some special cases however it is even sufficient to determine the relative time of flight between recoil ion and electron. The data for linear polarized light shown below are taken with a reference signal from a pulsed beam (see [15] for details on the timing), those for circular polarized light by only measuring the relative TOF (see [16]).

In the following we first briefly discuss the experimental technique and restrict ourself to discussing a few recent applications to low energy photo double ionization studies.

II EXPERIMENTAL TECHNIQUE

The basic principles for high resolution 4π spectrometers are identical for ion and electron detection. They are based on a small reaction volume (typically below 1 mm^3) from which the fragments are guided by electric and magnetic fields to large area position sensitive detectors. The momenta of electron and ion can then be calculated from the time of flight and the position where the particles hit the detectors. The ion momenta resulting from atomic reactions are typically in the range of a few atomic units (a.u.), their energies in the μeV - meV regime. This is comparable or even smaller than the thermal motion of the atoms at room temperature (4.6 a.u. for He). Thus, one has to provide an internally cold atomic target for the collision. This is presently achieved by using supersonic gas jet targets. A further improvement in resolution is envisaged by the future use of laser cooled targets [17].

In all high resolution collision COLTRIMS studies supersonic gas jet targets have been used. The gas jet is formed by expanding the gas through a 30 μm hole. In some cases the nozzle and the gas are precooled to 15-30 K to further improve the resolution. From this expansion a supersonic jet is formed and the He atoms have a mean speed proportional to the square root of the nozzle temperature. The momentum distribution around the mean value is given by the speed ratio of the expansion [19]. The precooling helps to achieve a narrow momentum distribution with small turbo pumps. About 1 cm above the nozzle the inner part of the preformed gas jet enters the scattering chamber through a skimmer of 0.3 mm diameter. A typical operating condition for precooled one stage jets with small turbo pumps is a driving pressure of 400 mbar. This results in 5×10^{-4} mbar in

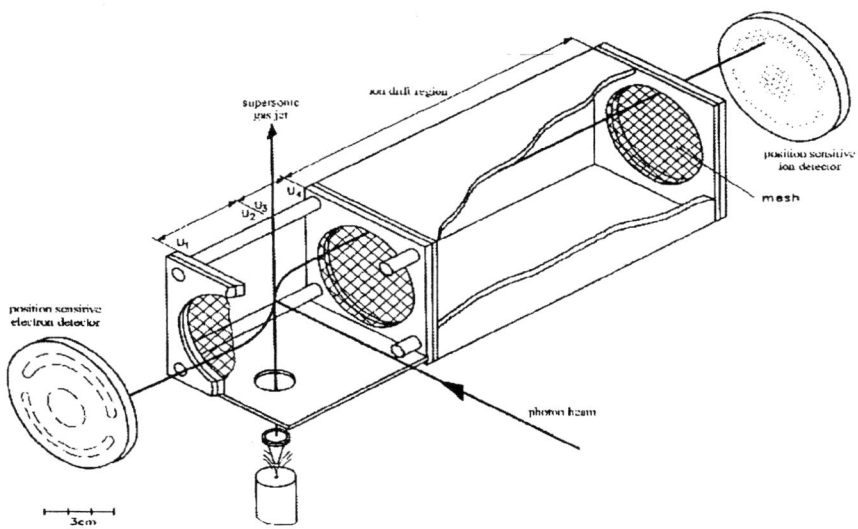

FIGURE 1. Typical COLTRIMS setup. The gas nozzle is cooled to 15-30 K, the super sonic gas jet has a diameter of 1.1 mm at the intersection with the photon (or charged particle) beam. The electron detector is located on the left side of the spectrometer and the ion detector on the right side. (from [18])

the source chamber and a local target pressure of a few 10^{-5} mbar at the target region, 3 cm above the skimmer. The gas jet leaves the scattering chamber into a separately pumped jet dump. In other experiments two stage jets [20,21] or jets backed with 8000 l/s diffusion pumps [22,23] are successfully used.

The ions are created in the overlap volume of the gas jet with the projectile beam. Different designs for recoil ion and electron spectrometers have been used. In a first version a homogeneous electric field followed by a drift tube guided the ions to a position sensitive channel plate [24–29]. With this homogeneous field spectrometer Mergel and coworkers reported a resolution of 0.26 a.u. [25]. A very flexible combination of electric fields for ion detection and magnetic fields for guiding the electrons has been used at GSI [30–34]. This 'momentum microscope' is discussed in detail in [21]. In all spectrometers with homogeneous fields the ion momentum resolution is restricted by the extension of the overlap volume. To circumvent this restriction Mergel [18] has designed a field geometry which is focussing in three dimensions. An electrostatic lens in the extraction region focuses different starting positions perpendicular to the extraction field onto one point on the detector. In the third direction different starting points along the field lead to the same time of flight. Thus, a high resolution can be achieved even with a gas target extended over several mm. With this spectrometer a resolution of 0.07 a.u FWHM, which is close to the internal temperature of the gas jet, has been reached [35]. Details on the field geometry can be found in [15,18]. A typical COLTRIMS setup is shown in figure 1.

Due to the low energy of the recoiling ions a moderate field of a few V/cm is sufficient to collect all ions onto the detector. The same field is used to guide the electrons in the opposite direction. If one chooses a distance of 2 cm from the target region to the electron detector a 4π collection efficiency can be reached only for very low energy electrons (1-5 eV). In fast particle collisions however often higher energy electrons are created. To guide such electrons to the detector Moshammer and coworkers have superimposed on the electric field a homogenous magnetic field yielding 4π detection efficiency up to 30 eV electron energy [21]. Higher electron energies can be accessed by increasing the magnetic and electric fields. Such electron imaging spectrometers with magnetic confinement are used with great success in ion impact [36,33,32,21,34] and in photo ionization studies [37,38].

III APPLICATION TO PHOTO IONIZATION STUDIES

A particular interesting case of an atomic few body reaction is double ionization of He by one single photon. This process is a detailed probe of the effects of dynamic electron correlation, one of the hottest topics in todays atomic collision physics. The most simple and widely studied quantity for this reaction is the ratio of total cross sections for double to single ionization (see eg the recent experimental papers [39–42]). For a detailed understanding of the dynamics however much higher differential cross sections are needed.

Starting with the pioneering work of Schwarzkopf and coworkers in 1993 [2] several groups have reported electron coincidence studies for He double double ionization [3–9]. These experiments have covered several angular settings between the electron spectrometers as well as equal and unequal sharing of the excess energy available to the system. They have however been restricted to the coplanar geometry, where the two electrons move in a plane containing the polarization vector of the linear polarized light.

As outlined in the introduction, detecting the momentum vector of the He^{2+} ion and one of the electrons is equivalent to detecting both electrons in coincidence. Several studies [43,44,37,15,16] reported on such experiments which mostly cover the full momentum space of all particles. One of the major advantages of such a comprehensive data set for double ionization is the possibility to display the data in any set of coordinates. Thus not only the traditional angles and energies of the two electrons (or momenta k_1, k_2) can be chosen but also collective coordinates like Jacobi coordinates ($\mathbf{k_r} = \mathbf{k_1} + \mathbf{k_2}$ and $\mathbf{k_R} = 1/2(\mathbf{k_1} - \mathbf{k_2})$) or hyperspherical coordinates. In such momentum space images the characteristics of the photo double ionization process become directly visible. Figure 2 compares a two dimensional

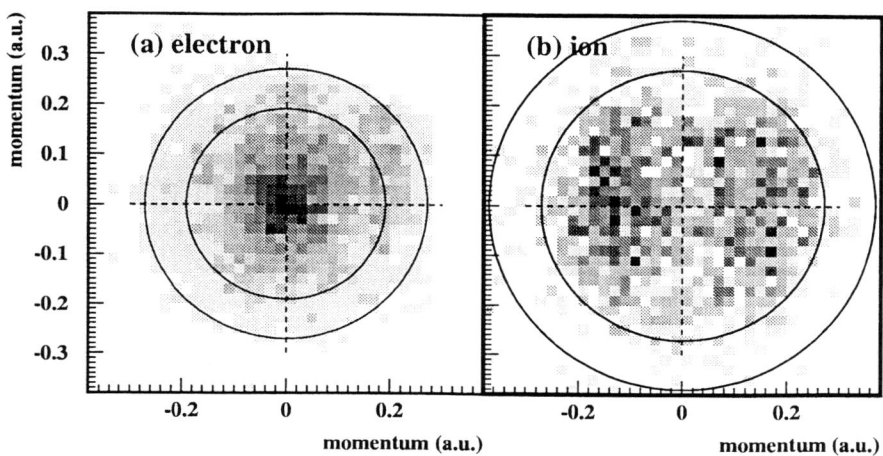

FIGURE 2. Density plots of projections of the momentum distribution from double ionization of He by 80.1 eV linear polarized photons. The z and y components of the momentum are plotted on the horizontal and vertical axes, respectively. The polarization vector of the photon is in the z direction and the photon propagates in the x direction. Only events with $-0.1 < k_{rx} < 0.1$ a.u. are projected onto the plane. a) The distribution of single electron momenta (k_1 or k_2). The circle locates the momentum of an electron which carries the full excess energy. (b) The recoil (or $-k_r$) momentum distribution. The outer circle indicates the maximum calculated recoil momentum, and the inner circle is the locus of events for which the k_r motion has half of the excess energy (from [43]).

representation of the final state momentum distribution of the He^{2+} ion and one of the electrons. The nucleus clearly shows a dipolar emission pattern as a result of the absorption of the photon. This characteristics of the primary absorption process is completely washed out in the electronic momentum distribution. This highlights the fact, that the nucleus as the center of positive charge in the system always participates in the absorbtion of the photon. It is the electron electron interaction which always is required in order for double ionization which smears out this reminiscent of the photons angular momentum in the momentum distribution of one single electron. A more detailed discussion of this problem can be found in [43,45,46,37].

An overview of the three body continuum in the momenta of the two electrons is given in figure 3. It shows the momentum of one electron with respect to the other at 1 and 20 eV above the double ionization threshold. All three particles are necessarily in one plane (following from momentum conservation). This internal plane of the breakup has some orientation to the electric field vector ϵ of the linear polarized photon beam.

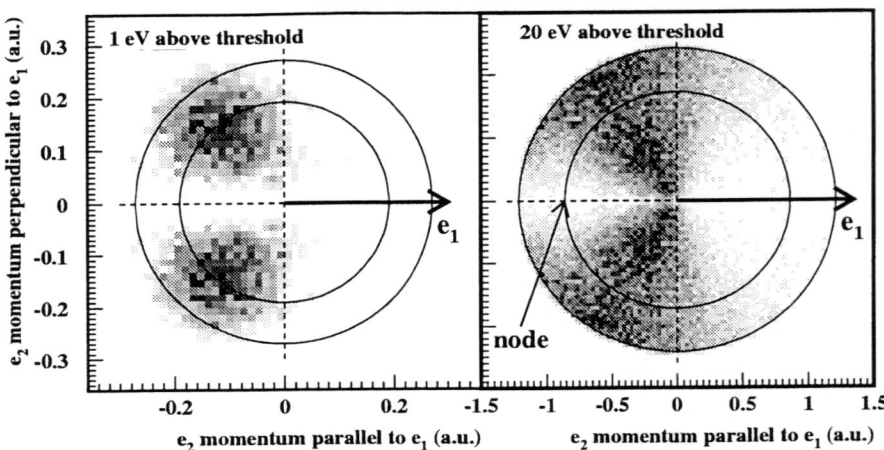

FIGURE 3. Photo double ionization of He at 1 and 20 eV above threshold by linear polarized light. Shown is the momentum distribution of electron 2 for fixed direction of electron 1 as indicated. The plane of the figure is the internal momentum plane of the three particles. The data are integrated over all orientations of the polarization axis with respect to this plane. The figure thus samples the full cross sections, all angular and energy distributions of the fragments. The outer circle corresponds to the maximum possible electron momentum, the inner one to the case of equal energy sharing. (see text for details)

In the figure we show the momenta of the electrons in this internal plane integrated over all directions of ϵ. Electron 1 is chosen along the positive x axis of the figure (shown by the arrow). The full cross section, all energy sharings and all angles, are sampled in this figure. Each circle on this plane corresponds to a particular energy sharing. The outer circle gives the maximum available momentum (i.e. extremely unequal energy sharing), the inner circle corresponds to equal energy sharing. As has been shown already by the work using coincident electron detection (see e.g. citeSchwarzkopf93,Schwarzkopf94,Schwarzkopf95jpb,Huetz94, two main features determine the momentum distribution: the effect of electron repulsion and a selection rule from the $^1P^0$ symmetry of the final state. Electron repulsion at these low excess energies leads to an emission of the electrons to opposite half spheres. There are almost no events on the right half of the figure corresponding to two electrons going to the same internal half plane. As can be expected from the velocity of the particles, this backward emission effect is slightly more pronounced at 1 eV than at 20 eV. Second, for a $^1P^0$ two body state $\mathbf{k_1} = -\mathbf{k_2}$, i.e. back to back emission of equal energy electrons, is prohibited (see selection rule C in [47]). For 1 eV the data show that this node extends all the way along the x-axis. Thus at such low energies back to back emission is suppressed at all energy sharings, even so this is not a strict selection rule (see also [7]). At 20 eV the node is really centered at $\mathbf{k_1} = -\mathbf{k_2}$ (indicated by the arrow). This presentation shows strikingly that this node is internal to the three body system and has nothing to do with ϵ, since the data are integrated over all orientations of ϵ.

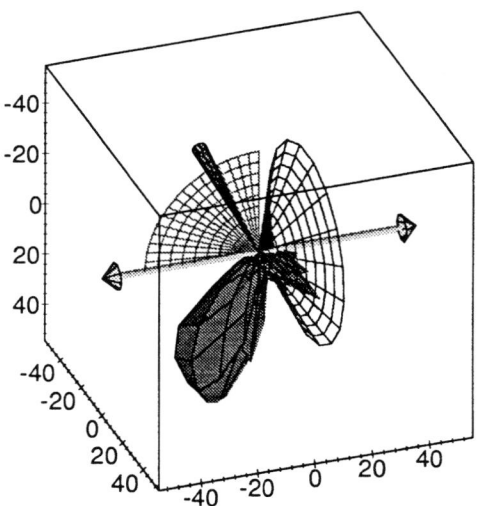

FIGURE 4. Four fold differential cross section $d^4\sigma/d\vartheta_1 d\vartheta_2 d\Phi_{1-2} dE_1$ The direction of the slow electron is indicated by the narrow dark cone (at ϑ_1 =65 deg), the polarization is horizontal, indicated by the double arrow. The excess energy is 6 eV, the fixed electron has 0.1-1 eV

An additional structure becomes visible if one avoids this integration and displays the data in the conventional polar coordinates. Figure 4 shows the fourfold differential cross section $d^4\sigma/d\vartheta_1 d\vartheta_2 d\Phi_{1-2} dE_1$ at 6 eV. Strictly speaking this is a fivefold differential cross section, since the are five linear independent observables. Within the dipol approximation, the additional rotational symmetry around the polarization axis reduces this to four dimension. The direction of the first electron is indicated by the narrow cone. The polarization axis is horizontal. The wide open cone on the right represented by the mesh indicates the locus of a two-dimensional node according to selection rule F of [47]. This selection rule holds strictly only for equal energy sharing. As the data show at 6 eV excess energy it prevails however for unequal energy sharing (see [15,6,7]).

An interesting twist is added to this three body breakup if one introduces a chirality in the initial state by inducing the transition with circular instead of linear polarized light. The question arises how or if at all the chirality of the photon is transferred to the three body continuum. It has been first pointed out by Berakdar and Klar [48] that such an effect, termed dichroism, might exist even for He double ionization. Viefhaus and coworkers [9] found the first experimental evidence for this effect. The two electrons and the photon axis can span a tripod which could have a handedness if its two legs defined by the electron momenta are distinguishable, i.e. the electrons have unequal energy. This shows up strongest if the three body plane (as it is shown in figure 3) is held fix perpendicular to the photon axis. At 20 eV above threshold, figure 5 shows the momentum distribution of the ion and electron 2 in this plane. The momentum of electron 1, which is chosen to be the faster one, is fixed along the x axis.

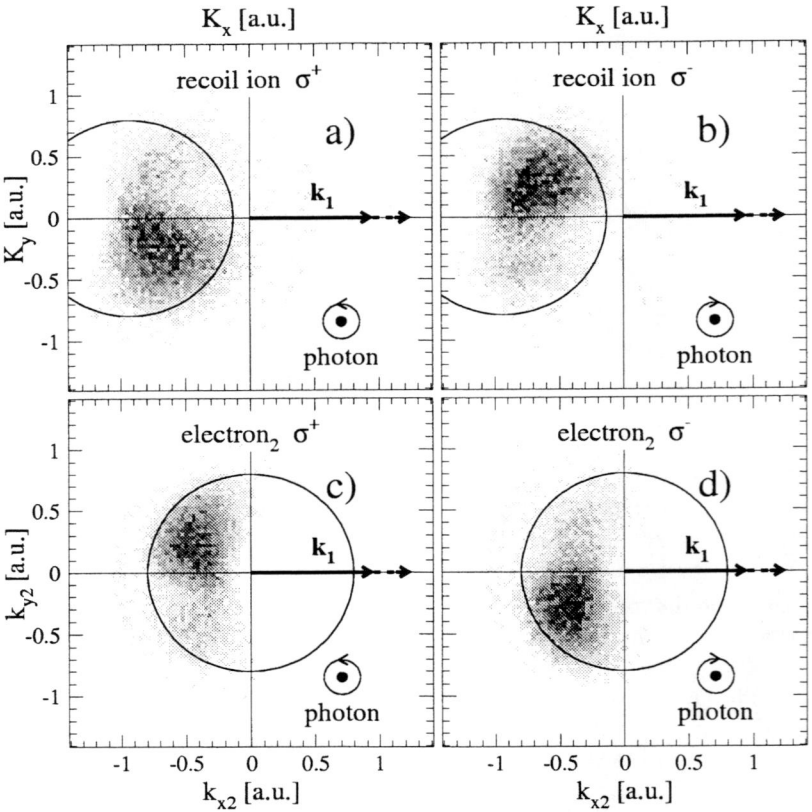

FIGURE 5. Recoil ion and electron momentum distribution in the x-y-plane both for σ^+ and σ^- as indicated. We fixed e_1 within 0.9a.u. $< k_{x1} < 1.2$a.u. and with $k_{y1} = k_{z1} = 0$. The wave vector $\mathbf{k}_\gamma$ points out of the plane of paper. Circles: maximum magnitude of $\mathbf{K}$ and $\mathbf{k}_2$. The grey scale represents the five fold differential cross section on a linear scale (from [16]).

Comparison with figure 3 as well as between left and right circular polarized light visualizes that dichroism is a huge effect in this system. While for linear polarized light upper and lower half of the figure are necessarily symmetric, this symmetry is broken for circular polarized light. A detailed comparison of these experimental results with 3C calculations can be found in [16]. In general the agreement between 3C theory and experiment is much worse for circular polarized [16] than for linear polarized light.

IV OUTLOOK

Multiparticle imaging with COLTRIMS provides new spectacular views on many body breakup of coulombic systems. It combines high resolution in momentum space (typically $<< 0.1 a.u.$) with 4π solid angle for all fragments. In many cases such kinematical complete pictures directly 'display' the processes responsible for the breakup of the atom or molecule. Thus some long standing puzzles in atomic collision physics were solved recently using this approach and many new questions and challenges to theory were raised. Due to limited space we had to restrict ourselves to a few selected results for photoionization at low energies. Another rich field of applications lies in the study of Compton scattering at high photon energies [23,42]. A great amount of important results could be obtained in particular for ion atom collisions. We name only a few of these: Momentum space of single and double ionization by relativistic heavy ions showed the explosion of the atom in the light of an attosecond pulse of virtual photons, stronger than any available laser [33,32,30]. A scattering of two electrons inside an atom in a transfer ionization process could be directly seen [49]. Two center electron-electron interaction has been separated experimentally from nuclear-electron scattering [50,27]. At slow collisions the promotion of one electron into the continuum via the saddle point mechanism has been imaged [51,52]. Besides this rapidly increasing progress in atomic and molecular [38] physics such imaging techniques can be expected to be particularly useful in material research and solid state physics in the future. On can envision even the imaging of the many electron emission from surfaces which will give insight in the correlated electron motion in solids.

V ACKNOWLEDGEMENTS

The work was financially supported by DFG, BMFT and by the Division of Chemical Sciences, Basic Energy Sciences, Office of Energy Research, U.S.Department of Energy, DOE grant 82ER53128 (KSU) and DOE contract No. DE-AC03-76SF00098 (LBNL). One of us (R.D.) was supported was supported by the Habilitanden Programm der DFG. H.B. and R.D. acknowledge support from the Alexander von Humboldt foundation. Kh. K. greatfully acknowledges support by the DAAD. We also acknowledge financial support from Max Planck Forschungspreis of the Humboldt foundation. The work presented for linear polarized light would not have been possible without the great support at the ALS in particular by R. Thatcher, T. Warwick, E. Rothenberg and J. Dennlinger. For the data with circular polarized light we acknowledge the help of Dr. T. Hatano and Prof. T. Miyahara and the excellent working conditions at the Photon Factory during the experiment (Proposal No. 96G160) and of T. Kambara, V. Zoran, Y. Awaya and B. Nystrom doe their help during the beamtime. The interpretation of the photo ionization data profited from many discussions with our friends and colleagues J. Feagin, S. Keller, J. Berakdar, V. Schmidt, B. Kässig, C. Wheelan and A. Huetz.

REFERENCES

1. O. Schwarzkopf, B. Krässig, J. Elmiger, and V. Schmidt. *Phys. Rev. Lett.*, 70:3008, 1993.
2. O. Schwarzkopf, B. Krässig, V. Schmidt, F. Maulbetsch, and J. Briggs. *J. Phys.*, B27:L347–50, 1994.
3. O. Schwarzkopf and V. Schmidt. *J. Phys.*, B28:2847, 1995.
4. A. Huetz, P. Lablanquie, L. Andric, P. Selles, and J. Mazeau. *J. Phys.*, B27:L13, 1994.
5. G. Dawber, L. Avaldi, A.G. McConkey, H. Rojas, M.A. MacDonald, and G.C. King. *J. Phys.*, B28:L271, 1995.
6. P. Lablanquie, J. Mazeau, L. Andric, P. Selles, and A. Huetz. *Phys. Rev. Lett.*, 74:2192, 1995.
7. J. Viefhaus, L. Avaldi, F. Heiser, R. Hentges, O. Gessner, A. Rüdel, M. Wiedenhöft, K. Wielczek, and U. Becker. *J. Phys.*, B29:L729, 1996.
8. J. Viefhaus, L. Avaldi, G. Snell, M. Wiedenhöft, R. Hentges, A. Rüdel, F. Schäfer, D. Menke, U. Heinzmann, A. Engelns, J. Berakdar, H. Klar, and U. Becker. *Phys. Rev. Lett.*, 77:3975, 1996.
9. M.A. Coplan et al. *Rev.Mod.Phys.*, 66:985, 1994, and References therein.
10. I.E. McCarthy and E. Weigold. *Rep. Prog. Phys.*, 54:789, 1991.
11. A. Lahmam-Bennani. *J. Phys*, B24:2401, 1991.
12. B. El Marji, A. Lahmam-Bennani, A. Duguet, and T.J. Reddish. *J. Phys*, B29:L1, 1996.
13. B. El Marji, C. Schröter, A. Duguet, A. Lahmam-Bennani, M. Lecas, and L. Spielberger. *J. Phys*, B, 1997. Zur Veröffentlichung eingereicht.
14. R. Dörner, H. Bräuning, J.M. Feagin, V. Mergel, O. Jagutzki, L. Spielberger, T. Vogt, H. Khemliche, M.H. Prior, J. Ullrich, C.L. Cocke, and H. Schmidt-Böcking. *Phys. Rev.*, 57:1074, 1998.
15. V.Mergel, M. Achler, R. Dörner, Kh. Khayyat, T. Kambara, Y. Awaya, V. Zoran, B. Nyström, L. Spielberger, J.H. McGuire, J. Feagin, J. Berakdar, and H. Schmidt-Böcking. . *Phys. Rev. Lett.*, :submitted for publication, 1998.
16. V. Mergel. PhD thesis, *Universität Frankfurt*. ISBN 3-8265-2067-X. Shaker Verlag, 1996.
17. G. Brusdeylins, J.P. Toennies, and R. Vollmer. page 98, Perugia, 1989.
18. P. Jardin, J.P. Grandin, A. Cassimi, J.P. Lemoigne, A. Gosseling, X. Husson, D. Hen-

necart, and A. Lepoutre. In *AIP Conference Proceedings No 274*, page 291. AIP, 1993.
19. O. Jagutzki, L. Spielberger, R. Dörner, S. Nüttgens, V. Mergel, H. Schmidt-Böcking, J. Ullrich, R.E. Olson, and U. Buck. *Zeitschrift für Physik*, D36:5, 1996.
20. L. Spielberger, O. Jagutzki, R. Dörner, J. Ullrich, U. Meyer, V. Mergel, M. Unverzagt, M. Damrau, T. Vogt, I. Ali, Kh. Khayyat, D. Bahr, H.G. Schmidt, R. Frahm, and H. Schmidt-Böcking. *Phys. Rev. Lett.*, 74:4615, 1995.
21. O. Jagutzki. *Dissertation Unversität Frankfurt*, unpublished, 1994.
22. V. Mergel, R. Dörner, J. Ullrich, O. Jagutzki, S. Lencinas, S. Nüttgens, L. Spielberger, M. Unverzagt, C.L. Cocke, R.E. Olson, M. Schulz, U. Buck, E. Zanger, W. Theisinger, M. Isser, S. Geis, and H. Schmidt-Böcking. *Phys. Rev. Lett*, 74, 1995.
23. V. Mergel. *Diploma thesis Universität Frankfurt*, unpublished, 1994.
24. R. Dörner, V. Mergel, R. Ali, U. Buck, C.L. Cocke, K. Froschauer, O. Jagutzki, S. Lencinas, W.E. Meyerhof, S. Nüttgens, R.E. Olson, H. Schmidt-Böcking, L. Spielberger, K. Tökesi, J. Ullrich, M. Unverzagt, and W. Wu. *Phys. Rev. Lett*, 72:3166, 1994.
25. R. Dörner, V. Mergel, L. Zhaoyuan, J. Ullrich, L. Spielberger, R.E. Olson, and H. Schmidt-Böcking. *J. Phys*, B28:435, 1995.
26. V. Mergel, R. Dörner, J. Ullrich, O. Jagutzki, S. Lencinas, S. Nüttgens, L. Spielberger, M. Unverzagt, C.L. Cocke, R.E. Olson, M. Schulz, U. Buck, and H. Schmidt-Böcking. *Nucl. Instr. Meth.*, B96:593, 1995.
27. R. Dörner, V. Mergel, L. Spielberger, O. Jagutzki, S. Nüttgens, M. Unverzagt, H. Schmidt-Böcking, J. Ullrich, R.E. Olson, K. Tökesi, W.E. Meyerhof, W. Wu, and C.L. Cocke. *AIP Conf. Proc. 360 (1995) Editors: L.J. Dube, J.B.A. Mitchell J.W. McConckey C.E. Brion*, AIP Press New York, 1995.
28. R. Moshammer, J. Ullrich, M. Unverzagt, W. Schmitt, P. Jardin, R.E. Olson, R. Mann, R. Dörner, V. Mergel, U. Buck, and H. Schmidt-Böcking. Low-energy electrons and their dynamical correlation with the recoil-ions for single ionisation of helium by fast ion impact. *Phys. Rev. Lett.*, , 1995.
29. H.P Bräuning, R. Dörner, C.L. Cocke, M.H. Prior, B. Krässig, A. Bräuning-Demian, K. Carnes, S. Dreuil, V. Mergel, P. Richard, J. Ullrich, and H. Schmidt-Böcking. *J. Phys.*, 30:L649, 1997.
30. R. Dörner, H. Bräuning, O. Jagutzki, V. Mergel, M. Achler, R. Moshammer, J. Feagin, A. Bräuning-Demian, L. Spielberger, J.H. McGuire, M.H. Prior, N. Berrah, J. Bozek, C.L. Cocke, and H. Schmidt-Böcking. *Phys. Rev. Lett.*, submitted for publication, 1998.
31. J.A.R. Samson, W.C. Stolte, Z.X. He, J.N. Cutler, Y. Lu, and R.J. Bartlett. *Phys. Rev.*, 57:1906, 1998.
32. R. Dörner, T. Vogt, V. Mergel, H. Khemliche, S. Kravis, C.L. Cocke, J. Ullrich, M. Unverzagt, L. Spielberger, M. Damrau, O. Jagutzki, I. Ali, B. Weaver, K. Ullmann, C.C. Hsu, M. Jung, E.P. Kanter, B. Sonntag, M.H. Prior, E. Rotenberg, J. Denlinger, T. Warwick, S.T. Manson, and H. Schmidt-Böcking. *Phys. Rev. Lett.*, 76:2654, 1996.
33. J. C. Levin, G. B. Armen, and I. A. Sellin. *Phys. Rev. Lett.*, 76:1220, 1996.
34. L. Spielberger, O. Jagutzki, B. Krässig, U. Meyer, Kh. Khayyat, V. Mergel, Th.

Tschentscher, Th. Buslaps, H. Bräuning, R. Dörner, T. Vogt, M. Achler, J. Ullrich, D.S. Gemmel, and H. Schmidt-Böcking. *Phys. Rev. Lett.*, 76:4685, 1996.

35. R. Dörner, J. Feagin, C.L. Cocke, H. Bräuning, O. Jagutzki, M. Jung, E.P. Kanter, H. Khemliche, S. Kravis, V. Mergel, M.H. Prior, H. Schmidt-Böcking, L. Spielberger, J. Ullrich, M. Unverzagt, and T. Vogt. *Phys. Rev. Lett.*, 77:1024, 1996. see also erratum in Phys. Rev. Lett. 78. 2031 (1997).
36. T. Vogt, R. Dörner, O. Jagutzki, C.L. Cocke, J. Feagin, M. Jung, E.P. Kanter, H. Khemliche, S. Kravis, V. Mergel, L. Spielberger, J. Ullrich, M. Unverzagt, H. Bräuning, U. Meyer, and H. Schmidt-Böcking. *in: Proceedings of the Euroconference Ionization and Coincidence Spectroscopy ed.: C.T. Whelan and H.R.J. Walters, Plenum, 1997.*
37. M. Pont and R. Shakeshaft. *Phys. Rev.*, A54:1448, 1996.
38. J.M. Feagin. *J. Phys.*, B29:1551, 1996.
39. F. Maulbetsch and J.S. Briggs. *J. Phys.*, B28:551, 1995.
40. J. Berakdar and H. Klar. *Phys. Rev. Lett.*, 69:1175, 1992.
41. V. Mergel, R. Dörner, M. Achler, Kh. Khayyat, S. Lencinas, J. Euler, O. Jagutzki, S. Nüttgens, M. Unverzagt, L. Spielberger, W. Wu, R. Ali, J. Ullrich, H. Cederquist, A. Salin, R.E. Olson, Dž. Belkić, C.L. Cocke, and H. Schmidt-Böcking. Intraatomic electron-electron scattering in p-He collisions (Thomas process) investigated by Cold Target Recoil Ion Momentum Spectroscopy (COLTRIMS). *Phys. Rev. Lett,* 79:387, 1997.
42. S.D. Kravis, M. Abdallah, C.L. Cocke, C.D. Lin, M. Stöckli, B. Walch, Y. Wang, R.E. Olson, V.D. Rodriguez, W. Wu, M. Pieksma, and N. Watanabe. *Phys. Rev.*, A54:1394, 1996.
43. R. Dörner, H. Khemliche, M.H. Prior, C.L. Cocke, J.A. Gary, R.E. Olson, V. Mergel, J. Ullrich, and H. Schmidt-Böcking. *Phys. Rev. Lett,* 77:4520, 1996.

Projectile and Recoil Ion Momentum Spectroscopy in p + He Collisions

M. Schulz, L. An, R.E. Olson, J. Ullrich[1], and H. Schmidt-Böcking[2]

University of Missouri-Rolla, Physics Department, Rolla, MO 65409, USA
[1]*Universität Freiburg, Fakultät für Physik, 79104 Freiburg, Germany*
[2]*Universität Frankfurt, Institut für Kernphysik, 60486 Frankfurt, Germany*

We have applied projectile momentum spectroscopy and cold target recoil ion momentum spectroscopy (COLTRIMS) to study single ionization in p + He collision at a projectile energy of 75 keV. From the projectile momentum measurements we obtained doubly differential single ionization cross section as a function of the longitudinal and transverse projectile momentum components. Our data demonstrate the importance of post-collision interactions between the projectile and the ionized electron even though the cross sections are integrated over all electron angles. As a result, theoretical models have to account for at least second order interactions in the projectile potential.

Introduction

Plasmas are of crucial importance in many areas of physics, such as astrophysics, nuclear fusion, electric discharges etc. The fundamental underlying process for formation of a plasma is atomic ionization. Therefore, a thorough understanding of ionization processes is of general importance extending well beyond the boundaries of atomic physics.

Our current understanding of ionization in ion-atom collisions is still rather limited even for the most simple collision systems. The reason is that any inelastic process in an ion-atom collision represents a many-body problem involving at least three particles (two nuclei and at least one electron) and it is well known that the Schrödinger equation is not solvable in closed form for more than two mutually interacting particles. As a result, theory has to heavily rely on approximations which in some cases can be quite severe. Nevertheless, significant progress has been achieved in describing ion-atom collisions using for example the Continuum Distorted Wave method (1,2), the Close Coupling approach (3), the Classical Trajectory Monte Carlo (CTMC) method (4), or the direct integration of the time-dependent Schrödinger equation (5).

In spite of the recent progress our understanding of atomic collision processes is not yet complete, especially for ionization processes. Ionization is particularly difficult to describe because of the long range nature of the Coulomb force acting between the nuclei and the ionized electron in the collision system. In other non-atomic many-body systems, the force between the mutually interacting particles is often a short range force, like for example in nuclei. The description of the time evolution under the influence of these forces then only involves integrations limited to finite space. In other atomic processes, like excitation and electron capture, the electron is spatially confined because it remains bound to one of the two collision partners again allowing for limited space integration. For ionization, in contrast, the electron will in general move to large distances from the collision partners, yet maintaining an interaction with both nuclei due to the long range Coulomb force which only vanishes asymptotically at infinity. Therefore, for ionization it is much more difficult to limit the integration to finite space.

Effects of the long range Coulomb interaction long after the actual ionization process (post-collision interaction, PCI) on the outgoing electron have been observed in many experiments (e.g. 6-10). Madison has demonstrated that much better agreement with experimental data can be achieved if PCI effects between the target nucleus and the ionized electron are properly accounted for in the calculation (1). The importance of PCI effects has also been demonstrated in CTMC calculations, which include the PCI to all orders (11). A satisfactory description of the PCI between the projectile and the ionized electron appears to be a significantly more difficult task than between the target nucleus and the electron. On the other hand, a good understanding of PCI effects is important to obtain a complete picture of many-body systems of particles mutually interacting through long range forces. For this reason, in the calculations reported here, we have used the three-body CTMC method. This is an "exact" classical description that include all pairwise Coulomb potentials throughout the collision. Hence, it is a kinematically complete calculation which yields longitudinal, transverse, and collision plane information.

Because of the problems theory is facing in describing ionization processes, it is critically important to obtain detailed experimental data in order to support theoretical modeling efforts. In the ideal case, a complete experiment would be performed, i.e. the complete final state of the entire collision system would be determined. In pure single ionization, the final state involves only free particles, for each of which the space state is completely determined by its momentum vector. Therefore, finding the final state of the entire collision system requires one to determine the three momentum vectors of the projectile, the recoiling target ion, and the ionized electron. This, in turn, requires one to measure the momentum vectors of two of these three particles. The momentum vector of the third particle can then be deduced from momentum conservation because the initial momentum of the collision system is known. A kinematically complete experiment has recently been performed for structured, heavy projectiles colliding with He by measuring the recoil ion and ionized electron momentum vectors (12). In this paper we report on experimental studies of single ionization in p + He collisions by measuring the projectile and the recoil ion momentum vectors.

Experiment

The schematic of the apparatus to measure the projectile momentum vector is shown in Fig. 1. A 2 keV proton beam with a very narrow energy spread (<< 1 eV) was produced with a hot cathode ion source and accelerated to an energy of 75 keV. The beam was collimated by a set of slits right in front of the target chamber to a size of 0.1 mm x 0.1 mm and steered through the target region. After the collision region, the protons, which were selected by a switching magnet, passed through a solid angle defining collimator and were decelerated to an energy of 2 keV. The projectile momentum vector was measured in polar coordinates, i.e. its magnitude and its direction were measured.

The magnitude of the momentum vector was measured by energy-analyzing the protons with an electrostatic 45° parallel plate analyzer (13). The analyzer was set to a fixed pass energy of 2 keV. A proton energy-loss spectrum was obtained by scanning an offset voltage on the accelerator relative to the decelerator such that only protons which suffered an energy-loss equal to the offset voltage would have the analyzer pass energy of 2 keV. The direction of the momentum vector was obtained by measuring the polar projectile scattering angle, which in turn is determined by the above mentioned collimators located before and after the target chamber. The azimuthal projectile scattering angle is fixed at 0°. The scattering angle could be scanned by rotating the accelerator about an axis going through the center of the target chamber. The angular resolution was about 0.15 mrad FWHM.

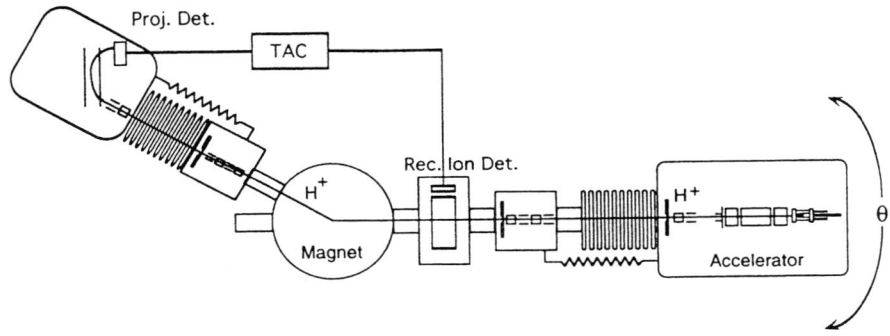

FIGURE 1. Schematic view of the experimental set-up.

In Fig. 2 we show the details of the target chamber. The initial direction of the proton beam is perpendicular to the plane of the figure, as indicated by the dot. The measurement of the recoil ion momentum vector involves two major components: the actual recoil ion momentum spectrometer and a supersonic gas jet. The latter is necessary because the energy transfer in an ion-atom collision is usually very small, typically 10 to 100 meV. On the other hand, thermal energy at room temperature is about 25 meV per particle. Therefore, the target gas has to be cooled to a temperature of about 1K or less (corresponding to a thermal energy of about 100μeV). The supersonic jet consists of a reservoir containing He gas at a pressure of 1-2 atm. The gas can escape through a very small nozzle (30 μm in diameter). The surrounding of the reservoir (first pumping stage) is kept at a pressure of 10^{-4} T by a turbopump. Because of the large pressure gradient between the reservoir and the first pumping stage, the gas expands adiabatically which leads to cooling in the direction of the pressure gradient. In the other two direction, the gas is cooled by using a hyperbolic skimmer with a diameter of 300 μm to collimate out any particles with a momentum component in these directions. The region after the skimmer (second pumping stage) is kept at a pressure of 10^{-7} T by a diffusion pump. The jet can be moved in all three directions by three micrometers for easy alignment of the target beam with the projectile beam.

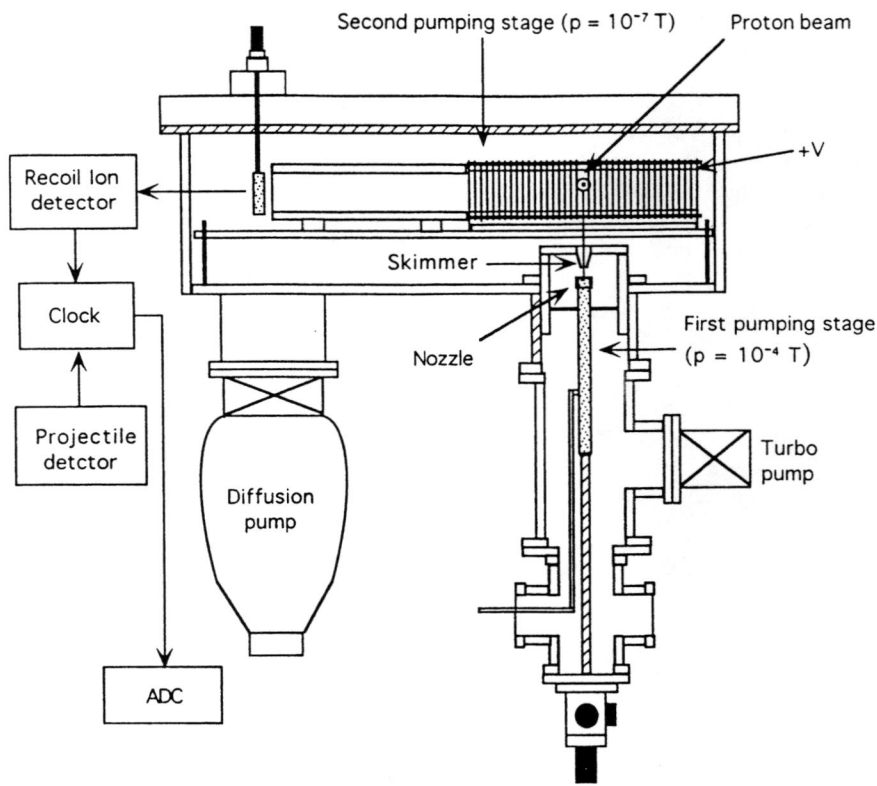

FIGURE 2. Schematic view of the target chamber with the recoil momentum spectrometer and the supersonic gas jet.

In the recoil ion momentum spectrometer a homogenous electric field of 1V/cm is generated by applying a voltage across a set of electrodes which are connected to each other by a resistor chain. After acceleration in this field over a distance of 10 cm, the recoil ions traverse a field free drift region 20 cm in length and are then detected by a two-dimensional position-sensitive channel plate detector. The recoil ion detector is set in coincidence with the projectile detector.

The recoil ion momentum distribution in the direction of the extraction field is determined from the time of flight of the recoil ions from the collision region to the detector, which, in turn, can be extracted from the coincidence time spectrum. The momentum components perpendicular to the extraction field are obtained from the position information from the detector anode. The momentum resolution in all three directions is about 0.2 a.u. FWHM.

Results and Discussion

In Fig. 3 we show on the left the singly differential ionization cross sections as a function of the longitudinal change of the projectile momentum p_z for 75 keV p + He collisions as the full circles. The data exhibit a threshold behavior at $p_z = -0.52$ a.u., which corresponds to the ionization energy of He of 24.6 eV. The cross sections then fall off rapidly with increasing negative p_z. At about $p_z = -1.4$ a.u. a pronounced shoulder structure is observed. A change of the longitudinal projectile momentum of -1.4 a.u. corresponds to an energy loss of 65 eV. This is the same energy for which a similar shoulder structure was observed in the energy-loss spectra for the same collision system (9,10). This behavior was interpreted as a PCI effect between the projectile and the ionized electron which leads to electron capture to a low lying continuum state of the projectile (ECC).

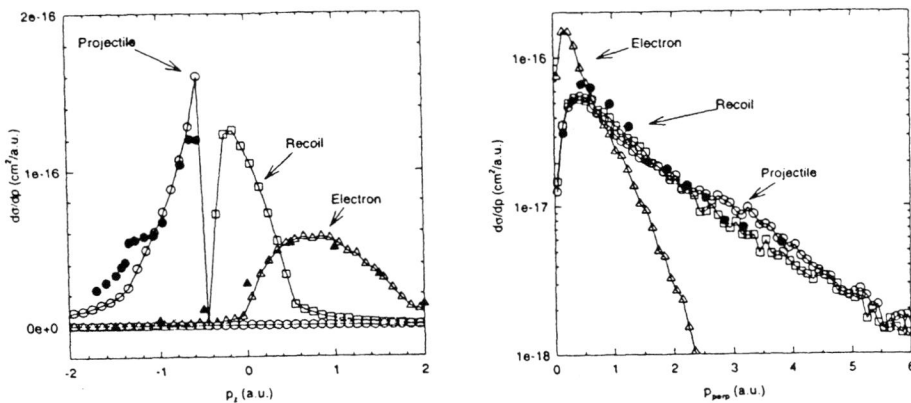

FIGURE 3. Longitudinal (left) and transverse momentum distributions. The full circles are the experimental data for the projectile, the full triangles the experimental data for the electrons (Kravis et al., 14) and the open symbols show the CTMC calculations.

For comparison, we show in the same figure as full triangles singly differential ionization cross sections as a function of the longitudinal ionized electron momentum obtained for the same collision system by Kravis et al. (14). The electrons are emitted almost exclusively in the forward direction (positive p_z) and the distribution peaks at about 0.8 a.u. The open symbols in Fig. 3 show our Classical Trajectory Monte Carlo calculation. The calculation is an excellent agreement with the data for the electrons, except for some discrepancies for negative p_z. For the projectiles the calculation is in good agreement with our data between the threshold and about -1.1 a.u. At more negative momenta the calculation underestimates the cross sections and the shoulder structure somewhat. However, overall there is good qualitative agreement. Because of momentum conservation this suggests that the calculation should provide a qualitatively correct behavior for the recoil ion longitudinal momentum distribution as well.

From the calculation it can be seen that the momentum distributions of all three particles have a significant width (0.5 a.u. for the projectiles, 0.7 a.u. for the recoil ions, and 1.4 a.u. for the electrons), i.e. there is a significant momentum transfer between all active particles in the collision. This is in sharp contrast to very fast collisions between heavy ions and He, where the width of the longitudinal projectile momentum distribution (0.2 a.u.) is small compared to the widths of the corresponding distributions for the recoil ion and the ionized electron (0.8 a.u. and 0.7 a.u, respectively). Therefore, in fast collisions the momentum transfer in the longitudinal direction appears to take place predominantly between the electron and the recoil ion. For the intermediate collision systems studied in this work, on the other hand, there is a significant momentum transfer between all active particles in the collision.

On the right of Fig. 3 we show the singly differential ionization cross section as a function of the projectile transverse momentum component as full circles. The open symbols are the CTMC calculation. The calculation is in very good agreement with the experimental data except for a slight underestimation of the cross sections at momenta near 1 a.u. Also, the calculated distribution is somewhat broader than in the data.

In Fig. 4 a coincidence time spectrum is shown for a projectile energy loss of 63 eV and a scattering angle of 0°. The coincidence time (time difference between the recoil ion signal and the projectile signal) is equivalent to the time of flight of the recoil ions because of the large speed of the projectiles. This time of flight, in turn, is inversely proportional to the transverse (perpendicular to the initial projectile beam direction) momentum component of the recoil ion. Therefore the shape of the time peak reflects the transverse recoil ion momentum distribution with the momentum axis running opposite to the time axis. Since the projectile energy loss and both the polar and azimuthal scattering angle, i.e. the projectile momentum vector, are fixed, the coincidence time spectrum is related to the quadruply differential ionization cross sections as a function of the transverse recoil ion momentum. It should be noted, however, that the conversion of the time spectrum to a differential cross section as a function of the transverse momentum component is itself time-dependent. Therefore, the shape of the time spectrum in Fig. 4 is very different from the calculated cross sections for the recoil ions shown in Fig. 3.

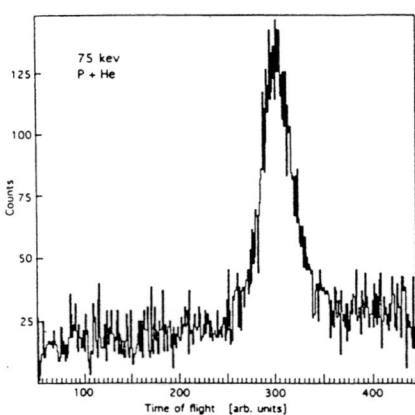

FIGURE 4. Coincidence time spectrum. This spectrum relates to the recoil ion transverse momentum distribution.

The longitudinal recoil ion momentum distribution is obtained by projecting the two-dimensional position spectrum of the recoil ions onto the longitudinal axis. Furthermore, collision events which led to specific projectile vectors can be selected by setting a condition on the time peak in the coincidence spectrum for the corresponding projectile energy loss and scattering angle. Such longitudinal recoil ion momentum spectra are shown in Fig. 5 for energy losses of 28, 48, and 63 eV and a scattering angle of 0° in each case. Again, these spectra are proportional to quadruply differential ionization cross sections. For comparison, we also show the total longitudinal recoil ion momentum distribution (no selection of any projectile parameter) in the top of Fig. 5.

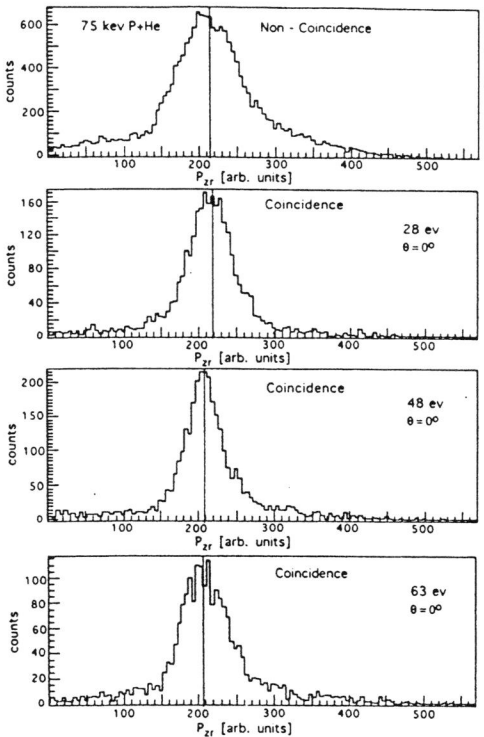

FIGURE 5. Longitudinal recoil ion momentum distributions without condition on the projectiles (top) and for energy losses of 28, 48, and 63 eV. The projectile scattering angle is 0 in each case.

Several effects can be observed in the comparison between the coincident spectra and the singles (non-coincident) spectrum: first, the momentum distributions in the coincident spectra are narrower than in the singles spectrum. Second, the centroid of the distributions in the coincident spectra (indicated by the vertical lines) are shifted compared to the centroid in the singles spectrum, where for an energy loss of 28 eV it is shifted to slightly more positive momenta and for the higher energy losses it is shifted to slightly more negative momenta. Third, there is a systematic trend of a shift toward more negative momenta with increasing energy loss. Finally, the low momenta wing of the distribution for 63 eV appears to be somewhat steeper than for 28 and 48 eV and than for the singles spectrum. All these observations can be explained by analyzing the kinematics of the collision. In the singles spectra, all projectiles regardless of energy loss and scattering angle are contributing, i.e. the longitudinal projectile momentum component is completely undetermined, which leaves a high degree of freedom for the longitudinal momentum components that the ionized electron and the recoil ion can have. This results in a relatively broad recoil momentum distribution. In the coincident spectra, in contrast, both the energy loss and the scattering angle, i.e. the complete momentum vector, of the projectile are fixed. This reduces the degree of freedom for the longitudinal momentum components of the electron and the recoil ion leading to a narrower recoil momentum distribution.

To understand the other points requires a more detailed analysis of the collision kinematics. By applying energy and momentum conservation and with the approximation of small projectile scattering angles (which is always the case for ion-atom collisions), it can be shown that for ionization the longitudinal recoil ion momentum component p_{zr} is given by (in atomic units):

$$p_{zr} = (-Q+K_e)/v_p - p_{ze} \qquad (1)$$

where the Q-value is defined as the difference in electron binding energy before and after the collision, K_e is the kinetic energy of the ionized electron, v_p is the initial projectile speed, and p_{ze} is the longitudinal electron momentum component. Since the Q-value is fixed for a given reaction (-24.6 eV for ionization), the longitudinal recoil momentum is determined by the kinetic energy and the longitudinal momentum of the ionized electron. Furthermore, the negative longitudinal recoil momentum maximizes for an ionized electron velocity equal to the projectile velocity (i.e. for cusp electrons), which at a projectile energy of 75 keV corresponds to a projectile energy loss of 65 eV. Therefore, as the energy loss approaches this energy, the longitudinal recoil momentum distribution should shift toward this maximum negative momentum, as observed in the data. Currently, no data are available yet for energy losses larger than 65 eV, however, we expect that there the spectrum will shift back to more positive momenta. The largest energy loss we studied so far (63 eV) is very close to 65 eV, which explains the steeper slope of the low momentum wing in the corresponding spectrum compared to those for the smaller energy losses. At 63 eV, the centroid is closer to the maximum negative momentum thus "squeezing" the spectrum at the low momentum wing.

Conclusions

We have measured projectile and recoil ion momentum vector distributions for ionizing 75 keV p + He collisions. From the projectile momentum measurements we found that the post-collision interaction (PCI) between the outgoing projectile and the ionized electron can be very important. This is a result of the long range nature of the Coulomb force. From a theoretical point of view this means that the asymptotic behavior of the collision long time after the ionization process has to be accurately described in terms of appropriate three-body Coulomb wavefunctions, which is a difficult task (15).

By measuring both the projectile and the recoil ion momentum vectors simultaneously in a coincidence experiment, we were able to completely fix the collision kinematics for each detected event since the ionized electron momentum vector is determined by momentum conservation. We observed a correlation between the kinematics of the projectile and the recoil ion for the longitudinal component. Both the shape and the position of the recoil momentum distribution depend on the projectile energy loss.

The results presented here only represent a small fraction of a large amount of information which can still be extracted from the data. For example, the longitudinal recoil ion and projectile momentum spectra can be created for a fixed transverse recoil momentum by setting a narrow condition on the coincidence time peak and the transverse position spectrum of the recoil ion detector. Furthermore, the electron momentum vector can be deduced from momentum conservation and the recoil and projectile momentum distributions can be studied as a function of the electron kinematics. As a result, theoretical models can be tested in unprecedented detail. For example, it will be possible to test the predictions of CTMC calculations, which analyzed the complete kinematics of ion-atom collisions (16,17).

References

1. Madison D.H., *Phys. Rev.* **A8**, 2449 (1973)
2. Gulyas L., Fainstein P.D., and Salin A., *J. Phys.* **B28**, 245 (1995)
3. Martin F. and Salin A., *J. Phys.* **B28**, 1985 (1995)
4. R.E. Olson, *Phys. Rev.* **A36**, 1519 (1987)
5. Wells J.C., Schultz D.R., Gavras P., and Pindzola M.S., *Phys. Rev.* **A54**, 593 (1996)
6. Crooks G.B. and Rudd M.E., *Phys. Rev. Lett.* **25**, 1599 (1970)
7. Harrison K.G. and Lucas M.W., *Physics Lett.* **33A**, 142 (1970)
8. Meckbach W., Focke P.J., Goni A.R., Suarez J, Macek J., and Menendez M.G., *Phys. Rev. Lett.* **57**, 1587 (1986)
9. Vajnai T., Gaus A.D., Brand J.A., Htwe W., Madison D.H., Olson R.E., and Schulz M., *Phys. Rev. Lett.* **74**, 3588 (1995)
10. Schulz M., Vajnai T., Gaus A.D., Htwe W., Madison D.H., and Olson R.E., *Phys. Rev.* **A54**, 2951 (1996)
11. Reinhold C.O. and Olson R.E., *Phys. Rev.* **A39**, 3861 (1989)
12. Moshammer R. et al., *Phys. Rev. Lett.* **73**, 3371 (1995)
13. Gaus A.D., Htwe W.T., Brand J.A., Gay T.J., and Schulz M., *Rev. Sci. Instrum.* **65**, 3739 (1994)
14. Kravis S.D., Abdallah M., Cocke C.L., Lin C.D., Stöckli M., Walch B., Wang Y.D., Olson R.E., *Phys. Rev.* **A54**, 1394 (1996)
15. McGuire J.H., Reeves T., Deb N.C., and Sil N.C., *Nucl. Instrum. Meth.* **B24/25**, 243 (1987)
16. Wood C.J., Feeler C.R., and Olson R.E., *Phys. Rev.* **A56**, 3701 (1997)
17. Wood C.J. and Olson R.E., *J. Phys.* **B29**, L257 (1996)

ATOMIC PROCESSES IN INDUSTRIAL PLASMAS

UV/VUV High Sensitivity Absorption Spectroscopy

A.N. Goyette, L.W. Anderson, K.L. Mullman and J.E. Lawler

Department of Physics, University of Wisconsin, 1150 University Avenue, Madison, WI 53706 USA

Abstract.
High sensitivity absorption spectroscopy is a powerful diagnostic technique for reactive glow discharges plasmas. Absolute column densities of many chemical radicals have been measured in both deposition and etching plasmas. Modern photodiode or charge-coupled device (CCD) detector arrays vastly increase the sensitivity of traditional absorption experiments enabling one to observe fractional absorptions of ultraviolet (UV) and vacuum ultraviolet (VUV) radiation less than 0.0001. Stable arc lamps provide a continuum source in some experiments, but experiments at very high spectral resolution or at VUV wavelengths require the greater spectral radiance of synchrotron radiation. High sensitivity absorption spectroscopy has been applied to intense glow discharges used for lighting, for diamond film deposition, and for both depositing and etching Si films. Absorption spectroscopy provides absolute column densities, is useful for transitions that do not fluoresce, and approaches the sensitivity of laser-induced fluorescence (LIF) in glow discharges under some conditions.

INTRODUCTION

Relatively few methods exist to determine absolute gas phase species densities in reactive glow discharges that do not perturb the species densities. High sensitivity absorption spectroscopy is one such method. It is easy to implement, reliable, and directly yields quantitative measurements provided that the absorption cross section is known. The densities of atoms, molecules, and transient radicals, both neutral and ionized in a glow discharge all may be probed with this technique. In addition, species inaccessible by other methods of measuring densities such as LIF, are accessible with high sensitivity absorption spectroscopy.

The basic elements of absorption spectroscopy are a source of continuum radiation, an absorbing gas phase sample, and a means of detecting light dispersed by a spectrometer. Stabilized deuterium or Xe arc lamps work well as sources of continuum radiation at UV wavelengths; experiments at VUV wavelengths require radiation from a storage ring. Traditionally, a single channel detector such as a photomultiplier tube has been used to detect the light dispersed by the spectrometer.

Much higher sensitivity, however, can be achieved by the use of either photodiode or CCD arrays as detection elements. A detector array provides two key advantages over a sequentially scanned single-channel experiment. The first advantage is that the detector array makes the experiment insensitive to low frequency noise or drift in the continuum source, since the spectral shape of the continuum from an arc lamp or storage ring is nearly constant even though the total intensity drifts. A single-channel sequentially scanned experiment maps the low frequency noise onto the absorption spectrum. Secondly, the multichannel detector compiles good photon statistics far more rapidly than a sequentially scanned single-channel experiment. Photon statistical noise is often the primary limit on the sensitivity of an absorption experiment after the effect of low frequency noise or drifts in the continuum source is suppressed. A digital technique is used to subtract the line emission from a glow discharge (the absorbing sample) and isolate the transmitted radiation from a continuum source with absorption features. As long as the spectral radiance of the continuum source is greater than, or similar to, that of the line emission from the glow discharge, the subtraction is possible. This relative spectral radiance criterion is generally met with reactive glow discharges when an arc lamp is used as a continuum source and is more than easily fulfilled when a storage ring is used [1].

APPLICATIONS TO REACTIVE GLOW DISCHARGES

In this section we discuss a few examples of the diverse applications in which high sensitivity absorption spectroscopy has been applied to study glow discharges.

Lighting Plasmas

Wamsley *et al.* measured excited Hg atom densities in a Hg-Ar positive column discharge with an estimated detection limit of 7×10^9 cm^{-2} [1]. These experiments used a stabilized Xe arc lamp as the source of continuum radiation. A gated, image-intensified CCD array used in conjunction with a 0.5m spectrometer equipped with an echelle grating and a premonochromator detected the dispersed spectra. Sensitivities of 0.001 were achieved, limited primarily by shot noise.

Diamond Film Deposition

High sensitivity absorption studies have provided a fairly complete picture of the gas phase chemistry in conventional (hydrogen-rich) low-pressure diamond synthesis. Absolute densities of CH_3, CH, C_2 and the hydrogen dissociation ratio have been measured in hydrocarbon glow discharges used for the chemical vapor deposition of diamond [2–5]. Additional species, C_2H_2 and C_6H_6, have been studied in diamond growth employing thermal, instead of plasma, activation of the gas chemistry.

The spectroscopic apparatus consists of a high pressure Xe arc lamp as a continuum source, a 0.5m focal length spectrometer with a 1200 line/mm grating, and a 1024-pixel photodiode array. A narrow bandpass filter in front of the entrance slit of the spectrometer rejects light not in the spectral region of interest, reducing the amount of stray light in the spectrometer. Figure 1 shows a diagram of this arrangement.

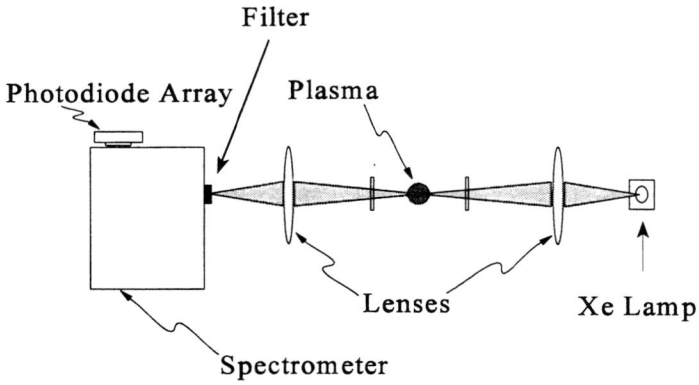

FIGURE 1. A schematic of the high sensitivity absorption spectroscopy system [5].

Column densities of diatomic species are calculated from the integrated equivalent width of the absorption band structure. The rotational partition function at a given discharge temperature can be calculated (the gas kinetic temperature is typically determined spectroscopically), and oscillator strengths for individual rotational transitions within the band can be calculated from the band oscillator strength. For an absorption signal on the linear part of the curve of growth, the integrated equivalent width is directly proportional to the column density. Careful curve of growth analysis can extend the dynamic range of these measurements beyond the optically thin limit. Polyatomic column densities are generally calculated from the absorption cross section and measured absorbance at line center.

Remarkable sensitivities have been achieved in these experiments. Figure 2 shows an absorption feature due to the CH radical in a hollow cathode discharge used for diamond deposition.

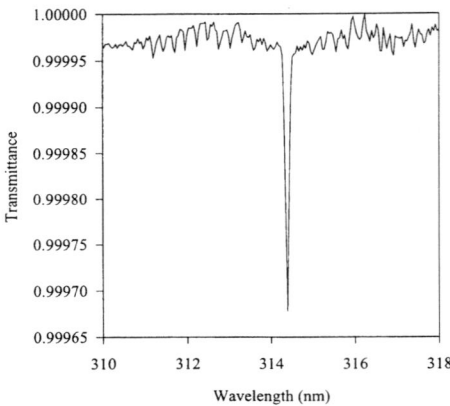

FIGURE 2. Absorption from the Q branch of the $C\,^2\Sigma^+ \leftarrow X\,^2\Pi$ (0,0) transition of CH near 314 nm. Immediately to the blue and to the red of the Q branch are the R and P branches of this band respectively. Fractional absorptions of ~ 0.00001 are observed in these branches [3].

Deposition and Etching of Si Films

Both SiF_2 and CF_2 have been detected using high sensitivity absorption spectroscopy in reactive ion etching plasmas [6]. The apparatus of Booth et al. is similar to that described in the previous section, with the exception that instead of a narrow bandpass filter, Booth et al. use a high quality fused silica prism as a premonochromator. Dielectric interference filters generally have a low peak transmittance far in the UV, typically less than 20%. The transmittance of optical grade fused silica in this wavelength range is considerably higher, allowing more rapid accumulation of photon statistics than with a filter.

The SiH_3 radical has been detected in H_2/SiH_4 rf plasmas used in the deposition of amorphous Si [7]. The absorption cross section for SiH_3 has not yet been measured, preventing determination of absolute column densities. Useful information can still be obtained, however, from observing the variation of relative radical column densities with processing conditions. Toyoda et al. also observed UV absorption caused by particulates in the plasma and thus were able to monitor the relative abundance of particulates with plasma conditions with high sensitivity absorption spectroscopy.

Laboratory Astrophysics

An experiment using a 3m focal length vacuum echelle spectrometer and a storage ring as a continuum source has been used to measure Fe^+ and Co^+ column densities as small as 3×10^8 cm^{-2} [8]. This experiment achieves both a high spectroscopic resolving power of 350,000 and sensitivity to small fractional absorptions at far UV and VUV wavelengths.

The white light beam line on the Aladdin Storage Ring at the Synchrotron Radiation Center provides a spectrally smooth and very stable source of continuum radiation across the VUV with a spectral radiance greater than 10^3 times that of an arc lamp. A hollow cathode discharge produces a sample of gas phase atoms and ions distributed across many metastable levels. Off-axis cylindrical mirrors reshape the light beam from the storage ring to match the slit size and angular acceptance of the spectrometer. The spectrometer is equipped with an echelle grating and a VUV sensitive CCD array. A premonochromator with 0.1 nm bandpass reduces stray light in the 3m focal length spectrometer and isolates a single order of the echelle grating. Stray light is typically 1% of the continuum flux at the peak of the premonochromator bandpass and absorption measurements are corrected for stray light levels in the spectrometer. Digital subtraction is used to eliminate line emission from the hollow cathode discharge and pixel-to-pixel variations in the quantum efficiency of the array can be divided out by using a dark-signal corrected, high signal-to-noise spectrum of the synchrotron radiation when the hollow cathode discharge is turned off. The resultant absorption measurements are limited in sensitivity by the Poisson statistical noise of the photoelectrons in the CCD array. Figure 3 shows a typical VUV absorption spectrum of the 160.845 nm transition in FeII.

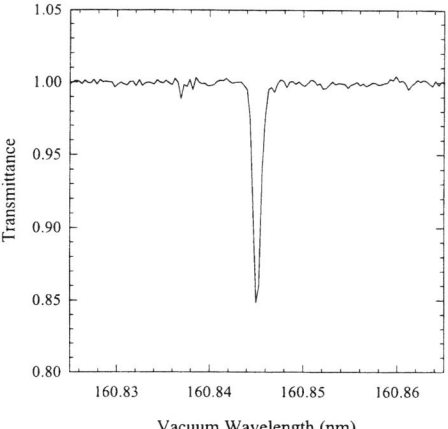

FIGURE 3. Absorption spectrum of the y $^6P^o_{7/2} \leftarrow$ a $^6D_{9/2}$ transition of Fe$^+$. The noise in the continuum is consistent with the Poisson statistical noise of photoelectrons in the CCD array [9].

These measurements are being used to make highly accurate measurements of the oscillator strengths of VUV transitions of atomic ions present in the interstellar medium. The oscillator strengths are necessary to interpret high quality absorption spectra from the Hubble Space Telescope and other orbiting observatories.

CONCLUSION

High sensitivity absorption spectroscopy is a simple, yet powerful technique for quantitative diagnosis of species densities in reactive glow discharges. Sensitivities to absorbances as small as 0.00001 are attainable even in the presence of intense line emission from the discharge. This technique has very broad applicability for reactive glow discharge studies.

ACKNOWLEDGMENTS

This work was supported by the USA Army Research Office under grant DAAH-04-96-0143, by NASA under grant NAGW-2908, and by the National Science Foundation under grant DMR-9212658 to the Synchrotron Radiation Center.

REFERENCES

1. Wamsley, R.C., Mitsuhashi, K., and Lawler, J.E., *Rev. Sci. Instrum.* **64**, 45-48 (1993).
2. Childs, M.A., Menningen, K.L., Toyoda, H., Anderson, L.W., and Lawler, J.E., *Europhys. Lett.* **25**, 729-734 (1994).
3. Childs, M.A., Menningen, K.L., Toyoda, H., Ueda, Y., Anderson, L.W., and Lawler J.E., *Phys. Lett. A* **194**, 119-123 (1994).
4. Erickson, C.J., Jameson, W.B., Watts-Cain, J., Menningen, K.L., Childs, M.A., Anderson, L.W., and Lawler, J.E., *Plasma Sources Sci. and Technol.* **5**, 761-764 (1996).
5. Goyette, A.N., Matsuda, Y., Anderson, L.W., and Lawler, J.E., *J. Vac. Sci. and Technol.* **16**, 337-340 (1998).
6. Booth, J.P., Cunge, G., Sadeghi, N., Chabert, P., and Neuilly, F., *Bull. Am. Phys. Soc.* **42**, 1756 (1997).
7. Toyoda, H., Goto, M., Kitagawa, M., Hirao, T., and Sugai, H., *Jap. J. Appl. Phys.* **34**, L448-L451 (1995).
8. Bergeson, S.D., Mullman, K.L., and Lawler, J.E., *Astrophys. J.* **464**, 1050-1053 (1996).
9. Mullman, K.L., Sakai, M., and Lawler, J.E., *Astron. Astrophys. Suppl. Ser.* **122**, 157-161 (1997).

ELECTRON COLLISION FREQUENCY IN INDUCTIVELY COUPLED PLASMAS

V. A. Godyak
OSRAM SYLVANIA Development Inc.
71 Cherry Hill Drive, Beverly, MA 01915

Inductively coupled plasma (ICP) sources are increasingly used in semiconductor manufacturing and lighting technology because they can provide a large plasma density at low gas pressure with no need for electrodes. These desirable performance features have spawned a new generation of plasma processing reactors and electrodeless light sources based on rf inductive plasma. The macroscopic analysis of ICP sources involves electrodynamic characteristics such as the real and imaginary part of the plasma conductivity. In the largest ICP applications (plasma processing and lighting) rf electron transport occurs in Ramsauer-like gases with a strong dependence of the electron-atom collision frequency on the electron energy. At such conditions the electron transport collision frequency in the rf field ν, and both components of the plasma conductivity are essential functions of the rf frequency ω and the electron energy distribution function EEDF which always is non-Maxwellian. Moreover, at low gas pressures, due to electron thermal motion in the inhomogeneous rf field, the electron drift is non-locally coupled with the rf field, resulting in a collisionless drag force and an additional collisionless rf power absorption. Such a situation is associated with an effective collisional frequency ν_{eff} accounting for both collisional and collisionless power absorption ($\nu_{eff} > \nu$). Langmuir probe measurement of the EEDF and magnetic probe measurement of the rf electric field and current density in ICP sources have permitted us to infer the values of ν and ν_{eff} over a wide range of gas pressure covering collisionless and collisional ICP regimes. Some theoretical aspects of ν and ν_{eff} estimations and their comparison with the experiment will be given in this lecture.

Spectroscopic Modeling and Interpretation of Industrial Plasmas in Materials Processing

J.L. Giuliani[1], J.P. Apruzese[1], A. Dasgupta[1], P. Kepple[1]
J. Rogerson[1], V. Shamamian[2], D. Counts[2]
R. Bicknell-Tassius[3] and F.J. Grunthaner[3]

[1] *Plasma Physics Division, Naval Research Laboratory, Washington DC*
[2] *Chemistry Division, Naval Research Laboratory, Washington DC*
[3] *Jet Propulsion Laboratory, Pasadena CA*

Plasma used in industrial processing of materials are characterized by their electron temperature and density. These properties control the plasma chemistry and hence the composition of particle flux to the working surface. Optical spectroscopy offers the industrial physicist an inexpensive and non-invasive diagnostic of such plasma conditions. However proper interpretation of the observed spectrum with atomic modeling is necessary in the process regime of a few Torr pressure and above where radiation trapping can become important. This presentation will focus on the analysis of nitrogen based industrial plasmas covering a broad pressure range.

The high pressure example is a 100kW atmospheric DC arc torch under investigation for shipboard waste remediation. Analysis is presented of the temperature profile in long arcs seeded with hydrogen as a diagnostic. A collisional radiative equilibrium (CRE) model with full radiative transfer for H and N was developed to calculate synthetic spectra. Comparison with the data using various temperature profiles indicates that photo-pumping from the hot core leads to non-LTE populations and an inversion in the emission ratio in the mantle, similar to the observed data.

A low pressure industrial system is a few 100W diffuse arc used as a clean source of nitrogen atoms for the growth of gallium-nitride films by molecular beam epitaxy. Optical emission spectra show the presence of both atomic and molecular nitrogen. Even at a few hundred Torr pressure the emission features indicate the plasma is neither coronal nor in equilibrium. The efficiency of cracking the N_2 molecule is therefore difficult to assess. Actinometry with argon can be used to estimate the dissociation fraction in conjunction with synthetic spectra. The data demonstrate that the diffuse arc jet is capable of supplying atomic nitrogen fluxes consistent with growth rates of several monolayers per second.

Finally, two theoretical issues will be addressed in regards to spectroscopic analysis of industrial plasmas. First, we note that excitation rates from metastable levels of argon are needed for actinometry, but calculated values differ significantly from recent data. Second, a complete CRE model of a typical plasma used in materials processing requires the coupling of atomic and molecular excitations. Initial results from such a model for N and N_2 will be presented.

Work supported by BMDO/IST.

List of Attendees

Abdallah Jr., Joseph
Los Alamos National Laboratory
Atomic and Optical Phys. Gr. T-4
Los Alamos, NM 87545
phone: (505) 667-7388
FAX: (505) 665-6229
email: abd@lanl.gov

Anderson, Harvey
University of Strathclyde
Department of Physics and
Applied Physics
Glasgow G4 0NG UK
phone: 44-014-155-341-96
FAX: 44-014-155-228-91
email:anderson@barwani.phys.
strath.ac.uk

Back, Christina
Lawrence Livermore National
Laboratory
700 East Avenue, L-488
Livermore, CA 94450
email: tinaback@llnl.gov

Bannister, Mark
Oak Ridge National Laboratory
Building 6003, MS-6372
PO Box 2008
Oak Ridge, TN 37831-6372
phone: (423) 574-4700

Bartschat, Klaus
Drake University
Physics Department
Des Moines, IA 53011
phone: (515) 271-3750
email: kb0001r@acad.drake.edu

Bauche, Jacques
Laboratoire Aime Colton
Bat. 505
91405 Orsay FRANCE
phone: (33) 169-352-106
FAX: (33) 169-352-100
email: claire.bauche@lac.u-psud.fr

Behar, Ehud
The Hebrew University of Jerusalem
Rahrah Institute of Physics
91904 Jerusalem ISRAEL
phone: 912-2-658-5179
FAX: 912-2-643-5034
email: behar@vms.huji.ac.id

Beiersdorfer, Peter
Lawrence Livermore National
Laboratory
PO Box 808, L-421
Livermore, CA 94551
phone: (510) 422-3985
FAX: (510) 422-5940
email: beiersdorfer@llnl.gov

Brown, Greg
Lawrence Livermore National
Laboratory
PO Box 808, L-421
Livermore, CA 94551
phone: (510) 422-6879
email: brown86@llnl.gov

Busquet, Michel
CEA Bruyeres, BP12
F-9169-Bruyeres le Chatel
Bruyeres FRANCE
FAX: 33-169-267-074
email: busquet@bruyeres.cea.fr

Cauble, Robert
Lawrence Livermore National
Laboratory, L-51
Livermore, CA 84550
email: cauble@llnl.gov

Ceccotti, Tiberio
Universite Paris VI
4 Place Jussieu, 75262
Paris Cedex, Paris   FRANCE
phone: 33-144-272-82
FAX: 33- 144-277-537
email: ceccotti@moka.ccr.jussieu.fr

Chenais-Popovics, Claude
Laboratoire pour l'Utilisation
Ecole Polytecnique
91128 Palaiseau Cedex   FRANCE
phone: 33-169-333-257
FAX: 33- 169-333-009
email:claude@greco2.polytechnique.fr

Chutjian, Ara
Jet Propulsion Laboratory
Caltech-MS 121-114
4800 Oak Grove Drive
Pasadena, CA 91109
phone: (818) 354-7012
email: ara.chutjian@jpl.nasa.gov

Clothiaux, Eugene
Auburn University
Physics Department
206 Allison Street
Auburn, AL 36849-5311
phone: (334) 844-4253
FAX: (334) 844-4613
email: ejc@physics.auburn.edu

Cosse, Philippe
CEA Bruyeres, BP12
F-9169 Bruyeres le Chatel
Bruyeres FRANCE
FAX: 33-169-267-074
email: cosse@bruyeres.cea.fr

Davis, Jack
National Research Laboratory
Plasma Physics Division
Radiation Hydrodynamics
Washington, DC 20375
phone: (202) 767-3278
email: davisj@ppdu.nrl.navy.mil

Dörner, Reinhart
Institut für Kernphysik
Universität Frankfurt
August Euler Str. 6
D-60486   GERMANY
email: doerner@alpha.ikf.physik.uni-frankfurt.de

Doron, Rami
The Hebrew University of Jerusalem
Racah Institute of Physics
Girat Ram 91904 Jerusalem ISRAEL
phone: (972) 658-5386
FAX: (972) 643-5034
email: doron@vms.haji.ac.id

Dunn, James
Lawrence Livermore National
Laboratory, L-251, PO Box 808
Livermore, CA 94550
phone: (510) 423-1557
FAX: (510) 423-2505
email: dunng@llnl.gov

Evans, Todd
General Atomic s
PO Box 85608
San Diego, CA 92186
phone: (619) 455-4269
FAX: (619) 455-4156
email: evans@gov.gat.com

Falcone, Roger
University of California
Department of Physics
Berkeley, CA 94720
email:chairman@physics.berkeley.edu

Fang, Zuyun
University of Nevada
Department of Physics
Las Vegas, NV 89154-4002
phone: (702) 895-1733
FAX: (702) 895-0804
email: js7s@mail.iapcm.ac.cn

Ferland, Gary
University of Kentucky
Department of Physics and Astronomy
Lexington, KY 40501
email: gary@cloud9.pa.uky.edu

Flannery, Raymond
Georgia Institute of Technology
School of Physics
Atlanta, GA 03332-0439
phone: (404) 894-5263
FAX: (404) 894-1101
email:ray.flannery@physics.gataech.edu

Fontes, Chris
Los Alamos National Laboratory
MS F663
Los Alamos, NM 87545
phone: (505) 665-7676
FAX: (505) 665-5553
email: cjf@lanl.gov

Fournier, Kevin
Lawrence Livermore National Laboratory
PO Box 808
Livermore, CA 94550
email: fournier2@llnl.gov

Gauthier, Jean-Claude
Ecole Polytechnique, Palaiseau
Laboratory LULI
91128 Palaisean FRANCE
phone: 33-169-333-269
FAX: 33-169-333-009
email: gauthier@grid-polytechnique.fr

Giuliani, Jr., John
National Research Laboratory
Plasma Physics Division
Code 6720
Washington, DC 20375
email: giul@ppdu.nrl.navy.mil

Godyak, Valery
OSRAM-Sylvania
71 Cherry Hill Drive
Beverly, MA 01915
phone: (058) 750-1510
email: godyak@rd.sylvania.com

Goldstein, William H.
Lawrence Livermore National Laboratory
8000 East Avenue
Livermore, CA 94550
phone: (510) 422-2215
FAX: (510) 422-2952
email: goldstein3@llnl.gov

Golovkin, Igor
University of Nevada
Physics Department/220
Reno, NV 89557
phone: (702) 784-4815
FAX: (702) 784-1398
email: golovkin@physics.unr.edu

Gorczyca, Tom
Western Michigan University
Department of Physics
Kalamazoo, MI 49001
email: gorczyca@umich.edu

Griesmann, Ulf
NIST/Atomic Physics Division
MS 221, Building A167
Gaithersburg, MD 20899
phone: (301) 975-4929
FAX: (301) 990-1350
email: ulf.griesmann@nist.gov

Griffin, Don
Rollins College
Physics Department
Winter Park, FL 37289
email: griffin@vanadium.rollins.edu

Gu, Pijun
Institute of Applied Physics and Mathematics
PO Box 8009-12
Beijina 100088   CHINA
phone: 86-10-620-144-11 ext. 3153
FAX: 86-10-620-101-08
email: js2s@mail.iapcm.ac.cn

Gunderson, Mark
Department of Physics
University of Florida
Gainesville, FL 32601

Hammel, Bruce A.
Lawrence Livermore National Laboratory
PO Box 808, L-473
Livermore, CA 94551
phone: (510) 422-3299
FAX: (510) 422-8395
email: bhammel@llnl.gov

Hanafi, Derfoul
Luli-Universite Paris VI
T22, 2 Etage, C22-23
4, 75252 Place Jussieu   FRANCE
phone: 33-144-274-282
FAX: 33-144-277-535
email: derfoul@noka.ccr.jussieu.fr

Havener, Charles C.
Oak Ridge National Laboratory
Physics Division
Building 6003; MS 6372
Oak Ridge, TN 37831
phone: (423) 574-4704
FAX: (423) 574-4745
email: havener@orph13.phy.ufl.edu

Haynes, Don
University of Florida
Department of Physics
Gainesville, FL 32611
phone: (352) 392-4036
FAX: (352) 392-3591
email: haynes@phys.ufl.edu

Hazi, Andrew
Lawrence Livermore National Laboratory
PO Box 808, L-051
Livermore, CA 94550
phone: (510) 422-4574
FAX: (510) 422-9523
email: hazi1@llnl.gov

Hooper, Charles
University of Florida
Physics Department
Gainesville, FL 32611
email: chooper@physics.ufl.edu

Iglesias, Carlos A.
Lawrence Livermore National Laboratory
PO Box 808, L-401
Livermore, CA 94550
phone: (510) 422-7252
FAX: (510) 422-7228
email: iglesias1@llnl.gov

Jacobs, Verne L.
Naval Research Laboratory
Condensed Matter and Radiation Science; Code 6693
Washington, DC 20375
phone: (202) 404-7147
FAX: (202) 404-7546
email: jacobs@dave.nrl.navy.mil

Johnson, Walter
University of Notre Dame
Department of Physics
Notre Dame, IN 46556
email: wrj@atomic3.phys.nd.edu

Junkel, Gwyneth
Department of Physics
University of Florida
Gainesville, FL 32601

Kaiser, Patricia
CEA Bruyeres, BP 12
F-9169-Bruyeres le Chatel
Bruyeres   FRANCE
FAX: 33-169-267-074
email: kaisert@bruyeres.cea.fr

Kalman, Gabor
Boston College
Physics Department
Chestnut Hill, MA 02167
email: kalman@bc.edu

Kantsyrev, Victor
University of Nevada
Department of Physics/220
Reno, NV 89501
phone: (702) 784-6809
FAX: (702) 784-1398
email: victor@physics.unr.edu

Keane, Christopher
Department of Energy
Germantown, MD 20874
phone: (301) 903-3345
FAX: (301) 903-4096
email: chris.keane@dp.doe.gov

Key, Michael
Lawrence Livermore National Laboratory
7000 East Avenue, L-488
Livermore, CA 94550
phone: (510) 424-2175
FAX: (510) 423-1076
email: key@llnl.gov

Kielkopf, John
University of Louisville
Department of Physics
Louisville, KY 40292
phone: (510) 852-5990
FAX: (510) 852-0742
email:john@nimbus.physics.louisville. edu

Kilcrease, David
Los Alamos National Laboratory
T4 B212
Los Alamos, NM 87545

Kim, Dong-Eon
Postech, San 31, Hyoja-Dong
Namku Pohang, Hyungbak
790-784 KOREA
phone: (562) 279-2089
FAX: (562) 279-3099
email: kimd@vision.postech.ac.kr

Kim, Yong-Ki
National Institute of Standards and Technology
Building 221, Room A267
Gaithersburg, MD 20899-0001
phone: (301) 975- 3203
FAX: (301) 975- 1350
email: kim@dirac.phy.nist.gov

Klapisch, Marel
ARTEC- Naval Research Laboratory
4555 Overlook Avenue, SW
Washington, DC 20008
phone: (202) 206-7802
FAX: (202) 767-0046
email: klapisch@this.nrl.navy.mil

Kolakowska, Alice
Auburn University
Physics Department
206 Allison Laboratory
Auburn, AL 36849
phone: (334) 844-4264
FAX: (334) 844-4613
email: kalosa@physics.auburn.edu

Kucera, Terese
NASA/Goddard Space Flight Center
Code 682.3
Greenbelt, MD 20771
phone: (301) 286-0829
FAX: (301) 286-0264
email: terry.kucera@gsfc.nasa.gov

Kukk, Edwin
University of Western Michigan
Department of Physics
Kalamazoo, MI 49001
email: kukk@umich.edu

Kunze, Hans J.
Ruhr-University-Bochum
Institut fuer Experimentalphysik V
44780 Bochum GERMANY
phone: 49-234-700-578-5
FAX: 49-234-709-417-5
email: hansjoachim.kunze
@lep5.ruhr-unibochum.de

Lawler, James
University of Wisconsin
Department of Physics
1150 University Avenue
Madison, WI 53706
phone: (608) 262-2918
FAX: (608) 265-2334
email: jelawler@facstaff.wisc.edu

Leboucher-Dalimier, Elisabeth
Physique Atomique dans les
Plasmas Denses - LULI University
Paris VI - Ecole Polytechnique
75252 Paris Cedex 05 FRANCE
email: lebda@ccr.jussieu.fr

Lee, Richard
Lawrence Livermore Nationa
Laboratory, L-399, PO Box 808
Livermore, CA 94551
phone: (510) 422-7209
FAX: (510) 423-6172
email: dicklee@llnl.gov

Manson, Steven
Georgia State University
Department of Physics and
Astronomy
Atlanta, GA 30303
phone: (404) 651-3082
FAX: (404) 651-1427
email: smanson@gsu.edu

Mackinnon, Andrew
Imperial College
Plasma Group, Blackett Laboratory
Imperial College, London
SW7 2BZ UK
phone: 44-171-594-765-4
FAX: 44-171-594-765-8
email: a.j.mackinnon@ic.ac.uk

Maksimchuk, Anatoly
Department of Physics
University of Michigan
Ann Arbor, MI 48109

Mancini, Roberto C.
University of Nevada
Department of Physics
Reno, NV 89557-0058
phone: (702) 784-6595
FAX: (702) 784-1398
email: rcman@physics.unr.edu

Mandelbaum, P.
Rach Institute of Physics
The Hebrew University
91904 Jerusalem ISRAEL
email: pinchasm@cms.huji.ac.il

May, Mark
John Hopkins University
Bloomberg Building
34th North Charles Street
Baltimore, MD 21218
phone: (410) 516-7929
FAX: (410) 516-5494
email: may@pha.jhu.edu

Meyer, Fred W.
Oak Ridge Natonal Laboratory
PO Box 2008
Oak Ridge, TN 37831-6372
phone: (423) 574-4705
FAX: (423) 574-4745
email: meyerfw@ornl.gov

Mitnik, Dario
Auburn University
Physics Department
206 Allison Laboratory
Auburn, AL 36849-5311
email: dario@physics.auburn.edu

Mueller, Alfred
Institut fur Kernphysik
University of Giessen
Leihgesterrner Weg 217
D-35392 Giessen GERMANY
email: alfred.mueller@stra.
uni-giessen.de

Oks, Eugene
Auburn University
Physics Department
206 Allison Laboratory
Auburn, AL 36849
email: goks@physics.auburn.edu

Osterheld, Albert L.
Lawrence Livermore National
Laboratory, PO Box 808
Livermore, CA 94550
phone: (510) 423-7432
FAX: (510) 423-2260
email: aosterheld@llnl.gov

Peyrusse, Olivir
CEA-Bruyeres le Chatel
Centre d'Etundes de Bruyeres
FRANCE
email: peyrusse@bruyeres.cea.fr

Pindzola, Michael S.
Auburn University
Physics Department
206 Allison Laboratory
Auburn, AL 36849
email: pindzola@physics.auburn.edu

Plante, Daniel
Stetson University
Department of Mathematics
Deland, FL 32720
email: dplante@stetson.edu

Podder, Nirmol
Auburn University
Physics Department
206 Allison Laboratory
Auburn, AL 36849-5311
phone: (334) 844-5108
FAX: (334) 844-4613
email: podder@physics.auburn.edu

Pratt, Richard H.
University of Pittsburgh
Department of Physics
Pittsburgh, PA 15260
phone: (412) 624-9052
FAX: (412) 624-9163
email: pratt@vms.cis.pitt.edu

Reader, Joseph
National Institute of Standards and
Technology; Bldg 221, Room A-167
Gaithersburg, MD 20899
phone: (310) 975-3222
FAX: (310) 975-3038
email: joseph.reader@nist.gov

Rejoub, Riad
University of Nevada
Physics Department/220
Reno, NV 89577
phone: (702) 784-1362
FAX: (702) 784-1398
email: rejoub@rigel.physics.unr.edu

Robicheaux, Francis
Auburn University
Physics Department
206 Allison Laboratory
Auburn, AL 36849
email: francisr@physics.auburn.edu

Rogers, Forrest J.
Lawrence Livermore National
Laboratory; PO Box 808, MS-041
Livermore, CA 94550
phone: (510) 422-7351
FAX: (510) 423-7228
email: rogers4@llnl.gov

Rowan, William L.
Fusion Research Center
University of Texas
Austin, TX 78712
phone: (617) 253-8663
FAX: (617) 253-0627
email: rowan@cmod2.pfc.mit.edu

Rozsnyai, Balazs F.
Lawrence Livermore National
Laboratory; PO Box 808
Livermore, CA 94550
phone: (510) 422-4085
FAX: (510) 423-0925
email: rozsnyai@virgo.llnl.gov

Safronova, Ulyana
University of Notre Dame
Physics Department
Notre Dame, IN 46556
phone: (219) 631-4008
FAX: (219) 631-5952
email:usafrono@darwin.helios.nd.edu

Sasaki, Akira
Japan Atomic Energy Research Inst.
25-1 Miiminami-Cho, Neyagawa
Osaka, 572-0019 JAPAN
phone: 81-720-310-709
FAX: 81-720-310-596
email: sasaki@apr.jaeri.go.jp

Schultz, David R.
Oak Ridge National Laboratory
Building 6003, MS-6372
Oak Ridge, TN 37831-6372
phone: (423) 576-9461
FAX: (423) 574-4745
email: schultzd@ornl.gov

Schulz, Michael
University of Missouri-Rolla
110 Physics Department
Rolla, MO 65409
email: schulz@physics.umr.edu

Scott, Howard
Lawrence Livermore National
Laboratory; PO Box 808, L-18
Livermore, CA 94551
phone: (510) 423-1530
FAX: (510) 423-5112
email: hascott@llnl.gov

Shaw, John
Auburn University
Physics Department
206 Allison Laboratory
Auburn, AL 36849
email: jshaw@physics.auburn.edu

Shlyaptseva, Alla
University of Nevada
Department of Physics/220
Reno, NV 89557
phone: (702) 784-4734
FAX: (702) 784-1398
email: aslava@engr.colostate.edu

Shlyaptsev, V. N.
Lawrence Livermore National
Laboratory; PO Box 808
Livermore, CA 94551

Smith, Anthony C. H.
University College-London
Department of Physics and Astronomy
London, WC1E 6BT   UK
phone: 44-713-807-138
FAX: 44-713-807-145

Smith, Augustine
Morehouse College
Department of Physics
phone: (404) 215-2615
FAX: (404) 614-6032
email: asmith@morehouse.edu

Soukhanovskii, Vlad
John Hopkins University
Department of Physics and Astronomy
Baltimore, MD 21218
phone: (410) 516-8705
FAX: (410) 516-5494
email: vlad@eta.pha.jhu.edu

Stamm, Roland
Marseille University
Universite de Provence
Centre St. Jerome, Case 232
13397 Marseille Cedex 20 FRANCE
email: rstamm@piima1.univ-mrs.fr

Stancil, Phillip C.
Oak Ridge National Laboratory
Building 6003; MS-6372
PO Box 2008
Oak Ridge, TN 37831-6372
email: stancil@mail.phy.ornl.gov

Stone, Philip
National Institute of Standards and Technology - Atomic Phys. Division
Building 221, Room A167
Gaithersburg, MD 20899
phone: (301) 975-3211
FAX: (301) 990-1350
email: pstone@erols.com

Stutman, Dan
Princeton University
Plasma Physics Laboratory
Princeton, NJ 08543
phone: (609) 243-2560
FAX: (609) 243-2418
email: stutman@pppl.gov

Svidzinsky, Vladimir
Auburn University
Physics Department
206 Allison Laboratory
Auburn, AL 36849
email: svidz@physics.auburn.edu

Terry, Jim
Massachusetts Institute of Technology
Alcator C-Mod, NW17-176
175 Albany Street
Cambridge, MA 02139
email: terry@psfc.mit.edu

Tetsuo, Ozaki
National Institute for Fusion Science
Oroshi 322-6, Toki
Gifu, 509-5922   JAPAN
phone: 81-572-582-240
FAX: 81-572-582-624
email: ozaki@nifs.ac.jp

Toth, Gabor
Western Michigan University
Department of Physics
Kalamazoo, MI 49008
phone: (616) 387-5360
FAX: (616) 387-4939
email: toth@umich.edu

Touma, Jimmy
Auburn University
Physics Department
206 Allison Laboratory
Auburn, AL 36849
email: jtouma@physics.auburn.edu

Umstadter, Karl
Los Alamos National Laboratory
P-23; MS H803
Los Alamos, NM 87545
phone: (505) 665-1509
FAX: (505) 665-4027
email: kru@lanl.gov

Utter, Steven
Lawrence Livermore National
Laboratory; PO Box 808; L-421
Livermore, CA 94551
phone: (510) 422-6879
email: utter1@llnl.gov

Zavodszky, Peter
Kansas State University
Department of Physics/116 Cardwell
Manhattan, KS 55605
phone: (785) 532-2652
FAX: (785) 532-6806
email: paz@phys.ksu.edu

Author Index

A

Achler, M., 334
Ahmad, I., 287
Alexiou, S., 216, 313
Altun, Z., 19
An, L., 347
Anderson, L. W., 359
Angelo, P., 216
Antonetti, A., 92
Apruzese, J. P., 366
Audebert, P., 92
Azuma, Y., 334

B

Bach, C. A., 282
Badnell, N. R., 265
Bannister, M. E., 149
Bartsch, T., 241
Bartschat, K., 121
Bastiani, S., 92
Behar, E., 256
Berrah, N., 29
Bicknell-Tassius, R., 366
Borghesi, M., 229
Boswell, C., 43
Bozek, J. D., 29
Brandau, C., 241
Bräuning, H., 334
Büscher, S., 287

C

Cable, M. D., 283
Calisti, A., 216, 299
Ceccotti, T., 216
Chakraborty, H., 19
Chung, Y.-S., 149
Chutjian, A., 134
Clark, R. W., 199
Cocke, C. L., 334
Cohen, M., 256
Counts, D., 366
Cowan, T. E., 283

D

Dasgupta, A., 366
Davis, J., 199
Derevianko, A., 3
Derfoul, H., 216
Deshmukh, P. C., 19
Dias, E., 19
Djurić, N., 149
Dörner, R., 334
Doron, R., 256
Dunn, G. H., 149
Dunn, J., 106

E

Estabrook, K. G., 283
Evans, T. E., 58

F

Falcone, R. W., 91
Ferland, G. J., 163
Ferri, S., 299
Finkenthal, D. F., 58
Finkenthal, M., 73
Flannery, M. R., 317
Fournier, K. B., 73

G

Gauthier, J. C., 92
Gauthier, P., 216
Geindre, J. P., 92
Giuliani, Jr., J. L., 199, 366
Godyak, V. A., 365
Goldstein, W. H., 73
Gorczyca, T. W., 265
Goyette, A. N., 359
Greenwood, J. B., 134
Gregory, B. C., 73
Grunthaner, F. J., 366
Gwinner, G., 241

379

H

Hammel, B. A., 283
Hatchett, S. P., 283
Henry, E. A., 283
Hinkel, D. E., 283
Hoffknecht, A., 241
Hunter, J. R., 106

I

Iwase, A., 229

J

Jagutzki, O., 334
Johnson, W. R., 3
Jones, M. W., 229

K

Karsheninnikov, S. I., 43
Keane, C. J., 281
Kepple, P., 366
Key, M. H., 283
Khayyat, Kh., 334
Kilkenny, J. D., 283
Koch, J. A., 283
Koubiti, M., 299
Kozhuharov, C., 241
Krstić, P. S., 185
Kruer, W. L., 283
Kucera, T. A., 173
Kukk, E., 29
Kunze, H.-J., 287

L

LaBombard, B., 43
Landen, O. L., 282
Langdon, A. B., 283
Langer, B., 29
Lasinski, B. F., 283
Lawler, J. E., 359
Leboucher-Dalimier, E., 216
Lee, R. W., 282, 283

Libby, S. B., 282
Lipschultz, B., 43

M

MacGowan, B. J., 283
Mackinnon, A. J., 229, 283
Manson, S. T., 19
May, M. J., 73
Meftah, T., 299
Mendelbaum, P., 256
Mergel, V., 334
Moody, J. D., 283
Moran, M. J., 283
Moshammer, R., 334
Mossé, C., 299
Mouret, L., 299
Müller, A., 241
Mullman, K. L., 359

O

Offenberger, A. A., 283
Olson, R. E., 347
Osterheld, A. L., 106

P

Pacella, D., 73
Pappas, D. A., 43
Peleg, A., 256
Pennington, D. M., 283
Perry, M. D., 283
Phillips, T. J., 283
Pigarov, A. Yu., 43
Pindzola, M. S., 265
Poquerusse, A., 216
Prior, M. H., 334

Q

Quoix, C., 92

R

Ralchenko, Y., 313
Reva, F., 299
Rice, J. E., 73
Robicheaux, F., 265
Rogerson, J., 366
Rousse, A., 92

S

Safronova, M. S., 3
Saghiri, A. A., 241
Sangster, T. C., 283
Sauvan, P., 216
Schippers, S., 241
Schlegel, Th., 92
Schmidt-Böcking, H., 334, 347
Schmitt, M., 241
Schultz, D. R., 185
Schulz, M., 347
Schwob, J. L., 256
Shamamian, V., 366
Shepherd, R., 106
Shlyaptsev, V. N., 106
Singh, M. S., 283
Smith, A. C. H., 149
Smith, S. J., 134
Spielberger, L., 334
Stamm, R., 299
Stancil, P. C., 185
Stewart, R. E., 106
Stoyer, M. A., 283

T

Tabak, M., 283
Talin, B., 299
Terry, J. L., 43, 73
Thornhill, J. W., 199
Tietbohl, G. L., 283
Tsukamoto, M., 283
Turner, C. S., 19

U

Ullrich, J., 334, 347

V

Verner, D. A., 163
Vrinceanu, D., 317

W

Wallbank, B., 149
Weber, T., 334
Wharton, K., 283
White, W. E., 106
Wilks, S. C., 283
Willi, O., 229
Wills, A. A., 29
Woitre, O., 149
Wolf, A., 241
Woolsey, N. C., 282
Wrubel, Th., 287